Selected Papers from 2017 International Conference on Micro/Nanomachines

Selected Papers from 2017 International Conference on Micro/Nanomachines

Special Issue Editors

Jianguo Guan
Fangzhi Mou

MDPI • Basel • Beijing • Wuhan • Barcelona • Belgrade

MDPI

Special Issue Editors

Jianguo Guan
Wuhan University of Technology
China

Fangzhi Mou
Wuhan University of Technology
China

Editorial Office
MDPI
St. Alban-Anlage 66
Basel, Switzerland

This is a reprint of articles from the Special Issue published online in the open access journal *Micromachines* (ISSN 2072-666X) from 2016 to 2017 (available at: http://www.mdpi.com/journal/micromachines/special_issues/ICMNM)

For citation purposes, cite each article independently as indicated on the article page online and as indicated below:

LastName, A.A.; LastName, B.B.; LastName, C.C. Article Title. *Journal Name* **Year**, *Article Number, Page Range.*

ISBN 978-3-03897-081-1 (Pbk)
ISBN 978-3-03897-082-8 (PDF)

Contents

About the Special Issue Editors

Jianguo Guan, Changjiang Scholar Distinguished Professor of MOE, Chief Professor of Wuhan University of Technology (WUT), China. He is also Dean for Education at the International School of Materials Science and Engineering, WUT. After receiving his BS (1990) and PhD degree (1995), he joined the State Key Laboratory of Advanced Technology for Materials Synthesis and Processing (SKLARMSP), WUT. He became a full professor in 1998. He worked as a Research Fellow (2000–01) in Nanyang Technological University, Singapore, and as a senior visiting scholar (2008–09) in Washington University in St Louis, USA. He has co-authored more than 210 SCI journal papers, published in various journals, including *Chem Soc Rev, Adv Mater, Angew Chem Int Ed, Nano Lett, Adv Funct Mater, ACS Nano* etc., as well as 34 authorized patents. He has been awarded several recognized Science and Technology Awards from the Chinese government. His current research interests focus on electromagnetic metamaterials, micro-, nanomotors, and photonic crystals.

Fangzhi Mou received his PhD in materials science from the State Key Laboratory of Advanced Technology for Materials Synthesis and Processing (SKLATMSP), Wuhan University of Technology in 2012. Between 2010 and 2011, he was a visiting PhD student at Washington University in St. Louis, USA. He then joined SKLATMSP at Wuhan University of Technology as an assistant professor and became an associate professor in 2013. He was elected as a Top Youth Talent (Second Class) supported by 15551 Talents Project of Wuhan University of Technology in 2016. He has published over 30 papers in leading journals, including Advanced Materials, Chemical Society Reviews, Angewandte Chemie International Edition, *Advanced Functional Materials, Small, Nanoscale, ACS Applied Materials & Interfaces*, etc. His current research interests include intelligent micro-, nanomotors and biomimic micro-, nanorobot swarms and systems.

micromachines

MDPI

Editorial

Selected Papers from the 2017 International Conference on Micro/Nanomachines

Fangzhi Mou and Jianguo Guan *

State Key Laboratory of Advanced Technology for Materials Synthesis and Processing, International School of
Materials Science and Engineering, Wuhan University of Technology, Wuhan 430070, China;
moufz@whut.edu.cn
* Correspondence: guanjg@whut.edu.cn; Tel.: +86-27-8721-8832

Received: 29 May 2018; Accepted: 31 May 2018; Published: 4 June 2018

Thanks to their capabilities of converting various energy into motions, micro/nanomachines are believed to bring about revolutionary changes in many fields, such as biomedicine, micro/nanoengineering, and environmental monitoring and remediation. Even though remarkable progress has been made in the last few decades, there are still some great challenges, such as toxic fuels, low controllability, poor intelligence, single functions, etc., that need to be overcome so that micro/nanomachines can work safely and effectively in living bodies, natural water systems, and micro/nanofactories. In addition, swarmings and assemblies of micro/nanomachines which have collective behaviors may cooperatively perform complex biological or engineering tasks that cannot be completed by single ones. In this present Special Issue, we have published 10 papers, covering topics ranging from the design strategies, motion control, and applications of single micro/nanomachines [1–7], to the collective behaviors of micro/nanomachines [1,8–10].

Among the 10 published papers in this Special Issue, there are three reviews and four research papers that concern single micro/nanomachines. Ning et al. introduced the design from aspects of materials, geometries and fuels, the motion control strategies, and the potential applications of micromotors, and they outlined some future research directions [1]. Considering the unique advantages of tubular micro/nanomotors in driving force and surface functionalization, Zha et al. overviewed in detail the propulsion mechanisms, fabrication techniques, and applications, and pointed out some challenging problems of the existing tubular micro/nanomotors, as well as possible solutions to be explored in the near future [2]. Chen et al. summarized the recent advances in the design, manufacture, structural features, motion performance, and motion manipulation of light-powered micro/nanomotors and came up with some of their challenges and opportunities [3]. Jiao et al. reported a magnetic and fluorescent hybrid Janus micromotor by embedding magnetic nanoparticles and fluorescent dyes into the microparticles in a one-step process [4]. Sun et al. investigated the motion behaviors of self-powered liquid metal droplet machines under an external electric field, and put forward two non-dimensional parameters ($\ddot{A}$ and $\ddot{O}$) to evaluate the ratio of the forces that resulted from the electric field to the fluidic viscous force, and the ratio of the friction force to the fluidic viscous force [5]. Li et al. proposed a self-propelled Janus foam motor, which can effectively integrate intriguing behaviors of the self-propulsion, efficient oil capture, and spontaneous self-assembly [6]. Feng et al. demonstrated three-dimensional (3D) control of the microrobot within a microfluidic chip using balanced magnetic and buoyancy forces, and the microrobot could grip particles (200 μm) and deliver it in a 3D space [7].

Motivated by the intriguing collective behaviors and the swarm intelligence of lives in nature, researchers are also dedicated to the understanding, construction, and manipulation of micro/nanomachine swarms and assemblies in this Special Issue. Shi et al. theoretically studied the pair dynamics of two self-propelled sphere dimers in the chemically active medium, which may shed light on the understanding of the collective dynamics of synthetic micro/nanomachines, as pair dynamics are the basic elements of the larger scale systems [8]. Ning et al. briefly introduced swarming, collective, and adaptive behaviors of interactive micromotors in the view of dynamic interactions between them [1]. Liu et al. summarized and compared the assembly and swarming of synthetic micro/nanomachines by the fuel induced methods (enzyme, hydrogen peroxide, hydrazine, etc.) and fuel-free induced approaches (electric, ultrasound, light, and magnetic) [9]. Zhang et al. streamlined the recent developments in light-controlled swarming and the assembly of colloidal particles based on the interactions that have arisen from optical forces, photochemical reactions, photothermal effects, and photoisomerizations, and also discussed the potential applications, challenges, and future prospects [10].

We wish to express our gratitude to all of the authors who submitted their papers to this Special Issue. We are also very grateful to all of the reviewers who helped us in an attentive and timely manner to improve the quality of this Special Issue.

Conflicts of Interest: The authors declare no conflict of interest.

References

1. Ning, H.; Zhang, Y.; Zhu, H.; Ingham, A.; Huang, G.; Mei, Y.; Solovev, A.A. Geometry design, principles and assembly of micromotors. *Micromachines* **2018**, *9*, 75. [CrossRef]
2. Zha, F.; Wang, T.; Luo, M.; Guan, J. Tubular micro/nanomotors: Propulsion mechanisms, fabrication techniques and applications. *Micromachines* **2018**, *9*, 78. [CrossRef]
3. Chen, H.; Zhao, Q.; Du, X. Light-powered micro/nanomotors. *Micromachines* **2018**, *9*, 41. [CrossRef]
4. Jiao, J.; Xu, D.; Liu, Y.; Zhao, W.; Zhang, J.; Zheng, T.; Feng, H.; Ma, X. Mini-emulsionfabricated magnetic and fluorescent hybrid janus micro-motors. *Micromachines* **2018**, *9*, 83. [CrossRef]
5. Sun, Y.; Xu, S.; Tan, S.; Liu, J. Multiple electrohydrodynamic effects on the morphology and running behavior of tiny liquid metal motors. *Micromachines* **2018**, *9*, 192. [CrossRef]
6. Li, X.; Mou, F.; Guo, J.; Deng, Z.; Chen, C.; Xu, L.; Luo, M.; Guan, J. Hydrophobic janus foam motors: Self-propulsion and on-the-fly oil absorption. *Micromachines* **2018**, *9*, 23. [CrossRef]
7. Feng, L.; Wu, X.; Jiang, Y.; Zhang, D.; Arai, F. Manipulating microrobots using balanced magnetic and buoyancy forces. *Micromachines* **2018**, *9*, 50. [CrossRef]
8. Shi, J.-M.; Cui, R.-F.; Xiao, J.; Qiao, L.-Y.; Mao, J.-W.; Chen, J.-X. Pair interaction of catalytical sphere dimers in chemically active media. *Micromachines* **2018**, *9*, 35. [CrossRef]
9. Liu, C.; Xu, T.; Xu, L.P.; Zhang, X. Controllable swarming and assembly of micro/nanomachines. *Micromachines* **2017**, *9*, 10. [CrossRef]
10. Zhang, J.; Guo, J.; Mou, F.; Guan, J. Light-controlled swarming and assembly of colloidal particles. *Micromachines* **2018**, *9*, 88. [CrossRef]

micromachines

MDPI

Review

Geometry Design, Principles and Assembly of Micromotors

Huanpo Ning [1], Yan Zhang [1], Hong Zhu [1], Andreas Ingham [2], Gaoshan Huang [1,*], Yongfeng Mei [1] and Alexander A. Solovev [1,*]

[1] Department of Materials Science, Fudan University, 220 Handan Road, 200433 Shanghai, China; h.ning@dhu.edu.cn (H.N.); zhangyan19861103@163.com (Y.Z.); 17110300005@fudan.edu.cn (H.Z.); yongfeng.mei@gmail.com (Y.M.)
[2] Department of Biology, University of Copenhagen, 5 Ole Maaløes Vej, DK-2200, 1165 København, Denmark; andreas.ingham@cpr.ku.dk
* Correspondence: gaoshan.huang@gmail.com or gshuang@fudan.edu.cn (G.H.); solovevlab@gmail.com (A.A.S.)

Received: 18 January 2018; Accepted: 7 February 2018; Published: 11 February 2018

Abstract: Discovery of bio-inspired, self-propelled and externally-powered nano-/micro-motors, rotors and engines (micromachines) is considered a potentially revolutionary paradigm in nanoscience. Nature knows how to combine different elements together in a fluidic state for intelligent design of nano-/micro-machines, which operate by pumping, stirring, and diffusion of their internal components. Taking inspirations from nature, scientists endeavor to develop the best materials, geometries, and conditions for self-propelled motion, and to better understand their mechanisms of motion and interactions. Today, microfluidic technology offers considerable advantages for the next generation of biomimetic particles, droplets and capsules. This review summarizes recent achievements in the field of nano-/micromotors, and methods of their external control and collective behaviors, which may stimulate new ideas for a broad range of applications.

Keywords: micromotor; microengine; micropump; catalyst; complex; self-assembly; fluidic; collective; sub-system; chemoton

---

1. Introduction

1.1. Mechanical Machines: Simple, Complex, Chaotic, and Fluidic

Beginning with illustrative examples in mechanical engineering of simple, complex, predictable, chaotic machines and extending to miniature systems on the scale of single atoms, this review and the cited references contained within stand alone as a comprehensive introduction to nano-/micro-motors (NMs), intended to answer questions about possible future design directions and considerations required for the next generation of nano-/micro-motors. This review discusses (i) new NMs' geometries, surfaces, fabrication, fuels, efficiency, applications; (ii) external control of NMs using magnetic, acoustic, electrical fields, light for cargo delivery and on-chip integration; (iii) swarming, collective, adaptive behaviors using chemically and external field triggered NMs; and (iv) bio-soft-hybrid NMs for roving sensors, drug delivery applications. We also propose a new system by integrating sub-systems for an assembly of "synthetic cells" using microfluidics. Our aim is to spark new collaborations across disciplines and open new exciting horizons in the field of nano-/micro-machines, which is currently in its infancy, yet in high demand.

The genesis of the man-made NMs (nano-/micro-engines, motors rotors, pumps) research field began about 15 years ago [1–7]. Inspired by the shape and geometry of biological motors like proteins, flagellum, and related cellular motility, small machines are also generated to duplicate the capabilities of large-scale motors, engines and rotors. The tremendous promise of NMs is to have a profound influence on the environment, biomedicine, clean energy, cleaning of the environment, and on-chip integration. Analogous to the manufacture of an automobile in an assembly line process of the automotive industry, which began around 300 years ago, small machines require the integration of many innovations, such as high efficiency motors, geometrical optimization, reduced time of fabrication, improved methods of fabrication, and precision assembly of components in the assembly line.

Moreover, designed micro-/nano- motors, rotors, engines and pumps represent a microscopic analogy of large-scale man-made machines [1–7]. Classical mechanics provides very accurate results, when describing the motion of macroscopic bodies, including joint machine parts, which are not extremely heavy nor very small with much lower speeds than the speed of light. According to the definition in classical mechanics, a simple machine consists of only one part and a complex machine consists of at least two or more machine parts working together. For instance, a car, or a clockwork mechanism, (Figure 1a,b) are classical examples of complex machines. Another illustrative example of a complex machine is the recently fabricated octobot, as shown in Figure 1d. This soft robot mimics movements of an octopus with embedded integrated fluidic logic circuits, which are powered by chemical decomposition of hydrogen peroxide into oxygen and water [8].

We currently live in an era of reductionism that provides many new opportunities in developing complex machines with deterministic behavior. In reductionism, a complex machine's parts are broken down into smaller components and, with these simple individual parts, forces and interactions can be understood using classical equations of motion, thus the motion of the composite complex system can be understood. It is a revolutionary mode for understanding complex systems, and thus very useful for having complex machines with easily predictable functions. However, it is already known that biological systems cannot be explained using this reductionist approach. According to Sapolsky [9], bio-systems are non-additive and nonlinear bifurcating systems. Additionally, there are not a sufficient number of individual components to explain their collective function (e.g., not enough neurons to recognize faces; not enough genes to program capillaries in a body). Moreover, the system itself can be a variability rather than the accompanied noise, which often needs to be avoided for obtaining reliable data. Thus, while knowing the initial state and conditions, there exists no predictability about determining the final state [9]. Luckily, the discovery of chaotic systems by Lorenz in 1962 led to a better understanding of new dynamic systems for preparing weather forecasts, understanding the combined motion of double pendulum and knowing the chaotic rotation of a water wheel (Figure 1d), all of which are examples of unpredictability of nonlinear dynamic systems [10]. It shows remarkable examples that not every dynamic state is linear and predictable as was previously accepted before the discovery of chaos. To illustrate the chaotic nature of nonlinear dynamical systems, let's consider the chaotic Lorenz wheel, which is driven by both gravity and the pouring of water into buckets as shown in Figure 1d. As water pours into the buckets at a steady rate, the wheel rotates. Complicating matters, each bucket has a small hole at its bottom. As water leaks from the bottom of each bucket, energy is removed from the system, making the system oscillate. However, if only a small amount of water/energy is added, the wheel rotates and stops after some time, leading to a system's static state (Figure 1e). When larger amounts of water/energy is added, the wheel starts to oscillate forth and back, leading to an appearance of unpredictable dynamic states (Figure 1f). In other words, although the entire structure of the machine can be understood, it is not possible to predict the dynamic behavior from its initial conditions, linear interactions and energy flow. According to Sapolsky, there is no real "static answer" in chaotic systems, noise does not depend on the quality of our reductive tools or if we observe the system closer—variability is the phenomenon at any scale. In other words, chaotic machines can escape from the control of an operator which is, obviously, not desirable for utility. Biological systems are examples of dynamic systems, which operate on different and far more complex principles than a mechanical clock. It remains a speculative question, if principles of more complex NMs can be based on purely predictable classical mechanics or we should search for spontaneous behaviors with so called "emergent properties", nonlinear interactions and self-organization in hierarchical structures with feedback loops between different levels [11].

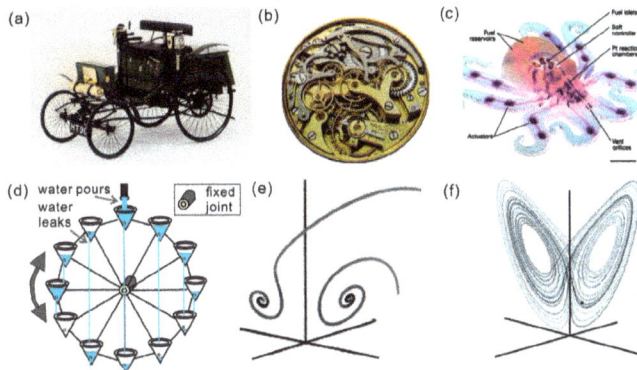

Figure 1. Mechanical, soft, complex (predictable) and complex (aperiodic, chaotic) macroscale machines. (**a**) one of the first modern car models (credit: Arnold Benz Motor Carriage, 1896). Reproduced from [12]; (**b**) a clockwork mechanism consisted of multiple mechanical elements with well understood forces, interactions and dynamics between mechanical components using reductionism approach. Reproduced from [13]; (**c**) biomimetic octobot that includes fluidic logic gates, powered by decomposition of chemical fuel. Reproduced with permission from [8], copyright (2016) Springer Nature; (**d**) chaotic Lorenz water wheel, which represents a classical example of the long-term unpredictability of a deterministic nonlinear dynamic system. Here, water pours into the buckets at a steady rate and gives the wheel energy, while gravity and water leakage out of each bucket removes energy from the system; (**e**) phase portrait of a typical system that reaches its equilibrium state. Reproduced from [14]; (**f**) a constant pouring of water leads to aperiodic oscillations of the wheel forth and back and an appearance of aperiodic/chaotic motion according to the Lorenz system. Reproduced from [15].

1.2. Small Machines from Atoms to Nano-, Micro-, and Meso-Particles

Technology has already advanced in both top-down and bottom-up fabrication methods. Atoms are the smallest stable building blocks and are known constituents for construction of machines. According to classifications based on size, nanoscale (<100 nm), microscale (100 nm–100 μm) and mesoscale (>100 μm) particles contain increasingly larger number of atoms, respectively. Small machines with atomic or molecular size can cross the border between classical and quantum mechanical regimes and include many biological cases. Figure 2a shows an image from a movie created by International Business Machines Corporation (IBM), entitled "a boy and his atoms", where a scanning tunneling microscopy was used to position individual carbon monoxide (CO) atoms on the surface of copper at low temperature, creating the smallest known robot [16]. When well-controlled conditions such as temperature, vibrations and pressure are achieved, atoms can be moved, assembled on the surface causing ripples or disturbances in the electronic density. Although IBM's "nanobot" cannot exist (yet) in ambient conditions, it demonstrates inspiring capability of new technology based on the smallest building blocks. Our bodies consist of atoms, ions and molecules and perhaps, nano- and Angstrom-scale "motors" made of individual atoms, ions and molecules are the most interesting to consider, not only because of their huge applications, but because of their unique properties, where fundamental laws of quantum mechanics dominate. For instance, wave-particle duality, uncertainty principle, quantization of energy levels, molecules and emission–absorption, i.e., "communication" using photons, and phonons become apparent. For instance, according to Heisenberg's uncertainty principle, we are not able to measure simultaneously, with high precision, a particle's position and momentum. This inability to know a particle's coordinates and momentum simultaneously remains a problem based on the wave-like nature of a particle and not a problem with today's measurement instruments, neither it is a reflection of the equality of experimental methods. Today, man-made nano-/micromachines are relatively large (10 nm–100 μm) in comparison to atomic structures. The largest known atom (cesium) has the diameter 0.53 nm that is approximately 20 times smaller than the diameter of smallest nanomotor, reported so far. When accepting matter as a continuum approximation, scaling laws can provide useful calculations for engineering of micromachines [17]. However, for truly nanoscale, as well as for molecular and atomic dimensions, scaling laws can be misleading due to an appearance of quantum size-confinement effects, where the mean free path, uncertainty principle, quantization of energy and radiation become significant. Remarkably, many tunable nanomaterials properties can be integrated into small machines/motors, tailored by changing both the dimension of NMs' layers, segments and their chemical compositions, leading to tunable reactivity, optical, magnetic, electronic, thermodynamic and mechanical properties [18,19].

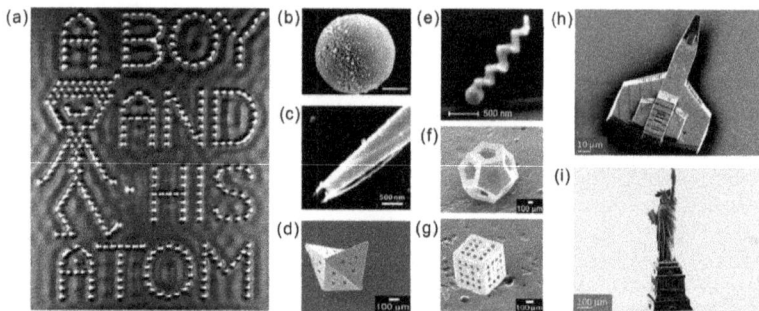

Figure 2. Scanning Tunneling Microscopy and Scanning Electron Microscopy images of fabricated small machines ranging from atoms to nano-, micro- and meso-particles. (**a**) the smallest movie created by International Business Machines Corporation (IBM), entitled "A boy and his atoms", it is based on nanomanipulation of atoms using scanning tunneling microscopy in ultra-high vacuum and low temperature conditions. Reproduced with permission from [16]. Reprint Courtesy of International Business Machines Corporation (Armonk, NY, USA) © 2013 International Business Machines Corporation. (**b**–**g**) three-dimensional micro-/nanoarchitectures with different shapes: (**b**) spherical Janus microparticle (reproduced from [20]). This work is licensed under the Creative Commons Attribution 4.0 International License [21]. (**c**) a microtube based on rolled-up nanomembrane (image courtesy of D. Gracias); (**d**) self-assembled irregular octahedron using surface tension (Reproduced with permission from [22], copyright (2010) Jon Wiley and Sons); (**e**) individual screw with nanostructured helicity, prepared by glancing angle deposition method (Reproduced with permission from [23], copyright (2009) American Chemical Society); (**f**) self-assembled dodecahedron (Reproduced with permission from [22], copyright (2010) Jon Wiley and Sons); (**g**) fabricated cubic container with holes (Reproduced with permission from [22], copyright (2010) Jon Wiley and Sons); (**h**,**i**) fabrication of complex shapes, micro-airplane and micro-statue of Liberty, enabled by advanced laser photolithography. Images are courtesy of Nanoscribe GmbH (Eggenstein-Leopoldshafen Germany).

Early research in the field of NMs began with NMs having characteristic dimensions that were several orders of magnitude larger than atoms—nano-/microparticles (Figure 2b–g). Self-electrophoretic bimetallic nanomotors [2,3], tubular micromotors made of rolled-up nanomembranes were fabricated using different combinations of inorganic catalytic/non-catalytic materials [5,6]. Gracias and co-workers demonstrated more complex 3D shapes using self-folding 2D layers into 3D architectures [22]. Recently, a new two-photon laser lithography enabled fabrication of almost any complex shapes/geometries of objects at the microscale (Figure 2h,i). For example, Figure 2h,i shows the smallest space ship and the Statue of Liberty. In the next step, these complex geometries can be coated with catalytic layers and placed in a fuel solution to achieve self-propelled motion. However, although complex shapes are achievable, it is unlikely that more complex NMs consisted of many mechanical moving parts will be utilized in the near future due to well-known problems of high viscosity of fluids at low Reynolds number, high surface tension and stiction between different parts due to short range forces (e.g., hydrophobic, electrostatic, Casimir).

The development of micro-/nano- motors, rotors, engines and pumps was inspired by biological motor proteins, jet propelled bacteria, flagella, cilia, shape changing and other impressive motile principles used in biology. In living systems, real chemical nanomachines are large macromolecules (enzymes), which catalyze many processes in organisms. Miniaturization in the design of man-made machines lead to multiple advantages, including (1) low weight; (2) fast performance; (3) less inertia, less mass; (4) less energy required to function; (5) increased strength to weight ratio; (6) increased power density, small power consumption and high energy conversion efficiency (chemo-mechanical coupling); (7) precise control of movement at the nano-/microscale; and (8) integration of a large number of devices in a small volume of space, which can enable breakthrough applications. Versatile nanodevices are not perturbed by thermal vibrations due to their high resonant frequencies (the resonant vibration is inversely proportional to the object mass) [23]. For instance, while enzymes can accomplish millions of operations per second, which is not possible by macroscopic robots. According to scaling laws, small machines are relatively stronger than larger machines (strength/weight = $1/D$, where D is a characteristic dimension), such as S/W ratio of individual micromotor can be up to 10^6 times higher in comparison to a macroscale motor) [24].

Bio-nanomotors have been considered promising prototypes for constructing of man-made chemically-actuated micro-/nano- motors, engines, rotors, pumps, powered by catalytic reactions and controlled using external fields [25–31]. NMs with different aspect ratios, shapes, materials, and fuels can effectively overcome Brownian diffusion and high viscosity of fluids at very low Reynolds number [32,33]. Swarming and collective "chemo- and phototactic" motion of nano-/microparticles were observed in chemical fuels and external fields. Gracias and co-workers noted, that in comparison to already well-established field of Micro-Electro-Mechanical-Systems (MEMS), the main difference being that the Micro-Chemo-Mechanical-Systems (MCMS) are triggered directly by chemistry, similar to biological motor proteins, as opposed to electro-mechanical energy conversion typically used to actuate conventional MEMS [34]. Nano- and micromachines have bright potential in analytical chemistry, sensing, migration, capture, delivery, and separation [35]. Today, several groups demonstrated design of the next generation of NMs at the smallest scales. Fischer and co-workers demonstrated helical nanopropellers [36], and Mei's group set the next world record for the smallest jet engine consisting of TiO_2/Pt nanotubes with the diameter as small as 30 nm [37]. External methods to power and control motion of NMs using versatile fuel-free magnetic and electric fields can be used [38]. It is important to mention that catalytic and biocatalytic reactions can enable applications of simpler and more environmentally friendly fuels than used today in gasoline engines, fuels cells or batteries. The fundamental mechanism of energy transduction can shed light on discovery of clean energy nanogenerators.

In the final part of our review, we concentrate on future generations of NMs: bio-hybrid-NMs for biosensing, drug delivery and, importantly, their fabrication and assembly using microfluidic technology, inspired by fundamental concepts concerning microfluidic machines proposed by Ganti [39], who stated two important concepts: (1) human technology is still unable to manipulate energy pathways using chemical means, rather than mechanical or electrical means. This represents a major difference between biological living and synthetic man-made machines; and (2) fluidic space of biological cells is free of mechanical constraints. Next, microfluidic techniques can provide tremendous flexibility in designing soft and fluidic systems. For instance, operation of crawling cells is not disturbed by stirring and diffusion of their internal components. While approaching functionalities inherent to biosystems, an exceptional opportunity exists to realize integration of multifunctional sub-systems using established microfluidic techniques to explore, optimize and mass-produce customized nano-/micromotors or "synthetic cells" with desired motive power and specific utility in very short time.

2. New Materials, Geometries and Fuels for Autonomous Motion

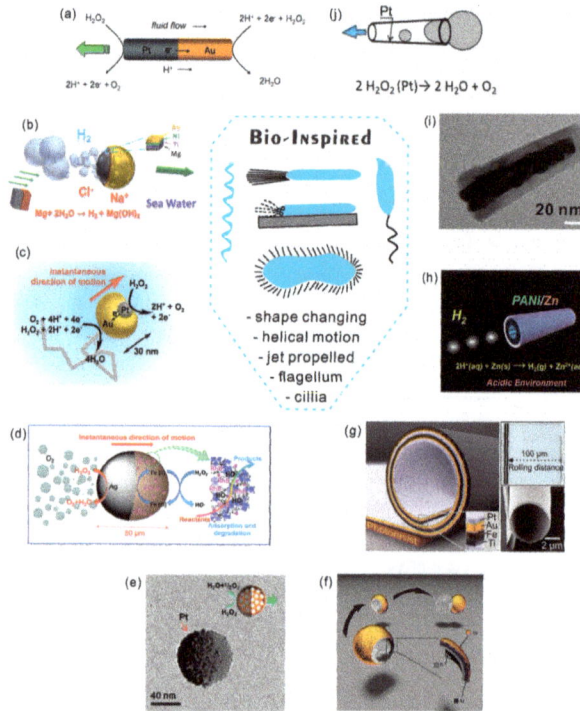

Figure 3. Several illustrative examples are shown: self-propelled nano-/micro-motors (NMs) with different materials, size, geometry, fuels and mechanisms of propulsion. (**a**) state-of-the-art bimetallic nanowire-based nanomotor driven by self-electrophoresis. Reproduced with permission from [40], copyright (2009) Royal Society of Chemistry; (**b**) hydrogen bubbles propelled magnesium spherical micromotor. Reproduced with permission from [41], copyright (2013) Royal Society of Chemistry; (**c**) one of the smallest reported Janus NM with diameter as small as 30 nm. Reproduced with permission from [42], copyright (2014) American Chemical Society; (**d**) metal-organic framework (MOF) based NM specifically designed for water purification [43]; (**e**) one of the smallest mesoporous NM with diameter around 80 nm. Reproduced with permission from [44], copyright (2015) American Chemical Society; (**f**) nanoshell-based NM with microcavity ideally suited for nucleation and generation of bubbles. Reproduced with permission from [45], copyright (2013) American Chemical Society; (**g**) the first example of fabricated tubular NM, made of rolled-up inorganic/catalytic nanomembranes. Reproduced with permission from [7], copyright (2009) Jon Wiley and Sons; (**h**) hydrogen bubbles driven NM in acidic environment. Reproduced with permission from [46], copyright (2012) American Chemical Society; (**i**) the smallest nanojet engine reported to date with diameter as small as 30 nm. Reproduced with permission from [37], copyright (2017) Jon Wiley and Sons; (**j**) schematic image of catalytic tubular microcavity/tube that is ideal for nucleation, growth and recoil of microbubbles, leading to effective bubble-induced pumping mechanism and ultra-high speeds of NMs.

Autonomous motion of NMs was initially inspired by motor proteins and self-propelled motile cells using cilia, flagellum, polymerization reactions, expelled slime from nanopores, helical motion and shape changing. This section discusses new materials, fuels, geometries, surfaces, fabrication methods and NMs' efficiencies recently demonstrated for autonomous propulsion of NMs (Figure 3a–j). Biological nanomachines are able to utilize extremely efficient chemo-mechanical energy coupling conversion mechanisms, providing a high specific power [47]. Moreover, a bacterium uses only 2% of its total energy for swimming [48]. A very large number of metals and enzymes can be used to catalyze reactions and power autonomous movement at the micro- and nanoscale. Depending on the shape of the object, the placement of the catalyst and balance of forces different kinds of movements and trajectories can be achieved. Several motive mechanisms have been previously reported: self-diffusiophoresis, self-electrophoresis, bubble recoil and surface/interfacial tension, which were previously not observed in biology.

2.1. Improvement of Nano-/Micromotors' Materials, Geometry, Surface and Motion

It is clear that previous problems of propulsion at a very low Reynolds number have been solved with micromotors, which now move faster than bacteria. Rapid progress in exploring new materials and improved geometries, surfaces and motion of nano-/micro-motors, engines, rotors and pumps began in the pioneering nanomotors research groups of Paxton, Sen, Mallouk and co-workers [2], Ozin and co-workers [3] (Figure 3a). Zhao and co-workers demonstrated applications of graphene oxide (GO/Ti/Pt) microjets by direct evaporation of metals on graphene surface, which self-assembled into tubes due to a weak bonding between GO layers [49]. In a similar study, Bandyopadhyay and co-workers presented glass beads as GO coated spherical motors [50]. Wang's group demonstrated electropolymerized outer layers, such as polypyrrole (PPy), poly (3,4-ethylenedioxythiophene) (PEDOT), and polyaniline (PANI) on various catalytic surfaces (Ag, Pt, Au, Ni-Pt alloy) [51]. Fischer's group experimentally proved self-propulsion of the smallest Janus particles with diameter as small as 50 nm (Figure 3c), which, however, were highly influenced by strong Brownian forces [42]. Mei and co-workers applied Atomic Layer Deposition of TiO_2/Pt Nanotubes to fabricate the smallest nanorockets with diameter around 30 nm (Figure 3i) [37]. Metal-polymer hybrid micromachines with bending and rotational motions based on stacking cationic poly (allylamine hydrochloride) (PAH) and anionic poly (acrylic acid) (PAA) were prepared by layer-by-layer technique [52]. Recently, new geometries and materials were used for catalytic locomotion, including core-shell nanowires, where it was found that self-diffusiophoresis could have a profound influence on the motion [53]. Interestingly, bubble-generation is not observed in smaller motors, where gaseous reaction products can diffuse out of surface quicker than bubble growth/nucleation can occur. Propelled NMs consisting of spherical Au/Ag/Pt nanoshell bubbles were realized by deposition of metals and a subsequent etching of spherical core particle (Figure 3f) [45]. Integration of several nanojets on a larger substrate was demonstrated using layer by layer deposition and under-etching techniques [54]. Theoretical models were developed to describe hydrodynamics of locomotion of both nanowires and engines by imbalance forces and to better understand propulsion mechanisms [55–57]. Recently, a low-surface-energy (LSE) layer combined with rough surfaces was fabricated on the outer surface of NMs for an efficient reduction of the fluidic drag force [58]. Au-mesoporous silica NMs powered by a hydrolysis reaction of aqueous $NaBH_4$ and KBH_4 and common H_2O_2 fuel [59] are particularly attractive for a micro/nano-carrier system [60]. Platinum-loaded NMs were prepared to power small under-water vehicles by deposition of platinum nanourchins (PNUs) onto cellulose (MFC) films via reduction of chloroplatinic acid (H_2PtCl_6) with formic acid (HCOOH) [61]. Metal-organic frameworks (MOF) was applied as motor's material [62]. It is known that MOF can have an ultrahigh surface area up to several thousand square meters per gram of material (m^2/g), which is particularly attractive for loading of a high surface area of catalysts or a reduction of consumption of expensive materials. Other impressive examples include tri-metallic microcaps [63], manganese oxide [64], carbon allotrope nanomaterials [65], copper-platinum segmented nanobattery [66] and nanoparticle-mediated motion [67]. Li et al. reported Au–Fe/Ni alloy hybrid nanowire motors, which can achieve speeds up to 850 μm·s^{-1} or 157 BL·s^{-1} [68].

2.2. Improvements of Fabrication Methods

Paradigm-shifting fabrication methods represent an important research direction for NMs, which include (i) improvement of fabrication methods and (ii) reduction of fabrication time. Usually, a standard photolithography procedure, followed by e-beam deposition of materials and optional supercritical point drying, was used at the beginning to prepare microjet engines based on strain-engineered rolled-up nanomembranes. However, this procedure takes many hours of fabrication. These methods can be contrasted with well-established fields of microfluidics and self-assembly, where NMs can be generated under ultra-fast fabrication conditions with desired properties and targeted for specific applications. Rapid progress of NMs fabrication has been discussed by Wang and Pumera, including electrochemical/electroless deposition, membrane template-assisted electrodeposition, asymmetric bipolar electrodeposition, physical vapor deposition, glancing angle deposition, self-scrolling method for helical NMs, three-dimensional direct laser writing and layer-by-layer assembly [69,70]. Other impressive fabrication methods include different components in oil-in-water droplets, followed by emulsification, solidification and direct assembly of asymmetric catalytic/magnetic NMs [71]. Examples of other methods are Co-Pt/Au motors with a three-step applied electrochemical potential process [72], shape-controlled fabrication of the polymer-based motors based on the polydimethylsiloxane template [73], template electrosynthesis of graphene microengines [74], layer-by-layer assembly technique in combination with micro-contact printing [75], evaporation-induced self-assembly using controllable crystals of ferrocene-based metal–organic (Fc-Ala-BCB) materials [76], and polymer NMs with doped Pt nanoparticles/carbon nanotubes [77].

2.3. New Fuels

Biology employs both "fuel free" Brownian motion and chemically powered motion to support viable functions of organisms. There is much to learn from biological motors, which often operate on very simple and environmentally clean fuels, such as ions and protons. Initially, only hydrogen peroxide was used as a fuel for NMs; however, recent progress in this area has provided a dramatic increase in the choice of a variety of fuels [78]. For instance, soft-oxometalates using dithionite as a fuel, where the redox active Mo^{VI} sites of soft-oxometalates (SOMs) were applied to oxidize dithionite and generate SO_2 for propulsion [79], magnesium-water reaction [80], motion of graphene swimmers in pure water [81], vapor-driven propulsion [82], hydrogen-bubble driven zinc material in strongly acidic media (Figure 3b) [46], moisture-activated torsional graphene-fiber motor [83], and water-powered cell-mimicking Janus motor [84]. Novel biocompatible fuels were reported using carbonate-based Janus particles propelling in acidic environments [85]. Biofuels are discussed in more detail in the biological section of this review. Not only materials, but effects of ionic screening, Debye lengths, dissociation in solvents [86] and different surfactants are crucial for the motion of NMs [87,88].

2.4. Efficiency

Similar to macroscale motors, an efficiency of NMs is highly important for achieving high speeds and long working time [89]. Biological nanomotors are very efficient in chemo-mechanical coupling without the need of an intermediate step for chemical-to-electrical-to-mechanical conversion. Biological nanomotors burn only a few fuel molecules to power mechanical actuation and thus self-propel almost at almost "no cost". Many other parameters also influence the efficiency of man-made NMs, including micromotors' geometry, size, materials, temperature, fuel composition and mechanisms of motion.

Valveless microbubble-driven micropumps without moving parts represent a distinct advantage over standard energetically costly pumps [90]. Pumping of fluids can be achieved simply by microbubbles driven by interfacial tension in microtubes. The addition of a surfactant is usually the only requirement for reducing the surface tension and stabilizing the microbubbles. Assuming incompressible, Newtonian and laminar fluid flows through a tube, a pressure drop can be calculated using the Hagen–Poiseuille equation, which states that the flow rate is proportional to the radius of the tube to the fourth power. In this case, a small decrease in the diameter of the tube yields a significant decrease in flow rate

$$\Delta P = 8\mu L Q / \pi R^4 \qquad (1)$$

where R is the tube radius, Q is the volumetric flow rate, L is the length of tube, μ is the dynamic viscosity and ΔP is the pressure difference between two ends. Equation (1) is also employed in better understanding of medical delivery and intravenous access of fluids.

Mallouk and co-workers recently discussed possible efficiency loss mechanisms of NMs with different shapes: wire, tubular and helical structure. So far, the efficiency of NMs was reported on order of 10^{-9} [91]. Solovev et al. found a factor of $\times 1000$ enhancement of microbubble nucleation and growth in tubular microcavity/pumps [92]. When fixed on the surface, microtubes with diameter 5–10 μm and length 30–1000 μm function as catalytic micropumps by decomposition of hydrogen peroxide into oxygen microbubbles and water. Micropump efficiency depends on the minimum fuel concentration for nucleation and stable generation of oxygen microbubbles in catalytic micropumps according to the reaction: $2H_2O_2 \rightarrow H_2O + 2O_2$. Catalytic microtubes played the role of tubular microcavity for gas collection, bubble nucleation, fluid pumping and ejection of oxygen microbubbles during the decomposition of hydrogen peroxide fuel into oxygen and water. This finding helped to reduce concentration of hydrogen peroxide fuel for microbubble-induced pumping of fluid to 0.009 vol % [92]. In comparison to shorter microtubes (diameter 5–10 μm, length 20–30 μm) by using longer microtubes (diameter 5–10 μm, length 100–1000 μm), it is possible to reduce the threshold hydrogen peroxide concentration for generation of microbubbles. Gao et al. reported efficient microengines fabricated using template electrosynthesis of polyaniline/platinum microtubes [93]. Microengines were 1–2 μm in diameter and 8 μm long self-propelled at 350 BL·s^{-1} in 0.2% of H_2O_2. This finding supports the statement that the aspect ratio of tubular micro-cavity is highly important for bubbles generation.

To better understand efficiency of microbubble-driven catalytic micropumps, we begin by simply estimating the time required for molecular diffusion over realistic distances intrinsic to micropumps. Catalytic micropumps operate using a chemical microreactor, where molecules react and reach opposite walls of the pump much faster than in larger pumps (mm–cm- scale). Typical mixing time in a micropump is a second or less, where gas/liquid can be supersaturated in a short period of time. By diffusion time, we can calculate how fast molecules can cross the diameter of pump according to the following equation, $t_{mix} = D^2/d$, where D—is the diameter of the pump, d—is the diffusion coefficient (for example, if $d = 2.1 \times 10^{-9}$ m^2/s for O_2 in water and $D = 10$ μm, $t = 50$ ms).

Other ideas to increase efficiency consist of new chemical fuels and higher catalyst turnover rates. For instance, Gao et al. experimentally proved that iridium-based Janus NMs could self-propel in ultralow levels of fuels [94]. Esplandiu and co-workers discussed fascinating new control parameters of electrochemical motors and pumps, where the surface potential of self-electrophoretic motors/pumps does not set only the electrochemical double layer, but the strength/direction of proton diffusion flux from the anode to cathode [95]. Increasing surface area is another effective strategy to reduce loading of expensive catalysts and increase efficiency of motors. Pumera's group used a high surface area iridium-based graphene motors at a low catalyst loading (0.54 at%) [96]. Mei's group demonstrated efficiency increase using a higher surface area by nanoparticle-decorated tubular microengines using atomic layer deposition, leading to ultrafast speeds up to 3200 μm·s^{-1} [97]. Mei's group designed a nanoporous reactor/microengine using nanoporous template to improve accessibility of reactants through the reactor walls and larger surface area [98]. Manjare et al. found that the hydrophobic surface of NMs is important to accelerate Janus particles due to possible depletion of water layer and interaction between generated reaction product (oxygen) and hydrophobic surface, leading to slip boundary condition and enhanced reaction rates [99].

Efficiency of electrophoretic nanowire-based nanomotors can be also increased. It is known that pumping of fluids in micro-/nanochannels is usually done not by using external pressure-driven flow, but by electrokinetic flows. For example, Rogers, Adams and Pennathur calculated difference in pressure driven flow necessary to drive water in a nanotube (100 nm in diameter) versus a macrotube

(1 m in diameter) at the same flow speed of 1 m·s^{-1} [24]. The reported difference in pressure per length was 14 orders of magnitude higher for a nanotube (2 × 10^{12} Pa/m for nanotube, 0.02 Pa/m macrotube). On the other hand, a very small electric field is needed to induce the flow of fluid through the nanotube (12 fA current for 100 nm diameter tube, accepting the double layer ~3 nm, the zeta potential ζ = −100 mV, the relative permittivity of water ε_r = 78.3) [24]. This can explain why an increase of conductivity of anode/cathode lead to high speeds of propulsion of bimetallic nanowire-based nanomotors due to efficient pumping of fluids induced by self-electrophoresis.

2.5. New Types of Motion and Trajectories

One very important research direction that needs further exploration is discovering and devising new types of motion by a comprehensive understanding of existing periodic/predictable trajectories and searching for aperiodic/unexpected dynamics for construction of more complicated micromachines. It is known that most NMs move in deterministic straight, helical and rotary trajectories [100,101]. Autonomous trajectories of micromotors can be controlled by the balance of motive–drag forces, shapes and geometrical asymmetry [102,103], effect of catalyst distribution [104], gravitaxis and separation phenomena for mass-anisotropic self-propelling colloids [105]. Interesting observations of periodic oscillatory motion driven by decomposition of H$_2$O$_2$ using catalase was reported [106]. Several groups observed that catalytic nanorods could spontaneously turn and tumble, which is similar to swimming of bacteria. Schmidt and co-workers showed unidirectional-overloaded transitions in microjet engines [107]. These observations and systematic designs can be the first steps towards efficient designs of more complex autonomous motion.

2.6. New Applications

As previously mentioned, NMs can be useful for biological and fluidic applications. Since NMs are coupled to their environment, their motion is influenced by Brownian diffusion, fuels and other processes. Widely used active transport mechanisms can be derived from biological systems albeit with some thoughtful consideration. For instance, Rogers, Adams and Pennathur proposed the idea that if a neuron cell is a meter long it can take thousands of years to deliver proteins from one end of neuron to the other [24]. Control of stepwise motion is another interesting prospect worthy of consideration. Molecules that move step by step while undergoing random collisions can be characterized by a mean free path. During the last few years, a possibility to construct nanoconfined Angstrom-size motors was considered, leading to unprecedented ways of understanding and controlling single molecules and ions [108,109]. It is perhaps most exciting to use NMs for multiple environmental cleaning tasks, where other types of human technology are inaccessible [110]. NMs were recently demonstrated for environmental remediation, pollutant removal and water cleaning application [111–116]. Since MOF motors often have a very high surface area, these new roles represent exciting prospects for water purification as shown by Wang et al. (Figure 3d) [43]. Moreover, charged molecules can be adsorbed, transported and separated by polymer motors [116,117]. Furthermore, it is known that micro-/nanobubbles can also be used for the degradation of organic molecules [118]. In this case, kinesin molecular motors can be used to transport proteins in about a week. Other applications include the use of assisted NMs to repair cracks [119], optical nanoscopy [120], pH sensing in motion [121], chemical sensing by quantum dots [122], threat detection [123] and gas sensing applications [124]. On the downside, another question that needs to be addressed is whether micro-/nanomotors themselves represent a new type of contamination or an undesirable byproduct after cleaning a river or lake, for example.

3. Motion Control and Externally Powered Micromotors

External control of autonomous nano-/micro-motors and engines (i.e., sphere, rod, tube and other shapes) can help to achieve fascinating applications such as delivery of micro-cargo, micromachine-enabled assembly of objects in desired configuration and on-chip integration. Followed by our discussion of advanced materials developed for NMs in the Section 2, here we highlight basic methods of NMs' external control. The first idea includes integration of an additional magnetic layer into NMs. For example, tubular microengines made of rolled-up nanomembranes can contain an additional ferromagnetic layer, leading to a controllable straight, helical, rotary movements of microengines. The second idea considers control of interface between the particle and fluid/fuel. For example, electrochemical modulation of NMs immersed in electrochemical cells can change speeds of NMs due to modification of oxidation and reduction processes, which are taking place on the anode and cathode of bimetallic nanorods. Similar, versatile wireless methods of motion control using light as a fuel source can be integrated, where semiconductor bandgap engineering, semiconductor-metal junctions, better control of electrons, holes, protons, electrons and reaction products are of paramount importance. External light control of motion enables new applications such as cleaning of water in rivers and lakes. In this case, NMs' knowledge can be often adopted from the well-known industrial methods related to research and discovery of new chemically relevant catalytic materials, processes and applications. Another exciting prospect of NMs is their small size that is particularly attractive for biomedical applications. Externally-powered and stimuli-responsive NMs have been shown, such as NMs driven by ultrasound waves. Multiple achievements of NMs' control are discussed below in more detail, including magnetic field, light, acoustic field and alternating current.

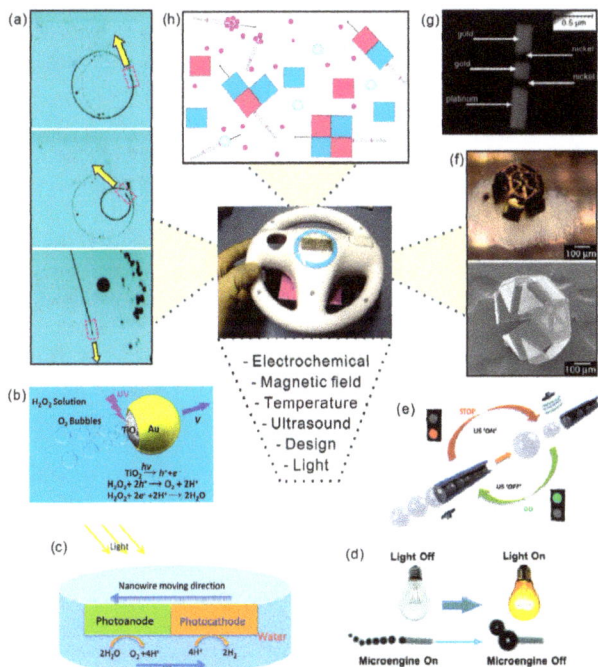

Figure 4. Externally powered "fuel-free" NMs and externally controlled motion of NMs driven by chemical fuels. (**a**) optical microscopy sequences of magnetic control of individual tubular Ti/Fe/Pt microjet in circular and straight motion; (**b**) Janus TiO_2/Au NM controlled by light. Reproduced with permission from [125], copyright (2016) Royal Society of Chemistry. (**c**) NM driven by photo-electrochemical reaction. Reproduced from [126]; (**d**) microengine switched "on" and "off" by white light by local degradation of hydrogen peroxide fuel above Pt-patterned silicon surface. Reproduced with permission from [127], copyright (2011) Jon Wiley and Sons; (**e**) ultrasound driven NMs.

Reproduced with permission from [128], copyright (2014) American Chemical Society; (**f**) magnetic stimuli-responsive microgrippers, design for sampling, analysis of tissue, biomedical minimally-invasive surgery and related operations. Reproduced from [22]; (**g**) the first example of bimetallic nanorods with integrated magnetic segments for external magnetic control of nanomotors. Reproduced with permission from [129], copyright (2005) Jon Wiley and Sons; (**h**) envisioning of a "microfactory", where externally controlled NMs are used for delivery and assembly of objects. This concept can be realized at the microscale using tubular microjet engines or similar micromotors with high motive power.

Dynamic nano-/microstructures, which can be controlled using external fields, already have real applications in separating and biosensing of molecules. For instance, "dynabeads" became a revolutionary technology for liquid separation of bio-materials using superparamagnetic polymer particles with bio-active surfaces to couple various cells and molecules. Therapeutic protocols and diagnostic assays include immunoprecipitation, cell isolation, cell activation and expansion, nucleic acid isolation, mRNA isolation, protein isolation and peptide purification, Streptavidin-coupling and in-vitro diagnostics (IVD) assay development [130]. Another example is biosensing, oscillating DNA helix bonded on surface of electrode enable measurements of affinity and kinetics of DNA-binding proteins, which is not possible to analyze by any other method [131]. Future and important applications of externally controlled NMs can be a "nano-micro-factories", where NMs can assemble building blocks from the bottom-up. Another envisioned application is chemotactic delivery of drugs.

3.1. Magnetically Powered/Controlled Motion

Artificial nano-/micro-motors can include an additional segment or layer for (i) external control of catalytic NMs or (ii) full power of NMs using magnetic, electric, acoustic fields, light source and related phoretic, osmotic, electrical, chemical and heat gradients [132–136]. External control helps to perform a task or series of tasks, such as transport, delivery and assembly of micro-cargo payloads and biosensing in motion (Figure 4a,f,g). Groups of Sen and Mallouk were among the first who demonstrated external magnetic control of catalytic nanorods with magnetic segments [129]. Later, Solovev et al. found that an incorporated ferromagnetic layer in rolled-up microtubes can be used to control the motion of microjet engines with magnetic fields [137]. Subsequently, magnetic control of microjets was improved [138–140] as well as other NMs including nanowire motors [141], Janus particles [142,143], paperbots [144], liquibots [145], freestyle nanoswimmer [146], fish-like nanoswimmers [147], liquid metal motors [148], flexible and linked superparamagnetic colloidal chains [149], chemo-magnetic structures [150], magneto-electric structures [151], magneto-acoustic structures [152] and chiral nanomagnets [153]. Fuel free NMs have advantages in biomedical applications, since there are no reaction products produced during the navigation [154]. Externally powered NMs are particularly attractive for delivery and assembly of microcargo payloads [155,156], as well as integration in lab-on-a-Chip devices [157,158]. Similar in scope to chemically-functionalized dynabeads, magnetic NMs can be used for cleaning operations and pollutant degradation [159]. Figure 4f shows self-assembling micro-grippers, which can be used to perform engineering functions on the micrometer scale, such as sampling, analysis of tissue, biomedical minimally-invasive surgery and related operations.

3.2. Light Powered Motion

Using light source as a fuel to power and control the motion of NMs is a very attractive due to the unique and already existing interactions light has with nano-/micromaterials [160]. However, many challenges and considerations remain such as how to choose and optimize elemental composition of nanomaterials, maximize light harvesting, optimize generation, separate and transport of electrons, holes and protons, match valence band, conduction band, energy band gap, Fermi level of photo-active cathode and anode (if two semiconductors are used), and match nanomaterial-molecule oxidation–reduction potentials, optimize temperature, light intensity and wavelength. Band gap engineering is needed to optimize the performance of light-powered NMs. Light absorption must be maximized while providing sufficient energy to facilitate reduction and oxidation. The thermodynamic potentials for molecules reduction products should fall within the band gap of the metal oxide in order for photo-reduction to occur (Figure 4b–d). Figure 4c shows a single nanomotor powered by photo-electrochemical reactions. Liu discussed an example of junctioned photochemical solar cells, where n-typed hetero-junctioned photoanode (TiO_2) and photocathode (Pt) can be used [126]. Figure 4d demonstrated light-control over the propulsion of microbubble-driven Ti/Cr/Pt catalytic microengines, where hydrogen peroxide fuel is degraded under local illumination of Si/Pt substrate [127]. Light powered/controlled motion can be related to several categories by using (i) semiconductor photocatalytic [161–167]; (ii) metallic plasmonic [168,169]; (iii) phototactic [170,171]; (iv) hybrid semiconductor-metallic swimmers [172–177]; (v) thermophoresis [178–180] and (vi) thermocapillary effects [181]. Other light controlled motors include: dual-light controlled [182] and a spectrally tunable light-driven silicon nanowires [183]. Zheng et al. showed a photo-electrochemically driven nanotree microswimmers with dyes, which can be coded with a distinct spectral response [184], structured light-enabled photoresponsive microstructures [185], chiral colloidal molecules [186], photochemically induced motion of liquid metal marbles [187], self-electrophoretic bimetallic nanomotors and micropumps in halogen media [188]. One of the most interesting applications includes light-driven micro- and nanomotors for environmental remediation of polluted waters [189–191] and transport of cargo [192].

3.3. Acoustically Powered Motion

Motion powered by acoustic waves became very popular because chemical fuels are not required and, thus, the system is fully biocompatible. Recent reports include artificial acoustically activated flagella [193], acousto-magnetic swimmers [194], metal nanoparticles for acoustic manipulation [195], nanorods trapped in an acoustic field [196], acoustic bubbles for microengines [197,198], acoustic microcannons [199], and nanoshells [200]. Figure 4e shows that ultrasound stimuli can be used to control the movement of bubble-propelled chemically powered PEDOT/Ni/Pt microengines. Applications include intracellular siRNA delivery [201], tissue welding [202] and holograms for acoustics [203]. Chemical and acoustic propulsion of bimetallic micromotors, moving up to speeds of $200 \ \mu m \cdot s^{-1}$, can be realized by ultrasonic standing waves at MHz frequency [204,205].

3.4. AC Field Powered Motion

One of the biggest advantages of alternating current (AC) powered motion consists of the possibility to modulate flow of electrons, pumping of fluids and control of speeds of motors/pumps without chemical fuels. Moreover, the action of charge separation using external fields is interesting for catalytic nano-/microparticles, where, for example, a catalytic powder can be suspended in aqueous solution for reaction, without physical connections to the surface of electrode. Several groups showed motion of conventional semiconductor diodes and pumps in external AC fields [206], electrochemical rotors [207], nanowire diodes [208], trajectory influenced by AC electrokinetics [209], characterized motion by particle–electrode impact voltammetry [210], and motion of liquid metal Al–Ga–In motors moving at high speed up to $43 \ cm \cdot s^{-1}$ under 20 V voltage [211].

3.5. Other Types of Motion Control

Other effects were observed, for example, motion can be influenced by thermal modulation [212–214], solutal and thermal buoyancy effects [215] and photochromic control of bubble-propelled motors by a spiropyran switch [216]. Huang et al. used grating-structured walls for guiding empennages, which improved linear motion of microengines [217]. Bimetallic motors can be accelerated in channels [218,219] and directed in teardrop-shaped posts [220], illustrating their integration into future Lab-on-a-Chip applications.

4. Interactive Micromotors: Swarming, Collective and Adaptive Behaviors

Emergent swarming, collective behaviors and Dynamic Self-Assembly (DySA) have been reported for NMs powered by chemical reactions and external fields [221,222]. Similar to biological motor proteins and cells, synthetic catalytic NMs operate by reaction, diffusion and motive forces that can move collectively. Communication by signals, spatiotemporal assembly, chemotactic response, motion either towards or away from chemical gradients, development of patterns and shapes are widely used in biological systems [223]. It is of fundamental importance to have a better understanding of dynamic forces and interactions in natural systems, which can be important for designing more complex NMs [224–226]. Short and long range static forces are influenced by (i) the direct motive power of NMs and (ii) released, i.e., "secreted", ions and molecules in solution, which can establish a long-term "communication" and change behavior of passive and active particles located nearby. Static forces and interactions between atoms, molecules, and micro-/nanoparticles are already relatively well understood (e.g., ionic, electrostatic, covalent, hydrophobic, Van der Waals, magnetic, capillary) [227]. It can explain, for example, why larger and "stickier" molecules have higher boiling points due to larger fluctuating dipoles in molecules. The motive power of synthetic NMs includes transport driven by (1) externally powered/controlled NMs; and (2) autonomous systems driven by diffusion and/or induced forces: self-electrophoresis, self-diffusiophoresis, gas recoil and surface tension driven.

4.1. Dynamic Self-Assembly Induced by External Fields and Chemical Reactions

External fields can be used to induce assembly of NMs, such as magnetic, acoustic, alternating electric fields, light (Figure 5a–i). At the same time, NMs can be passive or powered by chemical fuels. Guan's group reported light switchable colloidal TiO_2/Pt particles by modified electrostatic interactions [228,229]. Similar, light-induced clustering behavior has been reported by Singh et al. using SiO_2/TiO_2 [230], Hong et al. using TiO_2 [231], and Duan et al. using AgCl particles [232]. The TiO_2 NMs with a wide bandgap is effective material for water splitting using ultraviolet (UV) photons. Further studies performed by Zhou et al. demonstrated modification effect of Zeta potential due to pH and hydroxyl groups (OH), which influences aggregation of TiO_2 submicron particles [233]. Upon UV irradiation thermal energy of NMs is reduced from 118.2 kT to 33.6 kT for rutile and from 333.5 kT to 46.1 for anatase, respectively. Spiropyran functionalized SiO_2-Pt Janus particles in hydrogen peroxide and *N,N*-dimethylformamide (DMF) fuel mixture were demonstrated by Zhang et al. [234]. An assembly and transition into multiple motors was observed by electrostatic attraction and π–π stacking between molecules is induced by UV light irradiation (λ = 365 nm), disassembly in monomotors is realized by green light (λ = 520 nm). Wang's group showed Janus NMs with hydrophobic hemispheres [235]. Gibbs' group demonstrated contactless long range attractive, short range repulsive and mutual aligned interactions between swimming microparticles [236]. Dynamics of two interacting active Janus particles influence of hydrodynamic interaction due to the propagation of the ionic concentrations and flows influenced by electric force are also reviewed by Bayati's group demonstrating this principle [237]. Au-Pt catalytic pumps showed repulsion and attraction of silica colloids by the local change of the proton concentration and modification the colloid zeta potential and the electric force [238]. Other external fields used to induce assembly of NMs include Marangoni flow [239] magnetic [240,241], and acoustic fields [242].

For synthetic particles, "quorum sensing" micro- and nanoparticles can release ions, which can build an electric field around the particles due to different diffusion coefficients of anions and cations. Collective swarming behaviors and non-biological chemotaxis of catalytic nano-/microparticles and molecules are interesting fundamental phenomena. Groups of Sen and Mallouk showed biomimetic behavior of particles, such as predator–prey relationship, non-biological chemo- and photo-taxis [243,244]. It is known that biological chemotaxis depends on a temporal rather than on a spatial mechanism. For instance, immotile bacterial cells are too small to measure the difference in the chemical gradient. Bacteria solved this problem by swimming in short runs and sensing of changes in concentrations of attractants or repellents in time rather than in space [245]. When cells detect concentration gradient, they swim in that direction. Similar, synthetic nanomotors can elongate trajectories of their motion in higher concentrations of fuels and align their motion, according to ionic and molecular gradients. It is also known that, in nature, for instance, in stigmergy, intelligent structures can emerge without the direct awareness, communication, intelligence and memory, but due to traces left in the environment that stimulate next actions.

Altemose et al. reported spatiotemporal oscillatory behavior of silver orthophosphate particles under UV illumination in hydrogen peroxide fuel, where an electrostatic self-diffusiophoretic mechanism was connected to alternating electric fields by the reduction and oxidation of silver [246]. It was also shown that NMs with different stimuli might be employed as logic gates. *NOR* Gate with UV and addition of chemical fuel as inputs and collective behaviors as outputs: schooling and exclusion behaviors as *1* and *0*, respectively. From universal *NOR*, other logical gates can be constructed and their combinations, in principle, can be implemented in any digital component [247,248]. Groups of Krishna and Sen recently discovered that active biomolecules, such as enzymes, can increase their diffusion when located in chemical fuels. Followed by Michaelis–Menten kinetics, during substrate turnover, the increased reaction rate leads to the enhanced diffusion of enzymes [249,250]. However, no single conclusion exists about the molecular mechanism observed for enzymes. Groups of Grzybowski and Granick considered networking chemical systems [251] and active colloids with collective mobility [252], suggesting that dynamic interactions can also depend on time, history and feedback control in spatio-temporal scales.

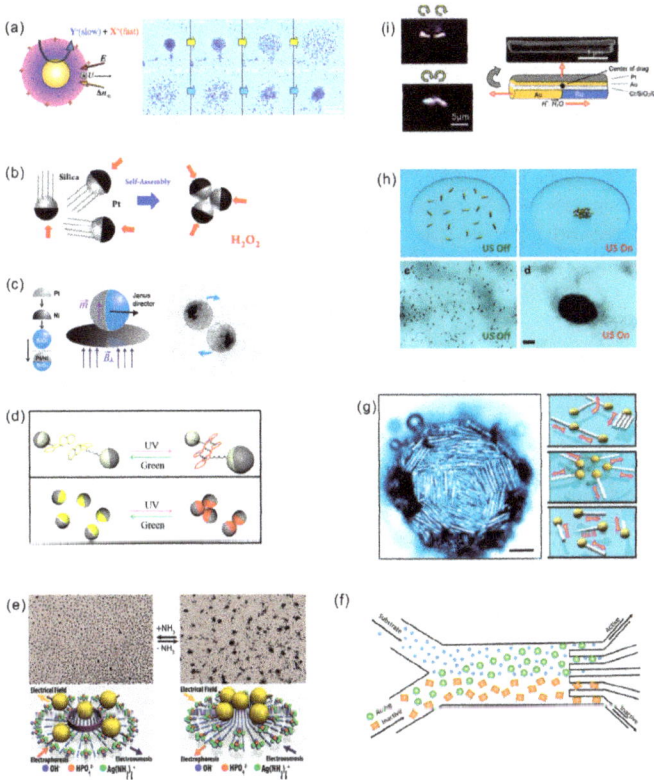

Figure 5. Dynamic self-assembly system based on externally triggered/controlled and autonomous micro-/nanomotors. (a) TiO$_2$ light driven reversible "microfireworks". Reproduced with permission

from [231], copyright (2010) Jon Wiley and Sons; (**b**) assembly of Janus motors with hydrophobic hemispheres. Reproduced with permission from [235], copyright (2013) American Chemical Society; (**c**) engineered contactless particle-particle interactions. Reproduced with permission from [236], copyright (2017) Jon Wiley and Sons; (**d**) light-induced assembly of spiropyran decorated SiO_2–Pt Janus particles. Reproduced with permission from [234], copyright (2015) American Chemical Society; (**e**) collective behaviors and response to different stimuli. Reproduced with permission from [247], copyright (2013) American Chemical Society; (**f**) a non-biological chemotaxis phenomena observed for catalytic enzymes in microfluidic channel. Reproduced with permission from [253], copyright (2014) American Chemical Society; (**g**) assembly and interactions between catalytic microtubes in chemical fuels of hydrogen peroxide. Reproduced with permission from [254], copyright (2013) Royal Society of Chemistry; (**h**) reversible swarms self-assembled under acoustic field. Reproduced with permission from [242], copyright (2015) American Chemical Society; (**i**) dynamic interactions between fast chemically-powered nanorotors. Reproduced with permission from [255], copyright (2009) American Chemical Society.

4.2. Dynamic Self-Assembly and Adaptive Systems

Dynamic self-assembly systems can be designed based on competing interactions between NMs. In this case, it is possible to estimate characteristic length scales at which collective and swarming interactions can occur. It was observed that catalytic microjet engines at the air–liquid interface of mixed fuels (hydrogen peroxide, propylene carbonate and water) can self-assemble by long lateral capillary forces [256]. Similar behavior was found with biological water striders, which self-organize in colonies and employ water menisci/capillary force to land on solid edges. If motive forces of microjets compete against lateral capillary forces $F_{motive} = F_{capillary}$, one can estimate a characteristic length at which swarming or collective behavior can occur [256]. Lateral capillary force can be simplified to the following equation $F_c \sim \gamma R^2/L$, where γ is the surface tension, R is the radius of generated bubbles, which shape meniscus and L is the center-to-center distance between particles. Microjet motive force can be described using Stokes law based on recoiling bubbles, $F_m = 6\pi\mu R\vartheta$, where μ is the fluid viscosity and ϑ is the speed of recoiling bubble. Characteristic length scale at which swarming behavior occurs can be determined, $L = \gamma R/6\pi\mu\vartheta$. The capillary force can provide long-range interaction up in mm–cm length scale, depending on parameters. It was observed that swarms of microengines decay when concentration of fuel leads to a decrease of bubble radius, thus decreasing the capillary attraction between the bubble-propelled microengines [256]. Similarly, Grzybowski and co-workers described the characteristic distance between the UV-light electrified spheres at the air–liquid interface [257]. In this case, particles attract each other by capillary forces and repel each other by the repulsive electrostatic forces, $F_q \sim kQ^2/L^2$, where k is the Coulomb constant and Q is the charge of each particle. A balance with capillary force leads to the following characteristic length, $L = kQ^2/R^2\gamma$, when the system can self-organize.

Prominent research challenges consist in better understanding of coupling between chemistry, mechanics, diffusion and active transport using collective micro-/nanomotors in biological and synthetic systems. In his seminal paper in 1952, Alan Turing theoretically predicted patterning by forming concentration gradients in biological systems, where reaction–diffusion driven substances called morphogens (e.g., proteins, small molecules) are non-uniformly distributed in space [258]. The generic mechanism is local activation and long-range inhibition or the diffusion coefficient of the activator (produces itself) must be much slower than the inhibitor. This process was found of paramount importance in development of shapes, forms and skin patterning of animals. Different patterns can be produced by modelling parameters, such as rate of reaction or diffusion. In other words, diffusion can be used to transport chemical signals and set the length scale of patterns. Howard, Grill and Bois suggested that the same holds true for biomolecular motors, which generate forces and the next Turing step could be the mechanochemical basis for morphogenesis [259]. In active transports, the Peclet number is important for both man-made micro-/nanomotors and biological motor proteins. The Peclet number measures ratio of advective to diffusive transport, $Pe = \vartheta l/D$, where l is the travelled distance and D is the coefficient of diffusion. The condition required to overcome diffusion is $l \geq D/\vartheta$. Usually, if Peclet number is smaller than 1, the diffusion dominates and, if Peclet number is larger than 1, the advective transport dominates. The transport driven only by diffusion leads to the following displacement, $d = \sqrt{Dt}$. If patterning of biological species is driven by reaction and diffusion, the characteristic scale of patterns can be determined, $\tau_{RD} = \sqrt{D/k}$, where k is the rate constant of degradation. However, in many cases, it is not practical to transport by diffusion over longer distances. Howard et al. indicated that chemical signals based only on RD process are often too slow and biology utilizes two strategies: (i) application of motor proteins capable to overcome diffusion and (ii) movement due to mechanical stress (forces exerted by proteins), which can travel 10^6 times faster than diffusion, i.e., at the speed of sound (1 m/s) [259]. In the case of an advective-diffusive transport by motor proteins, $\tau_{AD} = D/\vartheta$, where ϑ is the speed of motor protein or advective transport. If an advection-reaction only takes place, $\tau_{AR} = \vartheta/k$, where k is the rate constant of degradation. By taking into account viscosity and friction, $\tau_{VF} = \sqrt{\mu/\epsilon}$, where ϵ is the friction coefficient [259]. Figure 6a shows examples of well-known interactions, ranges and nano-/microparticles. Remarkably, long and short-range interactions are vital for dynamic self-assembly of both biological and synthetic nanomaterials.

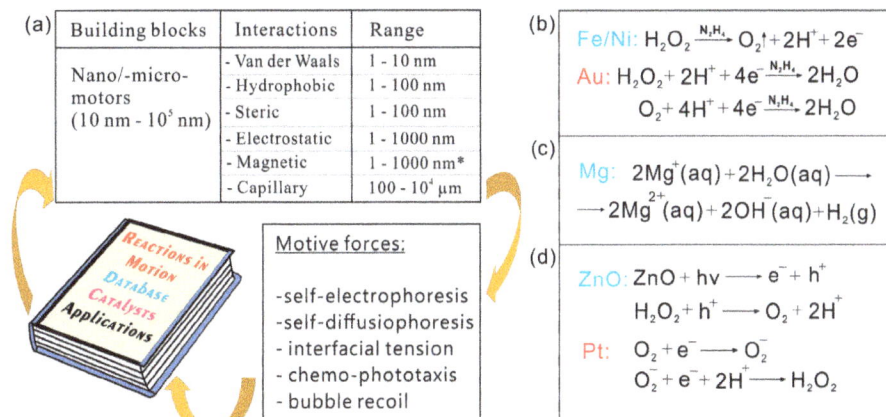

Figure 6. Static and dynamic interactions acting among synthetic nano-/micromotors. (**a**) well-known static interactions between particles (e.g., Van der Waals, hydrophobic, electrostatic, magnetic capillary), which depend on distances between particles, interaction type (* interaction range can vary by changing the size of particles) [251]. Interactions can be perturbed by adding motive force (self-elecrophoresis, self-diffusiophoresis, bubble recoil) to NMs. Databases of known reactions are particularly helpful when working with NMs; one can see a full list of potentially interesting and technologically relevant reactions, catalysts and applications; (**b–d**) examples of multi-electrons, protons and photons reactions, which power FeNi-Au, Mg and ZnO-Pt micro-/nanomotors. Released ions and molecules can lead to particle–particle interactions by chemo-and photo-taxis. Combination of individual NMs' reactions can lead to higher levels of applications, such as fuel cells.

Chemical databases are already full of innovative and transformative ideas, catalytic materials, tailored by compounds, elements, reactions and utilizations, which can be coupled to motion and assembly of NMs (Figure 6a). Better understanding of flow of electrons, protons, photons and ions during (photo-) catalytic reactions becomes a vital issue for future research directions. Many potentially useful reactions can be considered for design of "indirect" interactions and assemblies of different NMs'. For example, different nanomotors, if combined together, can achieve a higher level of functionality/utility. For example, Li et al. demonstrated application of reaction of hydrogen peroxide decomposition, where reaction products are water and oxygen, Figure 6b [68]. Gao et al. experimentally showed seawater magnesium driven NMs, where hydrogen gas was produced (Figure 6c) [41]. Wang's group showed symbiosis of two different NMs' types, production of oxygen and hydrogen fuels by NMs in different chambers for generation of electromotive force in a conventional fuel cells [260]. Dong et al. demonstrated a versatile control of speeds of micromotors using light source, where UV light is used to generate additional electron-hole pairs in ZnO materials for a subsequent enhanced reaction and increased speed (Figure 6) [167]. We believe that NMs can be potentially used for conversion of global warming CO_2 molecule into solar fuels and related useful by-products, such as methanol and plastics. For example, the following photo-catalytic reactions have been already demonstrated using WO_3 photoanode and layered double hydroxide (LDH) photocathode, respectively, in the "reversed" fuel cell configuration: $2H_2O \rightarrow O_2 + 4H^+ + 4e^-$ (WO_3 part), $CO_2 + 6H^+ + 6e^- \rightarrow CH_3OH + H_2O$ (LDH part), to achieve an overall reaction of methanol production using energy of light $CO_2 + 3H_2 \rightarrow CH_3OH + H_2O$ [261].

5. Towards Biomedical and Fluidic Micromachines

In recent years, much work on biomedical applications of micro-/nanomotors with high potential for prototype applications has been accomplished: (i) roving biosensors; (ii) drugs, cells delivery; (iii) isolation of pathogens, cancer cells by chemotactic microbots and (iv) cleaning of clogged arteries using microbots [262,263]. Magnetically actuated micro-/nanotools operate fuel-free, such as stimuli-responsive miniature grippers and [22] assisted fertilization by sperm-carrying externally powered microstructure [264]. For NMs powered by chemical fuels in living organisms, a crucial limitation exists: hydrogen peroxide is a highly cytotoxic fuel found in only very small quantities in the body. For this reason, novel biocompatible fuels are desperately needed. Furthermore, the reduction of threshold concentration of fuels does not solve the problem: ideally, the fuel must be a part of cellular metabolic pathways. There is already a well-established field of the U.S. Food and Drug Administration (FDA) approved biomedical drugs, micro-/nano-drops and capsules, which rely on flowing blood streams, passive diffusion, targeted delivery of drugs, nanoparticles and specific surface chemistry for interactions with cells. Weitz's group demonstrated that microfluidics can further strengthen fabrication of biomedical NMs, since the technology has enabled generation of microdrops at kHz rate, while screening for thousands of potentially useful chemical reactions and new drugs using very small volumes of reagents [265]. Comprehensive results obtained by microfluidics help to mass-produce lipid bilayers, permeable polymers and even nanoparticle-shelled capsules.

5.1. Biocompatible Fuels

A number of papers have recently been published on the topic of the motion of nanomotors both in vivo and ex vivo, which can be achieved by exploiting the energy from enzymatic catalysis [266–269]. Since enzymes normally function in a wide range of environments, both intra- and extracellular nanomotors can possibly be tailored to suit specific mediums with high biocompatibility. Propulsion of nanomotors has been demonstrated with enzymes such as urease and catalase with substrates urea and H_2O_2, respectively [270,271]. However, since the conversion of chemical energy into mechanical work is ubiquitous in the organic molecular world, there is a myriad of possibilities in novel nanobiology-related forward movement. Considering that enzymes have a significant turnover rate, even when low concentrations of substrate are available, it makes H_2O_2-powered nanomotors feasible in vivo.

The size, shape and material of NMs are of special importance when the motors need to enter the body. For instance, biocompatible nanotubes can be built from an outer coated PEDOT and an inner coated of Pt/Zn [77], making them biodegradable and non-toxic [272]. Besides nanotubes, hollow mesoporous silica Janus nanomotors pose an alternative to nanomotor-design. With their spherical design, Janus nanomotors have high drug-loading capacity as well as great biocompatibility both in vitro and in vivo due to silica being biodegradable, making them an exciting alternative to nanotubes. While nanotubes differ from less than 10 nm to 30 μm in diameter and even more so in length, which gives them a variety of functions, Janus nanomotors are typically 80–500 nm in diameter, which suits cells targeting.

Biofunctionalization and activity of enzymes operate within highly specific conditions. Activity of enzymes depends on their structure: amino-acid polypeptide chains fold in certain ways to form α-helixes, β-sheets or random coil structures as a consequence of hydrophobic interactions between water and the amino acid's side-chains. Since an attachment of enzyme to a nano-/microstructure requires linker molecules, the protein's tertiary or quaternary structure might be disrupted to such a degree that it inhibits catalysis. The orientation of the attached enzyme is also of importance, as it must be ensured that the nanomotor does not block the catalytic site or in other ways impose steric hindrance. This requires spatial awareness of the protein and must be accounted for in synthesis [273]. Note that NMs are not limited to nanotubes and Janus particles. For instance, another interesting approach is based on reactions of polymerization. Inspired by the *listeria monocytogenes* that moves by actin-tail polymerization, Pavlick et al. used Grubb's catalyst to emulate polymerization-movement [274].

5.2. Hybrid Bio-Micromotors for Drug-Delivery

Delivery of cells, drugs and proteins by NMs opened new perspectives in application of bio-nanomotors [275–280]. Drug delivery in vivo requires target recognition, uptake, movement and eventual release of cargo [281]. Both nanotubes and Janus NMs are implemented in drug-delivery studies. In nanotubes, cargo-uptake can be moderated by the hybridization of adhesive molecules or antibodies to the surface of the tube where molecules can recognize adhesive targets or specific antigens respectively on surface-cell areas, in extracellular matrix or in serum, shown in Figure 7a,b. In Figure 7a, one of the first examples of synthetic catalytic nanoengine assisted transport of yeast cells in hydrogen peroxide fuel is shown. The loading of nanotubes with cargo can also be achieved by pumping a fluid into the mouth of tubes using catalytic reactions [282]. Janus NMs are synthesized through an approach based on microfluidics, and therefore loading of the particles with proteins can be realized during synthesis [283]. Janus NMs can furthermore be filled with nanoparticles e.g., gold nanoaggregates that plays a role in enhancing photothermic tumor therapy [284]. Transport of NMs can be achieved using chemotaxis as previously considered, pH-taxis or external magnetic guidance [285]. Antibody-coatings may also enhance nanomotor-targeting [286]. Realization of cargo delivery is typically done as a consequence of a change in the environment of NMs or using external stimuli. This can be accomplished through chemotaxis or pH-change, where NMs consist of pH-sensitive polymer coatings, which can be dissolved by controllable pH of solution. NMs can also alter the pH locally by using acid as a fuel. Especially in the stomach region, NMs have been considered promising prototypes for drugs delivery and therapeutics [287,288]. Another mechanism of cargo release is through near-infrared (NIR) irradiation, when Au nanoparticles collapse due to photothermic effect and the nanomotors release drugs [289]. This principle is demonstrated in Figure 7c, where Janus particles are assembled with Au and Pt nanoparticles incorporated into the membrane of NMs. Delivery of interior cargo, molecules or particles can be triggered by NIR. A third drug-delivery system exploits endogenous glutathione to break down stomatocyte nanomotors by reduction of disulfide bonds to supply a redox-responsive drug-delivery mechanism [290].

Template-free Janus-like vesicles with micromotor function can act as motors for drug delivery; in particular, these NMs can be generated using a microfluidic-based approach. Careful injection of a lipid or block copolymers (BCPs) and inner content, such as inorganic nanoparticles, into a water-based medium through separate channels promotes formation of vesicles: amphiphilic BCP and lipids are self-assembling into vesicle-structures through hydrophobic interactions with surrounding water as it is thermodynamically favorable to minimize the surface-area of amphiphilic substances in a polar solution (Figure 7c). Typical diameters of these vesicles are around 1–2 μm, which depend on amphiphilic materials and the flow rate. For reference, a typical cell has a diameter of 20 μm.

Asymmetrical integration of Pt nanoparticles into the membrane can make the vesicles propel forward by H_2O_2 decomposition. Encapsulation of intravesicular material can be done co-synthetically through another fluidic channel or post-synthetically by treating the vesicles in the desired intravesicular material [291]. Then, the assembled vesicles can be used for both drug delivery and biosensors.

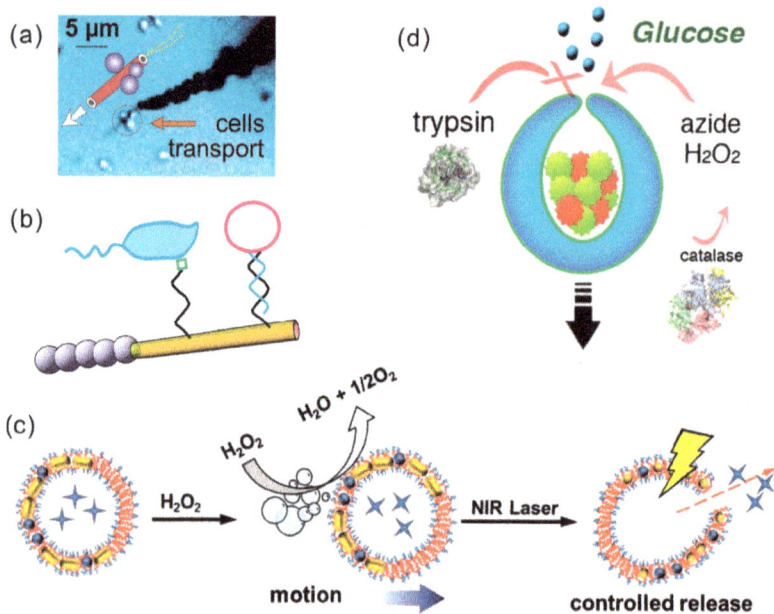

Figure 7. Recent review of bio-hybrid-fluidic NMs for biosensing, drugs delivery and assembly of new micromotors. (**a**) One of the first examples of catalytic nanojet engine transporting Yeast cells in hydrogen peroxide fuel. Reproduced with permission from [292], copyright (2016) American Chemical Society; (**b**) schematic image, biofunctionalized NMs capture and deliver cargo [281]; (**c**) novel architectures consisted of self-assembly of functional nanopraticles in the shell of lipid vesicles. Reproduced with permission from [291], copyright (2015) Jon Wiley and Sons; (**d**) schematic of stimuli-responsive glucose powered soft NM. Reproduced with permission from [263], copyright (2016) American Chemical Society.

5.3. Hybrid Bio-Motors for Biosensing

Micro- and nanomotors are already employed in the field of biosensors. The majority of biomolecules can be chemically-bonded to the surface of NMs, such as DNA-enabled environmental remediation [293] and small compounds like mercaptohexanol on a gold-surface to form self-assembled monolayers. Hybridization of various bioreceptors to the outer layer of tubular NMs forms a sensing interface on the surface of the motor that can detect biomolecules in situ with short assay time [294]. Information about chemical concentrations can be accessed via motor speed. An example is given from Wang's group: binding of toxins to the sensory unit of the nano- or micromotor can slow or inhibit catalase activity and thus reduce propulsion of the motor. NMs' speed is measured as a motion-based signal and chemical concentrations can be accordingly derived [295]. Detection of DNA and RNA can be achieved in a similar fashion. Duplex-formation with a thiolated DNA capture probe results in a binding of the duplex to the nanomotor-surface. When exposed to H_2O_2 the Ag-probes are dissolved, releasing Ag^+-ions in solution that increase the speed of nanotubes in peroxide solution (10% H_2O_2) and creates a motion-based DNA/RNA-detection mechanism [296]. In another example, nerve-agents were detected using similar methods [297,298]. Since all of these methods are ex vivo, diagnostics using a lab-on-a-chip devices and micro/nanomotors have great potential to achieve a rapid detection, cheap production and even naked-eye observation. Wang's group developed such a system, where cortisol, an important stress and clinical biomarker, was detected "on the fly" down to 0.1 µg·mL^{-1} using NMs and horseradish peroxidase [299]. There are still challenges that need to be overcome when applying micro/nanomotors in drug delivery such as poor tissue penetration, effective targeting with little to no off-target effects, and movement against blood flow and full biocompatibility. In biosensing, challenges include specificity and effective assay-application [300]. However, rapid advancements of the nano/microscale motors are expected to have a tremendous impact in the biomedical field.

5.4. Towards an Assembly of "Synthetic Cells" Using Microfluidics

Microfluidics can enable important breakthroughs in fabrication of programmable, evolvable, soft and fluidic micromachines. Since virtually any liquids, biocompatible, soft materials and drugs can be encapsulated in fluidic micromachines, microfluidic techniques hold great potential towards applications in medicine, the environment and on-chip technologies. Future challenges revolve around how to design more complex micromachines and "synthetic cells" from scratch. In 1971, Ganti attempted to characterize the fundamentals of life by conceiving a minimal system able to replicate. According to Ganti, the minimal requirements of a living system are (i) a metabolic chemical network that supplies energy to the collective system; (ii) template polymerization and replication, typically assessed through RNA-templates, and (iii) a lipid membrane enclosing the system. These three subsystems are coupled stoichiometrically, meaning that growth of one part leads to necessary, stoichiometric growth of the other parts though an interlinked, autocatalytic, cyclic system. In the latter case, if a chemical system contains three subsystems, it will be capable of growth and division and, thus, it can fulfill the criteria of life as we know it and observe biological systems [39]. Numerous systems can be designed to meet these criteria, but, in nature, this is observed as classical cell division. Through metabolism pathways, a cell can synthesize nucleotides de novo. High concentrations of nucleotides can contribute to ensuring successful DNA replication. The existence of a duplicate DNA-strand can trigger the expansion of the cell membrane through fatty acid synthesis and ultimately a contractile ring can divide the collective cell-system into two parts [301]. In this scenario, a chemical micromachine can represent a networking metabolic map that can be viewed as thousands of cog-wheels ordered in chemical rather than in geometrical fields. Chemical cycles do not couple with mechanical teeth, but through the fact that the product of one autocatalytic cycle is the "fuel" of another cycle, such as in the case of Belousov–Zhabotinskii reaction. Note that, even though this description is simplified, it demonstrates the principle of the chemoton theory very well. We propose the idea that the chemoton concept can be realized through microfluidics, as input energy can be harvested to synthesize multiple membrane-enclosed systems. Weitz's group showed that microfluidic devices can be used to prepare reaction vessels for biology, stimuli-responsive capsules, liposomes and polymersomes for diagnostics, drug delivery, design of new materials, isolation of cells and biomolecules. For example, a membrane can be made of lipids or permeable polymers for small molecules to enter and leave the capsule, while keeping larger molecules inside [302]. Figure 8 proposes a research plan, where advanced digital microfluidics can be used to generate multifunctional soft micromachines at a kHz rate.

Rogers, Adams and Pennathur discussed the traffic time, which provides capabilities for very precise and fast mixing of fluids and molecules in micro- drops and capsules [23]. If we estimate the time for two molecules inside the capsule, droplet or cell to meet, i.e., if molecules start off separated by the distance equal to the capsule's characteristic dimension, D

$$t_{traffic} = D^3/dr \qquad (2)$$

where d—sum of diffusion coefficients of both molecules, and r—the sum of the molecules radii. For example, a pair of enzymes, 12 and 10 nm in diameter, with diffusion coefficient 10×10^{-10} m^2/s located in 2 µm capsule can meet every 0.4 s. Both t_{mix} (Section 2.4) and $t_{traffic}$ are strongly dependent on a capsule characteristic dimension, scaling with D^2 and D^3 [21]—for example, for a capsule with diameter 1 µm $t_{traffic}$—is close to a second, for 10 µm—several minutes, for 100 µm—hours.

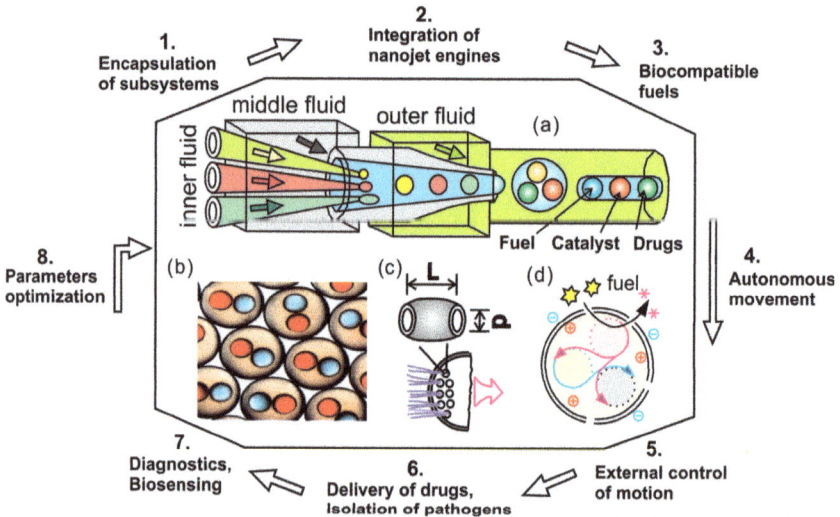

Figure 8. (**a**) A proposed plan to explore fabrication of programmable soft and fluidic NMs using microfluidics. Schematic shows glass multi-capillary device, where inner, middle and outer fluids are used to generate multicomponent drops, capsules for custom-designed NMs; (**b**) example of multicomponent drops, prepared using several syringe pumps. Reproduced with permission from [302], copyright (2012) Royal Society of Chemistry; (**c**) proposal of nanojet engines integration in multifunctional capsules to mimic nanojet powered cells; (**d**) schematic image: towards an assembly of "synthetic cells", containing coupled fluidic subsystems.

6. Conclusions

In summary, we highlighted recent significant results and achievements in the field of NMs, starting from historical overview of simple, complex machines and followed by recent rapid progress made in the field of nano-/micro- motors and engines. Multiple advantages exist for micromachines, which can be further explored and exploited for demanding applications. It includes advantages to overcome Brownian diffusion, outswim high viscosity of fluid at low Reynolds number, and construct better geometries, sizes, fuels for more energy efficient micro-motors and engines. Versatile control of NMs at the sub-microscale using chemical reactions, external fields as well as realization of delivery of objects on-chip and off-chip can be accomplished. Unlike large-scale motors, principles of micro-/nanomotors can go far beyond the classical mechanics analogies. Research progress in better understanding of reaction–diffusion processes, molecular systems, emergent behaviors on the level of individual/collective biomimetic particles and biological enzymes is unprecedented. So far, the only way to learn about NMs was to design, power and observe the motion of these bio-mimetic microparticles, and it remains unknown how many reactions must be tested and what we can really learn from the chemistry in motion. Outlining new horizons in applications, the next decade, in our opinion, is likely to see a considerable rise in application of drop-generating microfluidics to mass-produce fully biocompatible NMs and ultra-sensitive biosensors. A remarkable feature of small size NMs is a high potential for biomedical use, including minimally invasive surgery. In the future, we envision several research directions to advance the field of nano-/nanomachines: (i) better understanding of dynamics of NMs; (ii) understanding of scaling effects: reduction from mesoscale to molecular scale; (iii) application of drop-based microfluidics for fully biocompatible materials;

Micromachines **2018**, *9*, 75

(iv) design of reaction networks for design of more complex programmable and evolvable NMs. For example, if motion of NMs is usually recorded in straight, helical and circular trajectories, it can be related to well-understood periodic or predictable types of motion. Next, it is of high interest to explore transitional behaviors, overloaded and aperiodic motion. Biology is known to operate by nonlinear chemical dynamics, clocks, patterns, oscillations and chaos [303]. Until now, little progress has been given in design of more complex NMs. However, the whole picture of future research directions is still incomplete and NMs' research methods are often based on exploratory approaches, rather than on systematic designs—our research community has strong evidence that the era of micro-/nanomachinery industry has already started and evolved from its evolutionary stage to the revolutionary phase.

Acknowledgments: Alexander A. Solovev is grateful for the financial support from 1000 talent grant and Fudan University grant. The authors also acknowledge the financial support from the Natural Science Foundation of China (Grant No. 51475093). In addition, the authors thank Laura L.A. Adams for fruitful discussions.

Author Contributions: Alexander A. Solovev designed the structure, key concepts and wrote the paper. Gaoshan Huang and Yongfeng Mei conceived, wrote and revised the paper. Yan Zhang searched the literature and prepared figures. Huanpo Ning and Hong Zhu obtained permissions and copyrights of images, wrote and revised the paper. Andreas Ingham wrote and revised biological chapters.

Conflicts of Interest: The authors declare no conflict of interest.

References

1. Ismagilov, R.F.; Schwartz, A.; Bowden, N.; Whitesides, G.M. Autonomous movement and self-assembly. *Angew. Chem. Int. Ed.* **2002**, *41*, 652–654. [CrossRef]
2. Paxton, W.F.; Kistler, K.C.; Olmeda, C.C.; Sen, A.; Angelo, S.K.S.; Cao, Y.; Mallouk, T.E.; Lammert, P.E.; Crespi, V.H. Catalytic nanomotors: Autonomous movement of striped nanorods. *J. Am. Chem. Soc.* **2004**, *126*, 13424–13431. [CrossRef] [PubMed]
3. Ozin, G.A.; Manners, I.; Fournier-Bidoz, S.; Arsenault, A. Dream nanomachines. *Adv. Mater.* **2005**, *17*, 3011–3018. [CrossRef]
4. Xu, L.; Mou, F.; Gong, H.; Luo, M.; Guan, J. Light-driven micro/nanomotors: From fundamentals to applications. *Chem. Soc. Rev.* **2017**, *46*, 6905–6926. [CrossRef] [PubMed]
5. Li, J.; Rozen, I.; Wang, J. Rocket science at the nanoscale. *ACS Nano* **2016**, *10*, 5619–5634. [CrossRef] [PubMed]
6. Mei, Y.F.; Huang, G.S.; Solovev, A.A.; Ureña, E.B.; Mönch, I.; Ding, F.; Reindl, T.; Fu, R.K.; Chu, P.K.; Schmidt, O.G. Versatile approach for integrative and functionalized tubes by strain engineering of nanomembranes on polymers. *Adv. Mater.* **2008**, *20*, 4085–4090. [CrossRef]
7. Solovev, A.A.; Mei, Y.F.; Ureña, U.B.; Huang, G.S.; Schmidt, O.G. Catalytic microtubular jet engines self-propelled by accumulated gas bubbles. *Small* **2009**, *5*, 1688–1692. [CrossRef] [PubMed]
8. Wehner, M.; Truby, R.L.; Fitzgerald, D.J.; Mosadegh, B.; Whitesides, G.M.; Lewis, J.A.; Wood, R.J. An integrated design and fabrication strategy for entirely soft, autonomous robots. *Nat. Lett.* **2016**, *536*, 451–455. [CrossRef] [PubMed]
9. Robert, S. *Lecture: Chaos and Reductionism*; Stanford University: Stanford, CA, USA, 2018. Available online: www.robertsapolskyrocks.com (accessed on 10 January 2018).
10. Lorenz, E.N. Deterministic nonperiodic flow. *J. Atmos. Sci.* **1963**, *20*, 130–141. [CrossRef]
11. Bensaude-Vincent, B.; Guchet, X. One word for three different paradigms. *Tech. Res. Philos. Technol.* **2007**, *11*, 71–89.
12. Historical Motor Carriage (Year 1896). Available online: https://www.motorauthority.com/news/1110475_first-car-ever-caught-speeding-to-be-displayed-in-the-uk (accessed on 10 January 2018).
13. Mechanical Clock. Available online: https://www.pinterest.co.uk/pin/420664421428587728 (accessed on 10 January 2018).
14. Chaos Theory. Available online: http://universe-review.ca/R01-09-chaos.htm (accessed on 10 January 2018).
15. Lorenz System. Available online: https://en.wikipedia.org/wiki/Lorenz_system (accessed on 10 February 2018).
16. A Boy and His Atom: The World's Smallest Movie, IBM Company. Available online: http://www.research.ibm.com/articles/madewithatoms.shtml (accessed on 10 February 2018).

17. Reigh, S.Y.; Huang, M.J.; Schofield, J.; Kapral, R. Microscopic and continuum descriptions of Janus motor fluid flow fields. *Philos. Trans. A Math. Phys. Eng. Sci.* **2016**, *374*. [CrossRef] [PubMed]

18. Ozin, G.A.; Arsenault, A.; Cademartiri, L. *Nanochemistry: A Chemical Approach to Nanomaterials*, 1st ed.; Royal Society of Chemistry (RSC): London, UK, 2009; ISBN 9781847558954.

19. Cademartiri, L.; Ozin, G.A. *Concepts of Nanochemistry*, 1st ed.; Wiley-VCH Verlag GmbH: Weinheim, Germany, 2009; ISBN 978-3-527-32597-9.

20. Wu, Y.F.; Dong, R.F.; Zhang, Q.L.; Ren, B.Y. Dye-enhanced self-electrophoretic propulsion of light-driven TiO_2-Au Janus micromotors. *Nano-Micro Lett.* **2017**, *9*, 30. [CrossRef]

21. The Creative Commons Attribution 4.0 International License. Available online: http://creativecommons.org/licenses/by/4.0/ (accessed on 10 January 2018).

22. Leong, T.G.; Zarafshar, A.M.; Gracias, D.H. Three-dimensional fabrication at small size scales. *Small* **2010**, *6*, 792–806. [CrossRef] [PubMed]

23. Ghosh, A.; Fischer, P. Controlled propulsion of artificial magnetic nanostructured propellers. *Nano Lett.* **2009**, *9*, 2243–2245. [CrossRef] [PubMed]

24. Rogers, B.; Adams, J.; Pennathur, S. *Nanotechnology: Understanding Small Systems*, 2nd ed.; Taylor & Francis Inc.: Bosa Roca, FL, USA, 2011; ISBN-13 978-1439849200.

25. Alarcón-Correa, M.; Walker, D.; Qiu, T.; Fischer, P. Nanomotors. *Eur. Phys. J. Spec. Top.* **2016**, *225*, 2241–2254. [CrossRef]

26. Wong, F.; Dey, K.K.; Sen, A. Synthetic micro/nanomotors and pumps: Fabrication and applications. *Ann. Rev. Mater. Res.* **2016**, *46*, 407–432. [CrossRef]

27. Colberg, P.H.; Reigh, S.Y.; Robertson, B.; Kapral, R. Chemistry in motion: Tiny synthetic motors. *Acc. Chem. Res.* **2014**, *47*, 3504–3511. [CrossRef] [PubMed]

28. Wang, W.; Duan, W.; Ahmed, S.; Mallouk, T.E.; Sen, A. Small power: Autonomous nano- and micromotors propelled by self-generated gradients. *Nano Today* **2013**, *8*, 531–554. [CrossRef]

29. Yadav, V.; Duan, W.; Butler, P.J.; Sen, A. Anatomy of nanoscale propulsion. *Annu. Rev. Biophys.* **2015**, *44*, 77–100. [CrossRef] [PubMed]

30. Aubret, A.; Ramananarivo, S.; Palacci, J. Eppur si muove, and yet it moves: Patchy (phoretic) swimmers. *Curr. Opin. Colloid Interface Sci.* **2017**, *30*, 81–89. [CrossRef]

31. Ebbens, S.J. Active colloids: Progress and challenges towards realising autonomous applications. *Curr. Opin. Colloid Interface Sci.* **2016**, *21*, 14–23. [CrossRef]

32. Yamamoto, D.; Shioi, A. Self-propelled nano/micromotors with a chemical reaction: Underlying physics and strategies of motion control. *KONA Powder Part. J.* **2015**, *32*, 2–22. [CrossRef]

33. Ebbens, S.J.; Howse, J.R. In pursuit of propulsion at the nanoscale. *Soft Matter* **2010**, *6*, 726–738. [CrossRef]

34. Randhawa, J.S.; Lafl, K.E.; Seelam, N.; Gracias, D.H. Microchemomechanical systems. *Adv. Funct. Mater.* **2011**, *21*, 2395–2410. [CrossRef]

35. Duan, W.; Wang, W.; Das, S.; Yadav, V.; Mallouk, T.E.; Sen, A. Synthetic nano- and micromachines in analytical chemistry: Sensing, migration, capture, delivery, and separation. *Annu. Rev. Anal. Chem.* **2015**, *8*, 311–333. [CrossRef] [PubMed]

36. Qiu, T.; Lee, T.C.; Mark, A.G.; Morozov, K.I.; Munster, R.; Mierka, O.; Turek, S.; Leshansky, A.M.; Fischer, P. Swimming by reciprocal motion at low Reynolds number. *Nat. Commun.* **2014**, *5*, 5119. [CrossRef] [PubMed]

37. Li, J.; Liu, W.; Wang, J.; Rozen, I.; He, S.; Chen, C.; Kim, H.G.; Lee, H.-J.; Lee, H.-B.-R.; Kwon, S.-H.; et al. Nanoconfined atomic layer deposition of TiO_2/Pt nanotubes: Toward ultrasmall highly efficient catalytic nanorockets. *Adv. Funct. Mater.* **2017**, *27*, 1700598. [CrossRef]

38. Han, K.; Shields, C.W.; Velev, O.D. Engineering of self-propelling microbots and microdevices powered by magnetic and electric fields. *Adv. Funct. Mater.* **2018**, 1705953. [CrossRef]

39. Gánti, T. *The Principles of Life*, 1st ed.; Oxford University Press: Oxford, UK, 2003; ISBN 9780198507260.

40. Sen, A.; Ibele, M.; Hong, Y.; Velegol, D. Chemo and phototactic nano/microbots. *Faraday Discuss.* **2009**, *143*, 15–27. [CrossRef] [PubMed]

41. Gao, W.; Feng, X.; Pei, A.; Gu, Y.; Li, J.; Wang, J. Seawater-driven magnesium based Janus micromotors for environmental remediation. *Nanoscale* **2013**, *5*, 4696–4700. [CrossRef] [PubMed]

42. Lee, T.C.; Alarcon-Correa, M.; Miksch, C.; Hahn, K.; Gibbs, J.G.; Fischer, P. Self-propelling nanomotors in the presence of strong Brownian forces. *Nano Lett.* **2014**, *14*, 2407–2412. [CrossRef] [PubMed]

43. Wang, R.; Guo, W.; Li, X.; Liu, Z.; Liu, H.; Ding, S. Highly efficient MOF-based self-propelled micromotors for water purification. *RSC Adv.* **2017**, *7*, 42462–42467. [CrossRef]

44. Ma, X.; Hahn, K.; Sanchez, S. Catalytic mesoporous Janus nanomotors for active cargo delivery. *J. Am. Chem. Soc.* **2015**, *137*, 4976–4979. [CrossRef] [PubMed]

45. Huang, W.; Manjare, M.; Zhao, Y. Catalytic nanoshell micromotors. *J. Phys. Chem. C* **2013**, *117*, 21590–21596. [CrossRef]

46. Gao, W.; Uygun, A.; Wang, J. Hydrogen-bubble-propelled zinc-based microrockets in strongly acidic media. *J. Am. Chem. Soc.* **2012**, *134*, 897–900. [CrossRef] [PubMed]

47. Niemeyer, C.M.; Mirkin, C.A. *Nanobiotechnology, Concepts, Applications and Perspectives*, 1st ed.; Wiley-VCH: Weinheim, Germany, 2005; ISBN 978-3-527-30658-9.

48. Purcell, E.M. The efficiency of propulsion by a rotating flagellum. *Proc. Natl. Acad. Sci. USA* **1997**, *94*, 11307. [CrossRef] [PubMed]

49. Yao, K.; Manjare, M.; Barrett, C.A.; Yang, B.; Salguero, T.T.; Zhao, Y. Nanostructured Scrolls from Graphene Oxide for Microjet Engines. *J. Phys. Chem. Lett.* **2012**, *3*, 2204–2208. [CrossRef] [PubMed]

50. Kumar, S.; Singh, A.K.; Dasmahapatra, A.K.; Mandal, T.K.; Bandyopadhyay, D. Graphene based multifunctional superbots. *Carbon* **2015**, *89*, 31–40. [CrossRef]

51. Gao, W.; Sattayasamitsathit, S.; Uygun, A.; Pei, A.; Ponedal, A.; Wang, J. Polymer-based tubular microbots: Role of composition and preparation. *Nanoscale* **2012**, *4*, 2447–2453. [CrossRef] [PubMed]

52. Yoshizumi, Y.; Suzuki, H. Self-propelled metal-polymer hybrid micromachines with bending and rotational motions. *ACS Appl. Mater. Interfaces* **2017**, *9*, 21355–21361. [CrossRef] [PubMed]

53. Jang, B.; Wang, W.; Wiget, S.; Petruska, A.J.; Chen, X.; Hu, C.; Hong, A.; Folio, D.; Ferreira, A.; Pane, S.; et al. Catalytic locomotion of core-shell nanowire motors. *ACS Nano* **2016**, *10*, 9983–9991. [CrossRef] [PubMed]

54. Hu, L.; Miao, J.; Grüber, G. Disk-like nanojets with steerable trajectory using platinum nozzle nanoengines. *RSC Adv.* **2016**, *6*, 3399–3405. [CrossRef]

55. Manjare, M.; Yang, B.; Zhao, Y.P. Bubble-propelled microjets: Model and experiment. *J. Phys. Chem. C* **2013**, *117*, 4657–4665. [CrossRef]

56. Huang, G.S.; Wang, J.; Mei, Y.F. Material considerations and locomotive capability in catalytic tubular microengines. *J. Mater. Chem.* **2012**, *22*, 6519–6525. [CrossRef]

57. Wang, L.; Li, L.; Li, T.; Zhang, G.; Sun, Q. Locomotion of chemically powered autonomous nanowire motors. *Appl. Phys. Lett.* **2015**, *107*, 063102. [CrossRef]

58. Chang, X.; Li, L.; Li, T.; Zhou, D.; Zhang, G. Accelerated microrockets with a biomimetic hydrophobic surface. *RSC Adv.* **2016**, *6*, 87213–87220. [CrossRef]

59. Wang, Y.S.; Xia, H.; Lv, C.; Wang, L.; Dong, W.F.; Feng, J.; Sun, H.B. Self-propelled micromotors based on Au-mesoporous silica nanorods. *Nanoscale* **2015**, *7*, 11951–11955. [CrossRef] [PubMed]

60. Ma, X.; Feng, H.; Liang, C.; Liu, X.; Zeng, F.; Wang, Y. Mesoporous silica as micro/nano-carrier: From passive to active cargo delivery, a mini review. *J. Mater. Sci. Technol.* **2017**, *33*, 1067–1074. [CrossRef]

61. Claussen, J.C.; Daniele, M.A.; Geder, J.; Pruessner, M.; Makinen, A.J.; Melde, B.J.; Twigg, M.; Verbarg, J.M.; Medintz, I.L. Platinum-paper micromotors: An urchin-like nanohybrid catalyst for green monopropellant bubble-thrusters. *ACS Appl. Mater. Interfaces* **2014**, *6*, 17837–17847. [CrossRef] [PubMed]

62. Li, J.; Yu, X.; Xu, M.; Liu, W.; Sandraz, E.; Lan, H.; Wang, J.; Cohen, S.M. Metal-organic frameworks as micromotors with tunable engines and brakes. *J. Am. Chem. Soc.* **2017**, *139*, 611–614. [CrossRef] [PubMed]

63. Safdar, M.; Itkonen, T.; Jänis, J. Bubble-propelled trimetallic microcaps as functional catalytic micromotors. *RSC Adv.* **2015**, *5*, 13171–13174. [CrossRef]

64. Safdar, M.; Wani, O.M.; Janis, J. Manganese oxide-based chemically powered micromotors. *ACS Appl. Mater. Interfaces* **2015**, *7*, 25580–25585. [CrossRef] [PubMed]

65. Maria-Hormigos, R.; Jurado-Sanchez, B.; Vazquez, L.; Escarpa, A. Carbon allotrope nanomaterials based catalytic micromotors. *Chem. Mater.* **2016**, *28*, 8962–8970. [CrossRef]

66. Liu, R.; Sen, A. Autonomous nanomotor based on copper-platinum segmented nanobattery. *J. Am. Chem. Soc.* **2011**, *133*, 20064–20067. [CrossRef] [PubMed]

67. Liu, M.; Liu, L.; Gao, W.; Su, M.; Ge, Y.; Shi, L.; Zhang, H.; Dong, B.; Li, C.Y. Nanoparticle mediated micromotor motion. *Nanoscale* **2015**, *7*, 4949–4955. [CrossRef] [PubMed]

68. Li, J.; Xiao, Q.; Jiang, J.-Z.; Chen, G.-N.; Sun, J.-J. Au–Fe/Ni alloy hybrid nanowire motors with dramatic speed. *RSC Adv.* **2014**, *4*, 27522–27525. [CrossRef]

69. Wang, H.; Pumera, M. Fabrication of micro/nanoscale motors. *Chem. Rev.* **2015**, *115*, 8704–8735. [CrossRef] [PubMed]

70. Wang, H.; Pumera, M. Emerging materials for the fabrication of micro/nanomotors. *Nanoscale* **2017**, *9*, 2109–2116. [CrossRef] [PubMed]

71. Gao, W.; Liu, M.; Liu, L.; Zhang, H.; Dong, B.; Li, C.Y. One-step fabrication of multifunctional micromotors. *Nanoscale* **2015**, *7*, 13918–13923. [CrossRef] [PubMed]

72. Tabrizi, M.A.; Shamsipur, M. A simple method for the fabrication of nanomotors based on a gold nanosheet decorated with CoPt nanoparticles. *RSC Adv.* **2015**, *5*, 51508–51511. [CrossRef]

73. Su, M.; Liu, M.; Liu, L.; Sun, Y.; Li, M.; Wang, D.; Zhang, H.; Dong, B. Shape-controlled fabrication of the polymer-based micromotor based on the polydimethylsiloxane template. *Langmuir* **2015**, *31*, 11914–11920. [CrossRef] [PubMed]

74. Martin, A.; Jurado-Sanchez, B.; Escarpa, A.; Wang, J. Template electrosynthesis of high-performance graphene microengines. *Small* **2015**, *11*, 3568–3574. [CrossRef] [PubMed]

75. Gai, M.; Frueh, J.; Hu, N.; Si, T.; Sukhorukov, G.B.; He, Q. Self-propelled two dimensional polymer multilayer plate micromotors. *Phys. Chem. Chem. Phys.* **2016**, *18*, 3397–3401. [CrossRef] [PubMed]

76. He, T.; Wu, H.; Wang, X.; Zang, Q.; Xue, P.; Shen, R.; Dang, L.; Zhang, Y.; Xiang, J. Shape-controlled synthesis of organometallic microcrystal-based hollow hexagonal micromotors through evaporation-induced supramolecular self-assembly. *Cryst. Growth Des.* **2016**, *16*, 6239–6249. [CrossRef]

77. Li, Y.; Wu, J.; Xie, Y.; Ju, H. An efficient polymeric micromotor doped with Pt nanoparticle@carbon nanotubes for complex bio-media. *Chem. Commun.* **2015**, *51*, 6325–6328. [CrossRef] [PubMed]

78. Gao, W.; D'Agostino, M.; Garcia-Gradilla, V.; Orozco, J.; Wang, J. Multi-fuel driven Janus micromotors. *Small* **2013**, *9*, 467–471. [CrossRef] [PubMed]

79. Mallick, A.; Lai, D.; Roy, S. Autonomous movement induced in chemically powered active soft-oxometalates using dithionite as fuel. *New J. Chem.* **2016**, *40*, 1057–1062. [CrossRef]

80. Mou, F.; Chen, C.; Ma, H.; Yin, Y.; Wu, Q.; Guan, J. Self-propelled micromotors driven by the magnesium-water reaction and their hemolytic properties. *Angew. Chem. Int. Ed.* **2013**, *52*, 7208–7212. [CrossRef] [PubMed]

81. Akhavan, O.; Saadati, M.; Jannesari, M. Graphene jet nanomotors in remote controllable self-propulsion swimmers in pure water. *Nano Lett.* **2016**, *16*, 5619–5630. [CrossRef] [PubMed]

82. Dong, R.; Li, J.; Rozen, I.; Ezhilan, B.; Xu, T.; Christianson, C.; Gao, W.; Saintillan, D.; Ren, B.; Wang, J. Vapor-driven propulsion of catalytic micromotors. *Sci. Rep.* **2015**, *5*, 13226. [CrossRef] [PubMed]

83. Cheng, H.; Hu, Y.; Zhao, F.; Dong, Z.; Wang, Y.; Chen, N.; Zhang, Z.; Qu, L. Moisture-activated torsional graphene-fiber motor. *Adv. Mater.* **2014**, *26*, 2909–2913. [CrossRef] [PubMed]

84. Wu, Z.; Li, J.; de Ávila, B.E.-F.; Li, T.; Gao, W.; He, Q.; Zhang, L.; Wang, J. Water-powered cell-mimicking Janus micromotor. *Adv. Funct. Mater.* **2015**, *25*, 7497–7501. [CrossRef]

85. Guix, M.; Meyer, A.K.; Koch, B.; Schmidt, O.G. Carbonate-based Janus micromotors moving in ultra-light acidic environment generated by HeLa cells in situ. *Sci. Rep.* **2016**, *6*, 21701. [CrossRef] [PubMed]

86. Brown, A.T.; Poon, W.C.; Holm, C.; de Graaf, J. Ionic screening and dissociation are crucial for understanding chemical self-propulsion in polar solvents. *Soft Matter* **2017**, *13*, 1200–1222. [CrossRef] [PubMed]

87. Simmchen, J.; Magdanz, V.; Sanchez, S.; Chokmaviroj, S.; Ruiz-Molina, D.; Baeza, A.; Schmidt, O.G. Effect of surfactants on the performance of tubular and spherical micromotors—A comparative study. *RSC Adv.* **2014**, *4*, 20334–20340. [CrossRef] [PubMed]

88. Wang, H.; Zhao, G.; Pumera, M. Crucial role of surfactants in bubble-propelled microengines. *J. Phys. Chem. C* **2014**, *118*, 5268–5274. [CrossRef]

89. Liu, L.; Bai, T.; Chi, Q.; Wang, Z.; Xu, S.; Liu, Q.; Wang, Q. How to make a fast, efficient bubble-driven micromotor: A mechanical view. *Micromachines* **2017**, *8*, 267. [CrossRef]

90. Zhou, C.; Zhang, H.; Li, Z.; Wang, W. Chemistry pumps: A review of chemically powered micropumps. *Lab Chip* **2016**, *16*, 1797–1811. [CrossRef] [PubMed]

91. Wang, W.; Chiang, T.Y.; Velegol, D.; Mallouk, T.E. Understanding the efficiency of autonomous nano- and microscale motors. *J. Am. Chem. Soc.* **2013**, *135*, 10557–10565. [CrossRef] [PubMed]

92. Solovev, A.A.; Sanchez, S.; Mei, Y.F.; Schmidt, O.G. Tunable catalytic tubular micro-pumps operating at low concentrations of hydrogen peroxide. *Phys. Chem. Chem. Phys.* **2011**, *13*, 10131–10135. [CrossRef] [PubMed]

93. Gao, W.; Sattayasamitsathit, S.; Orozco, J.; Wang, J. Highly efficient catalytic microengines: Template electrosynthesis of polyaniline/platinum microtubes. *J. Am. Chem. Soc.* **2011**, *133*, 11862–11864. [CrossRef] [PubMed]
94. Gao, W.; Pei, A.; Dong, R.; Wang, J. Catalytic iridium-based Janus micromotors powered by ultralow levels of chemical fuels. *J. Am. Chem. Soc.* **2014**, *136*, 2276–2279. [CrossRef] [PubMed]
95. Esplandiu, M.J.; Afshar Farniya, A.; Reguera, D. Key parameters controlling the performance of catalytic motors. *J. Chem. Phys.* **2016**, *144*, 124702. [CrossRef] [PubMed]
96. Wang, H.; Sofer, Z.; Eng, A.Y.; Pumera, M. Iridium-catalyst-based autonomous bubble-propelled graphene micromotors with ultralow catalyst loading. *Chemistry* **2014**, *20*, 14946–14950. [CrossRef] [PubMed]
97. Jiang, C.; Huang, G.S.; Ding, S.J.; Dong, H.L.; Men, C.L.; Mei, Y.F. Atomic layer deposition of Pt nanoparticles for microengine with promoted catalytic motion. *Nanoscale Res. Lett.* **2016**, *11*. [CrossRef] [PubMed]
98. Li, J.X.; Liu, Z.Q.; Huang, G.S.; Mei, Y.F. Hierarchical nanoporous microtubes for high-speed catalytic microengines. *NPG Asia Mater.* **2014**, *6*, e94. [CrossRef]
99. Manjare, M.; Ting Wu, Y.; Yang, B.; Zhao, Y.P. Hydrophobic catalytic Janus motors: Slip boundary condition and enhanced catalytic reaction rate. *Appl. Phys. Lett.* **2014**, *104*, 054102. [CrossRef]
100. Kim, K.; Guo, J.; Liang, Z.X.; Zhu, F.Q.; Fan, D.L. Man-made rotary nanomotors: A review of recent developments. *Nanoscale* **2016**, *8*, 10471–10490. [CrossRef] [PubMed]
101. Fournier-Bidoz, S.; Arsenault, A.C.; Manners, I.; Ozin, G.A. Synthetic self-propelled nanorotors. *Chem. Commun.* **2005**, *28*, 441–443. [CrossRef] [PubMed]
102. Yang, M.; Ripoll, M.; Chen, K. Catalytic microrotor driven by geometrical asymmetry. *J. Chem. Phys.* **2015**, *142*, 054902. [CrossRef] [PubMed]
103. Hayakawa, M.; Onoe, H.; Nagai, K.; Takinoue, M. Influence of asymmetry and driving forces on the propulsion of bubble-propelled catalytic micromotors. *Micromachines* **2016**, *7*, 229. [CrossRef]
104. Gregory, D.A.; Campbell, A.I.; Ebbens, S.J. Effect of catalyst distribution on spherical bubble swimmer trajectories. *J. Phys. Chem. C* **2015**, *119*, 15339–15348. [CrossRef]
105. Campbell, A.I.; Wittkowski, R.; Ten Hagen, B.; Lowen, H.; Ebbens, S.J. Helical paths, gravitaxis, and separation phenomena for mass-anisotropic self-propelling colloids: Experiment versus theory. *J. Chem. Phys.* **2017**, *147*, 084905. [CrossRef] [PubMed]
106. Nakata, S.; Nomura, M.; Yamamoto, H.; Izumi, S.; Suematsu, N.J.; Ikura, Y.; Amemiya, T. Periodic oscillatory motion of a self-propelled motor driven by decomposition of H_2O_2 by catalase. *Angew. Chem. Int. Ed.* **2017**, *56*, 861–864. [CrossRef] [PubMed]
107. Klingner, A.; Khalil, I.S.M.; Magdanz, V.; Fomin, V.M.; Schmidt, O.G.; Misra, S. Modeling of unidirectional-overloaded transition in catalytic tubular microjets. *J. Phys. Chem. C* **2017**, *121*, 14854–14863. [CrossRef]
108. Colberg, P.H.; Kapral, R. Nanoconfined catalytic Angstrom-size motors. *J. Chem. Phys.* **2015**, *143*, 184906. [CrossRef] [PubMed]
109. Colberg, P.H.; Kapral, R. Ångström-scale chemically powered motors. *Europhys. Lett.* **2014**, *106*, 30004. [CrossRef]
110. Moo, J.G.; Pumera, M. Chemical energy powered nano/micro/macromotors and the environment. *Chemistry* **2015**, *21*, 58–72. [CrossRef] [PubMed]
111. Wang, H.; Gu, X.; Wang, C. Self-propelling hydrogel/emulsion-hydrogel soft motors for water purification. *ACS Appl. Mater. Interfaces* **2016**, *8*, 9413–9422. [CrossRef] [PubMed]
112. Wang, S.; Jiang, Z.; Ouyang, S.; Dai, Z.; Wang, T. Internally/externally bubble-propelled photocatalytic tubular nanomotors for efficient water cleaning. *ACS Appl. Mater. Interfaces* **2017**, *9*, 23974–23982. [CrossRef] [PubMed]
113. Eskandarloo, H.; Kierulf, A.; Abbaspourrad, A. Nano- and micromotors for cleaning polluted waters: Focused review on pollutant removal mechanisms. *Nanoscale* **2017**, *9*, 13850–13863. [CrossRef] [PubMed]
114. Soler, L.; Sanchez, S. Catalytic nanomotors for environmental monitoring and water remediation. *Nanoscale* **2014**, *6*, 7175–7182. [CrossRef] [PubMed]
115. Jurado-Sanchez, B.; Sattayasamitsathit, S.; Gao, W.; Santos, L.; Fedorak, Y.; Singh, V.V.; Orozco, J.; Galarnyk, M.; Wang, J. Self-propelled activated carbon Janus micromotors for efficient water purification. *Small* **2015**, *11*, 499–506. [CrossRef] [PubMed]

116. Lin, Z.; Wu, Z.; Lin, X.; He, Q. Catalytic polymer multilayer shell motors for separation of organics. *Chemistry* **2016**, *22*, 1587–1591. [CrossRef] [PubMed]
117. Xuan, M.; Lin, X.; Shao, J.; Dai, L.; He, Q. Motion-based, high-yielding, and fast separation of different charged organics in water. *Chem. Phys. Chem.* **2015**, *16*, 147–151. [CrossRef] [PubMed]
118. Wani, O.M.; Safdar, M.; Kinnunen, N.; Janis, J. Dual effect of manganese oxide micromotors: Catalytic degradation and adsorptive bubble separation of organic pollutants. *Chemistry* **2016**, *22*, 1244–1247. [CrossRef] [PubMed]
119. Li, J.; Shklyaev, O.E.; Li, T.; Liu, W.; Shum, H.; Rozen, I.; Balazs, A.C.; Wang, J. Self-propelled nanomotors autonomously seek and repair cracks. *Nano Lett.* **2015**, *15*, 7077–7085. [CrossRef] [PubMed]
120. Li, J.; Liu, W.; Li, T.; Rozen, I.; Zhao, J.; Bahari, B.; Kante, B.; Wang, J. Swimming microrobot optical nanoscopy. *Nano Lett.* **2016**, *16*, 6604–6609. [CrossRef] [PubMed]
121. Su, Y.; Ge, Y.; Liu, L.; Zhang, L.; Liu, M.; Sun, Y.; Zhang, H.; Dong, B. Motion-based pH sensing based on the cartridge-case-like micromotor. *ACS Appl. Mater. Interfaces* **2016**, *8*, 4250–4257. [CrossRef] [PubMed]
122. Jurado-Sanchez, B.; Escarpa, A.; Wang, J. Lighting up micromotors with quantum dots for smart chemical sensing. *Chem. Commun.* **2015**, *51*, 14088–14091. [CrossRef] [PubMed]
123. Ezhilan, B.; Gao, W.; Pei, A.; Rozen, I.; Dong, R.; Jurado-Sanchez, B.; Wang, J.; Saintillan, D. Motion-based threat detection using microrods: Experiments and numerical simulations. *Nanoscale* **2015**, *7*, 7833–7840. [CrossRef] [PubMed]
124. Liu, M.; Sun, Y.; Wang, T.; Ye, Z.; Zhang, H.; Dong, B.; Li, C.Y. A biodegradable, all-polymer micromotor for gas sensing applications. *J. Mater. Chem. C* **2016**, *4*, 5945–5952. [CrossRef]
125. Li, Y.; Mou, F.; Chen, C.; You, M.; Yin, Y.; Xu, L.; Guan, J. Light-controlled bubble propulsion of amorphous TiO_2/Au Janus micromotors. *RSC Adv.* **2016**, *6*, 10697–10703. [CrossRef]
126. Schematic Image of Photocatalytic Micromotor. Available online: http://research.chem.psu.edu/axsgroup/Ran/research/energyconversion.html (accessed on 15 January 2018).
127. Solovev, A.A.; Smith, E.J.; Bof Bufon, C.C.; Sanchez, S.; Schmidt, O.G. Light-controlled propulsion of catalytic microengines. *Angew. Chem.* **2011**, *50*, 10875–10878. [CrossRef] [PubMed]
128. Xu, T.; Soto, F.; Gao, W.; Garcia-Gradilla, V.; Li, J.; Zhang, X.; Wang, J. Ultrasound-modulated bubble propulsion of chemically powered microengines. *J. Am. Chem. Soc.* **2014**, *136*, 8552–8555. [CrossRef] [PubMed]
129. Kline, T.R.; Paxton, W.F.; Mallouk, T.E.; Sen, A. Catalytic nanomotors: Remote-controlled autonomous movement of striped metallic nanorods. *Angew. Chem.* **2005**, *117*, 754–756. [CrossRef]
130. Dynabeads, Commercial Product. Available online: www.thermofisher.com (accessed on 10 February 2018).
131. Dynamic Biosensors, Commercial Product. Available online: www.dynamic-biosensors.com (accessed on 10 February 2018).
132. Wang, J.; Manesh, K.M. Motion control at the nanoscale. *Small* **2010**, *6*, 338–345. [CrossRef] [PubMed]
133. Haeufle, D.F.; Bauerle, T.; Steiner, J.; Bremicker, L.; Schmitt, S.; Bechinger, C. External control strategies for self-propelled particles: Optimizing navigational efficiency in the presence of limited resources. *Phys. Rev. E* **2016**, *94*, 012617. [CrossRef] [PubMed]
134. Teo, W.Z.; Pumera, M. Motion control of micro-/nanomotors. *Chemistry* **2016**, *22*, 14796–14804. [CrossRef] [PubMed]
135. Tu, Y.; Peng, F.; Wilson, D.A. Motion manipulation of micro- and nanomotors. *Adv. Mater.* **2017**, *1701970*. [CrossRef] [PubMed]
136. Ariga, K.; Mori, T.; Ishihara, S.; Kawakami, K.; Hill, J.P. Bridging the difference to the billionth-of-a-meter length scale: How to operate nanoscopic machines and nanomaterials by using macroscopic actions. *Chem. Mater.* **2013**, *26*, 519–532. [CrossRef]
137. Solovev, A.A.; Sanchez, S.; Pumera, M.; Mei, Y.F.; Schmidt, O.G. Magnetic control of tubular catalytic microbots for the transport, assembly, and delivery of micro-objects. *Adv. Funct. Mater.* **2010**, *20*, 2430–2435. [CrossRef]
138. Karshalev, E.; Chen, C.; Marolt, G.; Martin, A.; Campos, I.; Castillo, R.; Wu, T.; Wang, J. Utilizing iron's attractive chemical and magnetic properties in microrocket design, extended motion, and unique performance. *Small* **2017**, *13*. [CrossRef] [PubMed]
139. Zhao, G.; Pumera, M. Magnetotactic artificial self-propelled nanojets. *Langmuir* **2013**, *29*, 7411–7415. [CrossRef] [PubMed]

140. Li, J.-X.; Lu, B.-R.; Shen, Z.; Xu, Z.; Li, H.; Wen, J.; Li, Z.; Qu, X.-P.; Chen, Y.-F.; Mei, Y.; et al. Magnetic and meniscus-effect control of catalytic rolled-up micromotors. *Microelectron. Eng.* **2011**, *88*, 1792–1794. [CrossRef]

141. Liu, M.; Wu, F.; Piao, H.; Huang, X.; Cong, J.; Luo, Z.; Pan, L.; Liu, Y. Rod-shaped nanomotor powered by magnetic field gradients and its application to surface-enhanced Raman-scattering-based detection. *Appl. Phys. Express* **2017**, *10*, 045202. [CrossRef]

142. Schattling, P.S.; Ramos-Docampo, M.A.; Salgueirino, V.; Stadler, B. Double-fueled Janus swimmers with magnetotactic behavior. *ACS Nano* **2017**, *11*, 3973–3983. [CrossRef] [PubMed]

143. Baraban, L.; Makarov, D.; Schmidt, O.G.; Cuniberti, G.; Leiderer, P.; Erbe, A. Control over Janus micromotors by the strength of a magnetic field. *Nanoscale* **2013**, *5*, 1332–1336. [CrossRef] [PubMed]

144. Singh, A.K.; Mandal, T.K.; Bandyopadhyay, D. Magnetically guided chemical locomotion of self-propelling paperbots. *RSC Adv.* **2015**, *5*, 64444–64449. [CrossRef]

145. Kumar, S.; Ali Faridi, M.R.; Dasmahapatra, A.K.; Bandyopadhyay, D. Magnetic field induced push–pull motility of liquibots. *RSC Adv.* **2016**, *6*, 107049–107056. [CrossRef]

146. Li, T.; Li, J.; Morozov, K.I.; Wu, Z.; Xu, T.; Rozen, I.; Leshansky, A.M.; Li, L.; Wang, J. Highly efficient freestyle magnetic nanoswimmer. *Nano Lett.* **2017**, *17*, 5092–5098. [CrossRef] [PubMed]

147. Li, T.; Li, J.; Zhang, H.; Chang, X.; Song, W.; Hu, Y.; Shao, G.; Sandraz, E.; Zhang, G.; Li, L.; et al. Magnetically propelled fish-like nanoswimmers. *Small* **2016**, *12*, 6098–6105. [CrossRef] [PubMed]

148. Tan, S.-C.; Gui, H.; Yuan, B.; Liu, J. Magnetic trap effect to restrict motion of self-powered tiny liquid metal motors. *Appl. Phys. Lett.* **2015**, *107*, 071904. [CrossRef]

149. Yang, T.; Tasci, T.O.; Neeves, K.B.; Wu, N.; Marr, D.W.M. Magnetic microlassos for reversible cargo capture, transport, and release. *Langmuir* **2017**, *33*, 5932–5937. [CrossRef] [PubMed]

150. Singh, A.K.; Dey, K.K.; Chattopadhyay, A.; Mandal, T.K.; Bandyopadhyay, D. Multimodal chemo-magnetic control of self-propelling microbots. *Nanoscale* **2014**, *6*, 1398–1405. [CrossRef] [PubMed]

151. Chen, X.-Z.; Shamsudhin, N.; Hoop, M.; Pieters, R.; Siringil, E.; Sakar, M.S.; Nelson, B.J.; Pané, S. Magnetoelectric micromachines with wirelessly controlled navigation and functionality. *Mater. Horiz.* **2016**, *3*, 113–118. [CrossRef]

152. Li, J.; Li, T.; Xu, T.; Kiristi, M.; Liu, W.; Wu, Z.; Wang, J. Magneto-acoustic hybrid nanomotor. *Nano Lett.* **2015**, *15*, 4814–4821. [CrossRef] [PubMed]

153. Eslami, S.; Gibbs, J.G.; Rechkemmer, Y.; Van Slageren, J.; Alarcón-Correa, M.; Lee, T.-C.; Mark, A.G.; Rikken, G.L.J.A.; Fischer, P. Chiral nanomagnets. *ACS Photonics* **2014**, *1*, 1231–1236. [CrossRef]

154. Xu, T.; Gao, W.; Xu, L.P.; Zhang, X.; Wang, S. Fuel-free synthetic micro-/nanomachines. *Adv. Mater.* **2017**, *29*, 1603250. [CrossRef] [PubMed]

155. Mou, F.; Pan, D.; Chen, C.; Gao, Y.; Xu, L.; Guan, J. Magnetically modulated pot-like $MnFe_2O_4$ micromotors: Nanoparticle assembly fabrication and their capability for direct oil removal. *Adv. Funct. Mater.* **2015**, *25*, 6173–6181. [CrossRef]

156. Sattayasamitsathit, S.; Kou, H.; Gao, W.; Thavarajah, W.; Kaufmann, K.; Zhang, L.; Wang, J. Fully loaded micromotors for combinatorial delivery and autonomous release of cargoes. *Small* **2014**, *10*, 2830–2833. [CrossRef] [PubMed]

157. Maria-Hormigos, R.; Jurado-Sanchez, B.; Escarpa, A. Labs-on-a-chip meet self-propelled micromotors. *Lab Chip* **2016**, *16*, 2397–2407. [CrossRef] [PubMed]

158. Maria-Hormigos, R.; Jurado-Sanchez, B.; Escarpa, A. Tailored magnetic carbon allotrope catalytic micromotors for 'on-chip' operations. *Nanoscale* **2017**, *9*, 6286–6290. [CrossRef] [PubMed]

159. Garcia-Torres, J.; Serra, A.; Tierno, P.; Alcobe, X.; Valles, E. Magnetic propulsion of recyclable catalytic nanocleaners for pollutant degradation. *ACS Appl. Mater. Interfaces* **2017**, *9*, 23859–23868. [CrossRef] [PubMed]

160. Eskandarloo, H.; Kierulf, A.; Abbaspourrad, A. Light-harvesting synthetic nano- and micromotors: A review. *Nanoscale* **2017**, *9*, 12218–12230. [CrossRef] [PubMed]

161. Dong, R.; Zhang, Q.; Gao, W.; Pei, A.; Ren, B. Highly efficient light-driven TiO_2-Au Janus micromotors. *ACS Nano* **2016**, *10*, 839–844. [CrossRef] [PubMed]

162. Enachi, M.; Guix, M.; Postolache, V.; Ciobanu, V.; Fomin, V.M.; Schmidt, O.G.; Tiginyanu, I. Light-induced motion of microengines based on microarrays of TiO_2 nanotubes. *Small* **2016**, *12*, 5497–5505. [CrossRef] [PubMed]

163. Mou, F.; Li, Y.; Chen, C.; Li, W.; Yin, Y.; Ma, H.; Guan, J. Single-component TiO$_2$ tubular microengines with motion controlled by light-induced bubbles. *Small* **2015**, *11*, 2564–2570. [CrossRef] [PubMed]

164. Lin, Z.; Si, T.; Wu, Z.; Gao, C.; Lin, X.; He, Q. Light-activated active colloid ribbons. *Angew. Chem. Int. Ed.* **2017**, *56*, 13517–13520. [CrossRef] [PubMed]

165. Palacci, J.; Sacanna, S.; Vatchinsky, A.; Chaikin, P.M.; Pine, D.J. Photoactivated colloidal dockers for cargo transportation. *J. Am. Chem. Soc.* **2013**, *135*, 15978–15981. [CrossRef] [PubMed]

166. Palacci, J.; Sacanna, S.; Kim, S.H.; Yi, G.R.; Pine, D.J.; Chaikin, P.M. Light-activated self-propelled colloids. *Philos. Trans. Soc. A* **2014**, *372*, 20130372. [CrossRef] [PubMed]

167. Dong, R.; Wang, C.; Wang, Q.; Pei, A.; She, X.; Zhang, Y.; Cai, Y. ZnO-based microrockets with light-enhanced propulsion. *Nanoscale* **2017**, *9*, 15027–15032. [CrossRef] [PubMed]

168. Meng, F.; Hao, W.; Yu, S.; Feng, R.; Liu, Y.; Yu, F.; Tao, P.; Shang, W.; Wu, J.; Song, C.; et al. Vapor-enabled propulsion for plasmonic photothermal motor at the liquid/air interface. *J. Am. Chem. Soc.* **2017**, *139*, 12362–12365. [CrossRef] [PubMed]

169. Jeong, H.H.; Mark, A.G.; Lee, T.C.; Alarcon-Correa, M.; Eslami, S.; Qiu, T.; Gibbs, J.G.; Fischer, P. Active nanorheology with plasmonics. *Nano Lett.* **2016**, *16*, 4887–4894. [CrossRef] [PubMed]

170. Dai, B.; Wang, J.; Xiong, Z.; Zhan, X.; Dai, W.; Li, C.C.; Feng, S.P.; Tang, J. Programmable artificial phototactic microswimmer. *Nat. Nanotechnol.* **2016**, *11*, 1087–1092. [CrossRef] [PubMed]

171. Grant, A. Micromotors swim toward and away from the light. *Phys. Today* **2016**, *69*, 25. [CrossRef]

172. Dong, R.; Hu, Y.; Wu, Y.; Gao, W.; Ren, B.; Wang, Q.; Cai, Y. Visible-light-driven BiOI-based Janus micromotor in pure water. *J. Am. Chem. Soc.* **2017**, *139*, 1722–1725. [CrossRef] [PubMed]

173. Jang, B.; Hong, A.; Kang, H.E.; Alcantara, C.; Charreyron, S.; Mushtaq, F.; Pellicer, E.; Buchel, R.; Sort, J.; Lee, S.S.; et al. Multiwavelength light-responsive Au/B-TiO$_2$ Janus micromotors. *ACS Nano* **2017**, *11*, 6146–6154. [CrossRef] [PubMed]

174. Zhou, D.; Ren, L.; Li, Y.C.; Xu, P.; Gao, Y.; Zhang, G.; Wang, W.; Mallouk, T.E.; Li, L. Visible light-driven, magnetically steerable gold/iron oxide nanomotors. *Chem. Commun.* **2017**, *53*, 11465–11468. [CrossRef] [PubMed]

175. Zhou, D.; Li, Y.C.; Xu, P.; Ren, L.; Zhang, G.; Mallouk, T.E.; Li, L. Visible-light driven Si-Au micromotors in water and organic solvents. *Nanoscale* **2017**, *9*, 11434–11438. [CrossRef] [PubMed]

176. Zhou, D.; Li, Y.C.; Xu, P.; McCool, N.S.; Li, L.; Wang, W.; Mallouk, T.E. Visible-light controlled catalytic Cu$_2$O-Au micromotors. *Nanoscale* **2017**, *9*, 75–78. [CrossRef] [PubMed]

177. Ye, Z.; Sun, Y.; Zhang, H.; Song, B.; Dong, B. A phototactic micromotor based on platinum nanoparticle decorated carbon nitride. *Nanoscale* **2017**, *9*, 18516–18522. [CrossRef] [PubMed]

178. Rao, Q.; Si, T.; Wu, Z.; Xuan, M.; He, Q. A light-activated explosive micropropeller. *Sci. Rep.* **2017**, *7*, 4621. [CrossRef] [PubMed]

179. Wu, Z.; Si, T.; Gao, W.; Lin, X.; Wang, J.; He, Q. Superfast near-infrared light-driven polymer multilayer rockets. *Small* **2016**, *12*, 577–582. [CrossRef] [PubMed]

180. Wu, Y.; Si, T.; Shao, J.; Wu, Z.; He, Q. Near-infrared light-driven Janus capsule motors: Fabrication, propulsion, and simulation. *Nano Res.* **2016**, *9*, 3747–3756. [CrossRef]

181. Maggi, C.; Saglimbeni, F.; Dipalo, M.; De Angelis, F.; Di Leonardo, R. Micromotors with asymmetric shape that efficiently convert light into work by thermocapillary effects. *Nat. Commun.* **2015**, *6*, 7855. [CrossRef] [PubMed]

182. Foy, J.T.; Li, Q.; Goujon, A.; Colard-Itte, J.R.; Fuks, G.; Moulin, E.; Schiffmann, O.; Dattler, D.; Funeriu, D.P.; Giuseppone, N. Dual-light control of nanomachines that integrate motor and modulator subunits. *Nat. Nanotechnol.* **2017**, *12*, 540–545. [CrossRef] [PubMed]

183. Wang, J.; Xiong, Z.; Zhan, X.; Dai, B.; Zheng, J.; Liu, J.; Tang, J. A silicon nanowire as a spectrally tunable light-driven nanomotor. *Adv. Mater.* **2017**, *29*, 1701451. [CrossRef] [PubMed]

184. Zheng, J.; Dai, B.; Wang, J.; Xiong, Z.; Yang, Y.; Liu, J.; Zhan, X.; Wan, Z.; Tang, J. Orthogonal navigation of multiple visible-light-driven artificial microswimmers. *Nat. Commun.* **2017**. [CrossRef] [PubMed]

185. Palagi, S.; Mark, A.G.; Reigh, S.Y.; Melde, K.; Qiu, T.; Zeng, H.; Parmeggiani, C.; Martella, D.; Sanchez-Castillo, A.; Kapernaum, N.; et al. Structured light enables biomimetic swimming and versatile locomotion of photoresponsive soft microrobots. *Nat. Mater.* **2016**, *15*, 647–653. [CrossRef] [PubMed]

186. Schamel, D.; Pfeifer, M.; Gibbs, J.G.; Miksch, B.; Mark, A.G.; Fischer, P. Chiral colloidal molecules and observation of the propeller effect. *J. Am. Chem. Soc.* **2013**, *135*, 12353–12359. [CrossRef] [PubMed]

187. Tang, X.; Tang, S.-Y.; Sivan, V.; Zhang, W.; Mitchell, A.; Kalantar-zadeh, K.; Khoshmanesh, K. Photochemically induced motion of liquid metal marbles. *Appl. Phys. Lett.* **2013**, *103*, 174104. [CrossRef]

188. Wong, F.; Sen, A. Progress toward light-harvesting self-electrophoretic motors: Highly efficient bimetallic nanomotors and micropumps in halogen media. *ACS Nano* **2016**, *10*, 7172–7179. [CrossRef] [PubMed]

189. Safdar, M.; Simmchen, J.; Jänis, J. Light-driven micro- and nanomotors for environmental remediation. *Environ. Sci. Nano* **2017**, *4*, 1602–1616. [CrossRef]

190. Zhang, Z.; Zhao, A.; Wang, F.; Ren, J.; Qu, X. Design of a plasmonic micromotor for enhanced photo-remediation of polluted anaerobic stagnant waters. *Chem. Commun.* **2016**, *52*, 5550–5553. [CrossRef] [PubMed]

191. Zhang, Q.; Dong, R.; Wu, Y.; Gao, W.; He, Z.; Ren, B. Light-driven Au-WO$_3$@C Janus micromotors for rapid photodegradation of dye pollutants. *ACS Appl. Mater. Interfaces* **2017**, *9*, 4674–4683. [CrossRef] [PubMed]

192. Martinez-Pedrero, F.; Massana-Cid, H.; Tierno, P. Assembly and transport of microscopic cargos via reconfigurable photoactivated magnetic microdockers. *Small* **2017**, *13*. [CrossRef] [PubMed]

193. Ahmed, D.; Baasch, T.; Jang, B.; Pane, S.; Dual, J.; Nelson, B.J. Artificial swimmers propelled by acoustically activated flagella. *Nano Lett.* **2016**, *16*, 4968–4974. [CrossRef] [PubMed]

194. Ahmed, D.; Dillinger, C.; Hong, A.; Nelson, B.J. Artificial acousto-magnetic soft microswimmers. *Adv. Mater. Technol.* **2017**, *2*, 1700050. [CrossRef]

195. Chen, M.; Cai, F.; Wang, C.; Wang, Z.; Meng, L.; Li, F.; Zhang, P.; Liu, X.; Zheng, H. Observation of metal nanoparticles for acoustic manipulation. *Adv. Sci.* **2017**, *4*, 1600447. [CrossRef] [PubMed]

196. Collis, J.F.; Chakraborty, D.; Sader, J.E. Autonomous propulsion of nanorods trapped in an acoustic field. *J. Fluid Mech.* **2017**, *825*, 29–48. [CrossRef]

197. Feng, J.; Yuan, J.; Cho, S.K. Micropropulsion by an acoustic bubble for navigating microfluidic spaces. *Lab Chip* **2015**, *15*, 1554–1562. [CrossRef] [PubMed]

198. Feng, J.; Yuan, J.; Cho, S.K. 2-D steering and propelling of acoustic bubble-powered microswimmers. *Lab Chip* **2016**, *16*, 2317–2325. [CrossRef] [PubMed]

199. Soto, F.; Martin, A.; Ibsen, S.; Vaidyanathan, M.; Garcia-Gradilla, V.; Levin, Y.; Escarpa, A.; Esener, S.C.; Wang, J. Acoustic microcannons: Toward advanced microballistics. *ACS Nano* **2016**, *10*, 1522–1528. [CrossRef] [PubMed]

200. Soto, F.; Wagner, G.L.; Garcia-Gradilla, V.; Gillespie, K.T.; Lakshmipathy, D.R.; Karshalev, E.; Angell, C.; Chen, Y.; Wang, J. Acoustically propelled nanoshells. *Nanoscale* **2016**, *8*, 17788–17793. [CrossRef] [PubMed]

201. Esteban-Fernandez de Avila, B.; Angell, C.; Soto, F.; Lopez-Ramirez, M.A.; Baez, D.F.; Xie, S.; Wang, J.; Chen, Y. Acoustically propelled nanomotors for intracellular siRNA delivery. *ACS Nano* **2016**, *10*, 4997–5005. [CrossRef] [PubMed]

202. He, W.; Frueh, J.; Hu, N.; Liu, L.; Gai, M.; He, Q. Guidable thermophoretic Janus micromotors containing gold nanocolorifiers for infrared laser assisted tissue welding. *Adv. Sci.* **2016**, *3*, 1600206. [CrossRef] [PubMed]

203. Melde, K.; Mark, A.G.; Qiu, T.; Fischer, P. Holograms for acoustics. *Nature* **2016**, *537*, 518–522. [CrossRef] [PubMed]

204. Rao, K.J.; Li, F.; Meng, L.; Zheng, H.; Cai, F.; Wang, W. A force to be reckoned with: A review of synthetic microswimmers powered by ultrasound. *Small* **2015**, *11*, 2836–2846. [CrossRef] [PubMed]

205. Wang, W.; Duan, W.; Zhang, Z.; Sun, M.; Sen, A.; Mallouk, T.E. A tale of two forces: Simultaneous chemical and acoustic propulsion of bimetallic micromotors. *Chem. Commun.* **2015**, *51*, 1020–1023. [CrossRef] [PubMed]

206. Chang, S.T.; Paunov, V.N.; Petsev, D.N.; Velev, O.D. Remotely powered self-propelling particles and micropumps based on miniature diodes. *Nat. Mater.* **2007**, *6*, 235–240. [CrossRef] [PubMed]

207. Essmann, V.; Voci, S.; Loget, G.; Sojic, N.; Schuhmann, W.; Kuhn, A. Wireless light-emitting electrochemical rotors. *J. Phys. Chem. Lett.* **2017**, *8*, 4930–4934. [CrossRef] [PubMed]

208. Calvo-Marzal, P.; Sattayasamitsathit, S.; Balasubramanian, S.; Windmiller, J.R.; Dao, C.; Wang, J. Propulsion of nanowire diodes. *Chem. Commun.* **2010**, *46*, 1623–1624. [CrossRef] [PubMed]

209. Yoshizumi, Y.; Honegger, T.; Berton, K.; Suzuki, H.; Peyrade, D. Trajectory control of self-propelled micromotors using AC electrokinetics. *Small* **2015**, *11*, 5630–5635. [CrossRef] [PubMed]

210. Moo, J.G.S.; Pumera, M. Self-propelled micromotors monitored by particle-electrode impact voltammetry. *ACS Sens.* **2016**, *1*, 949–957. [CrossRef]

211. Tan, S.-C.; Yuan, B.; Liu, J. Electrical method to control the running direction and speed of self-powered tiny liquid metal motors. *Proc. R. Soc. A* **2015**, *471*, 20150297. [CrossRef]

212. Balasubramanian, S.; Kagan, D.; Manesh, K.M.; Calvo-Marzal, P.; Flechsig, G.U.; Wang, J. Thermal modulation of nanomotor movement. *Small* **2009**, *5*, 1569–1574. [CrossRef] [PubMed]

213. Cai, K.; Yu, J.; Liu, L.; Shi, J.; Qin, Q.H. Rotation measurements of a thermally driven rotary nanomotor with a spring wing. *Phys. Chem. Chem. Phys.* **2016**, *18*, 22478–22486. [CrossRef] [PubMed]

214. Li, Q.; Liang, T.; Ye, W. Knudsen torque: A rotational mechanism driven by thermal force. *Phys. Rev. E Stat. Nonlin. Soft Matter Phys.* **2014**, *90*, 033009. [CrossRef] [PubMed]

215. Valdez, L.; Shum, H.; Ortiz-Rivera, I.; Balazs, A.C.; Sen, A. Solutal and thermal buoyancy effects in self-powered phosphatase micropumps. *Soft Matter* **2017**, *13*, 2800–2807. [CrossRef] [PubMed]

216. Moo, J.G.; Presolski, S.; Pumera, M. Photochromic spatiotemporal control of bubble-propelled micromotors by a spiropyran molecular switch. *ACS Nano* **2016**, *10*, 3543–3552. [CrossRef] [PubMed]

217. Huang, G.S.; Wang, J.Y.; Liu, Z.Q.; Zhou, D.K.; Tian, Z.A.; Xu, B.R.; Li, L.Q.; Mei, Y.F. Rocket-inspired tubular catalytic microjets with grating-structured walls as guiding empennages. *Nanoscale* **2017**, *47*, 18590–18596. [CrossRef] [PubMed]

218. Liu, C.; Zhou, C.; Wang, W.; Zhang, H.P. Bimetallic microswimmers speed up in confining channels. *Phys. Rev. Lett.* **2016**, *117*, 198001. [CrossRef] [PubMed]

219. Yang, F.; Qian, S.; Zhao, Y.; Qiao, R. Self-diffusiophoresis of Janus catalytic micromotors in confined geometries. *Langmuir* **2016**, *32*, 5580–5592. [CrossRef] [PubMed]

220. Davies Wykes, M.S.; Zhong, X.; Tong, J.; Adachi, T.; Liu, Y.; Ristroph, L.; Ward, M.D.; Shelley, M.J.; Zhang, J. Guiding microscale swimmers using teardrop-shaped posts. *Soft Matter* **2017**, *13*, 4681–4688. [CrossRef] [PubMed]

221. Illien, P.; Golestanian, R.; Sen, A. 'Fuelled' motion: Phoretic motility and collective behaviour of active colloids. *Chem. Soc. Rev.* **2017**, *46*, 5508–5518. [CrossRef] [PubMed]

222. Maggi, C.; Simmchen, J.; Saglimbeni, F.; Katuri, J.; Dipalo, M.; De Angelis, F.; Sanchez, S.; Di Leonardo, R. Self-assembly of micromachining systems powered by Janus micromotors. *Small* **2016**, *12*, 446–451. [CrossRef] [PubMed]

223. Ben-Jacob, E.; Cohen, I.; Levine, H. Cooperative self-organization of microorganisms. *Adv. Phys.* **2000**, *49*, 395–554. [CrossRef]

224. Kei Cheang, U.; Lee, K.; Julius, A.A.; Kim, M.J. Multiple-robot drug delivery strategy through coordinated teams of microswimmers. *Appl. Phys. Lett.* **2014**, *105*, 083705. [CrossRef]

225. Huang, M.J.; Kapral, R. Collective dynamics of diffusiophoretic motors on a filament. *Eur. Phys. J. E Soft Matter* **2016**, *39*, 36. [CrossRef] [PubMed]

226. Kolmakov, G.V.; Yashin, V.V.; Levitan, S.P.; Balazs, A.C. Designing communicating colonies of biomimetic microcapsules. *Proc. Natl. Acad. Sci. USA* **2010**, *107*, 12417–12422. [CrossRef] [PubMed]

227. Bishop, K.J.M.; Wilmer, C.E.; Soh, S.; Grzybowski, B.A. Nanoscale forces and their uses in self-assembly. *Small* **2009**, *5*, 1600–1630. [CrossRef] [PubMed]

228. Gao, Y.; Mou, F.; Feng, Y.; Che, S.; Li, W.; Xu, L.; Guan, J. Dynamic colloidal molecules maneuvered by light-controlled Janus micromotors. *ACS Appl. Mater. Interfaces* **2017**, *9*, 22704–22712. [CrossRef] [PubMed]

229. Mou, F.; Kong, L.; Chen, C.; Chen, Z.; Xu, L.; Guan, J. Light-controlled propulsion, aggregation and separation of water-fuelled TiO_2/Pt Janus submicromotors and their "on-the-fly" photocatalytic activities. *Nanoscale* **2016**, *8*, 4976–4983. [CrossRef] [PubMed]

230. Singh, D.P.; Choudhury, U.; Fischer, P.; Mark, A.G. Non-equilibrium assembly of light-activated colloidal mixtures. *Adv. Mater.* **2017**, 1701328. [CrossRef] [PubMed]

231. Hong, Y.; Diaz, M.; Córdova-Figueroa, U.M.; Sen, A. Light-driven titanium-dioxide-based reversible microfireworks and micromotor/micropump systems. *Adv. Funct. Mater.* **2010**, *20*, 1568–1576. [CrossRef]

232. Duan, W.; Ibele, M.; Liu, R.; Sen, A. Motion analysis of light-powered autonomous silver chloride nanomotors. *Eur. Phys. J. E Soft Matter* **2012**, *35*, 77. [CrossRef] [PubMed]

233. Zhou, C.; Bashirzadeh, Y.; Bernadowski, T.; Zhang, X. UV light–induced aggregation of Titania submicron particles. *Micromachines* **2016**, *7*, 203. [CrossRef]

234. Zhang, Q.; Dong, R.; Chang, X.; Ren, B.; Tong, Z. Spiropyran-decorated SiO_2-Pt Janus micromotor: Preparation and light-induced dynamic self-assembly and disassembly. *ACS Appl. Mater. Interfaces* **2015**, *7*, 24585–24591. [CrossRef] [PubMed]

235. Gao, W.; Pei, A.; Feng, X.; Hennessy, C.; Wang, J. Organized self-assembly of Janus micromotors with hydrophobic hemispheres. *J. Am. Chem. Soc.* **2013**, *135*, 998–1001. [CrossRef] [PubMed]

236. Nourhani, A.; Brown, D.; Pletzer, N.; Gibbs, J.G. Engineering contactless particle-particle interactions in active microswimmers. *Adv. Mater.* **2017**, *29*. [CrossRef] [PubMed]

237. Bayati, P.; Najafi, A. Dynamics of two interacting active Janus particles. *J. Chem. Phys.* **2016**, *144*, 134901. [CrossRef] [PubMed]

238. Afshar Farniya, A.; Esplandiu, M.J.; Bachtold, A. Sequential tasks performed by catalytic pumps for colloidal crystallization. *Langmuir* **2014**, *30*, 11841–11845. [CrossRef] [PubMed]

239. Manjare, M.; Yang, F.; Qiao, R.; Zhao, Y. Marangoni flow induced collective motion of catalytic micromotors. *J. Phys. Chem. C* **2015**, *119*, 28361–28367. [CrossRef]

240. Kokot, G.; Kolmakov, G.V.; Aranson, I.S.; Snezhko, A. Dynamic self-assembly and self-organized transport of magnetic micro-swimmers. *Sci. Rep.* **2017**, *7*, 14726. [CrossRef] [PubMed]

241. Vach, P.J.; Walker, D.; Fischer, P.; Fratzl, P.; Faivre, D. Pattern formation and collective effects in populations of magnetic microswimmers. *J. Phys. D* **2017**, *50*, 11LT03. [CrossRef]

242. Xu, T.; Soto, F.; Gao, W.; Dong, R.; Garcia-Gradilla, V.; Magana, E.; Zhang, X.; Wang, J. Reversible swarming and separation of self-propelled chemically powered nanomotors under acoustic fields. *J. Am. Chem. Soc.* **2015**, *137*, 2163–2166. [CrossRef] [PubMed]

243. Wang, W.; Duan, W.; Ahmed, S.; Sen, A.; Mallouk, T.E. From one to many: Dynamic assembly and collective behavior of self-propelled colloidal motors. *Acc. Chem. Res.* **2015**, *48*, 1938–1946. [CrossRef] [PubMed]

244. Hong, Y.; Velegol, D.; Chaturvedi, N.; Sen, A. Biomimetic behavior of synthetic particles: From microscopic randomness to macroscopic control. *Phys. Chem. Chem. Phys.* **2010**, *12*, 1423–1430. [CrossRef] [PubMed]

245. Purcell, E.M. Life at low Reynolds number. *Am. J. Phys.* **1977**, *45*, 3–11. [CrossRef]

246. Altemose, A.; Sanchez-Farran, M.A.; Duan, W.; Schulz, S.; Borhan, A.; Crespi, V.H.; Sen, A. Chemically controlled spatiotemporal oscillations of colloidal assemblies. *Angew. Chem. Int. Ed. Engl.* **2017**, *56*, 7817–7821. [CrossRef] [PubMed]

247. Duan, W.; Liu, R.; Sen, A. Transition between collective behaviors of micromotors in response to different stimuli. *J. Am. Chem. Soc.* **2013**, *135*, 1280–1283. [CrossRef] [PubMed]

248. Zhang, L.; Zhang, H.; Liu, M.; Dong, B. Reprogrammable logic gate and logic circuit based on multistimuli-responsive raspberry-like micromotors. *ACS Appl. Mater. Interfaces* **2016**, *8*, 15654–15660. [CrossRef] [PubMed]

249. Dey, K.K.; Wong, F.; Altemose, A.; Sen, A. Catalytic motors—Quo vadimus? *Curr. Opin. Colloid Interface Sci.* **2016**, *21*, 4–13. [CrossRef]

250. Dey, K.K.; Sen, A. Chemically propelled molecules and machines. *J. Am. Chem. Soc.* **2017**, *139*, 7666–7676. [CrossRef] [PubMed]

251. Grzybowski, B.A.; Fitzner, K.; Paczesny, J.; Granick, S. From dynamic self-assembly to networked chemical systems. *Chem. Soc. Rev.* **2017**, *46*, 5647–5678. [CrossRef] [PubMed]

252. Zhang, J.; Luijten, E.; Grzybowski, B.A.; Granick, S. Active colloids with collective mobility status and research opportunities. *Chem. Soc. Rev.* **2017**, *46*, 5551–5569. [CrossRef] [PubMed]

253. Dey, K.K.; Das, S.; Poython, M.F.; Sengupta, S.; Butler, P.J.; Cremer, P.S; Sen, A. Chemotactic separation of enzymes. *ACS Nano* **2014**, *8*, 11941–11949. [CrossRef] [PubMed]

254. Solovev, A.A.; Mei, Y.; Schmidt, O.G. Catalytic microstrider at the air-liquid interface. *Adv. Mater.* **2010**, *22*, 4340–4344. [CrossRef] [PubMed]

255. Wang, Y.; Fei, S.; Byun, W.M.; Lammert, P.E.; Crespi, V.H.; Sen, A.; Mallouk, T.E. Dynamic interactions between fast microscale rotors. *J. Am. Chem. Soc.* **2009**, *131*, 9926–9927. [CrossRef] [PubMed]

256. Solovev, A.A.; Sanchez, S.; Schmidt, O.G. Collective behaviour of self-propelled catalytic micromotors. *Nanoscale* **2013**, *5*, 1284–1293. [CrossRef] [PubMed]

257. Fialkowski, M.; Bishop, K.J.M.; Klajn, R.; Smoukov, S.K.; Campbell, C.J.; Grzybowski, B.A. Principles and implementations of dissipative (dynamic) self-assembly. *J. Phys. Chem. B* **2006**, *110*, 2482–2496. [CrossRef] [PubMed]

258. Turing, A.M. The chemical basis of morphogenesis. *Philos. Trans. R. Soc. B Biol. Sci.* **1952**, *237*, 37–72. [CrossRef]

259. Howard, J.; Grill, S.W.; Bois, J.S. Turing's next steps: The mechanochemical basis of morphogenesis. *Nat. Rev. Mol. Cell Biol.* **2011**, *12*, 392–398. [CrossRef] [PubMed]

260. Singh, V.V.; Soto, F.; Kaufmann, K.; Wang, J. Micromotor-based energy generation. *Angew. Chem. Int. Ed.* **2015**, *54*, 6896–6899. [CrossRef] [PubMed]

261. Morikawa, M.; Ogura, Y.; Ahmed, N.; Kawamura, S.; Mikami, G.; Okamotoband, S.; Izumi, Y. Photocatalytic conversion of carbon dioxide intomethanol in reverse fuel cells with tungsten oxide and layered double hydroxide photocatalysts forsolar fuel generation. *Catal. Sci. Technol.* **2014**, *4*, 1644–1651. [CrossRef]
262. Peng, F.; Tu, Y.; Wilson, D.A. Micro/nanomotors towards in vivo application: Cell, tissue and biofluid. *Chem. Soc. Rev.* **2017**, *46*, 5289–5310. [CrossRef] [PubMed]
263. Abdelmohsen, L.K.E.A.; Nijemeisland, M.; Pawar, G.M.; Janssen, G.-J.A.; Nolte, R.J.M.; van Hest, J.C.M.; Wilson, D.A. Dynamic loading and unloading of proteins in polymeric stomatocytes: Formation of an enzyme-loaded supramolecular nanomotor. *ACS Nano* **2016**, *10*, 2652–2660. [CrossRef] [PubMed]
264. Medina-Sanchez, M.; Schwarz, L.; Meyer, A.K.; Hebenstreit, F.; Schmidt, O.G. Cellular cargo delivery: Toward assisted fertilization by sperm-carrying micromotors. *Nano Lett.* **2016**, *16*, 555–561. [CrossRef] [PubMed]
265. Shah, R.K.; Shum, H.C.; Rowat, A.C.; Lee, D.; Agresti, J.J.; Utada, A.S. Designer emulsions using microfluidics. *Mater. Today* **2008**, *11*, 18–27. [CrossRef]
266. Gaspar, S. Enzymatically induced motion at nano- and micro-scales. *Nanoscale* **2014**, *6*, 7757–7763. [CrossRef] [PubMed]
267. Ma, X.; Hortelao, A.C.; Patiño, T.; Sanchez, S. Enzyme catalysis to power micro/nanomachines. *ACS Nano* **2016**, *10*, 9111–9122. [CrossRef] [PubMed]
268. Lin, X.; Wu, Z.; Wu, Y.; Xuan, M.; He, Q. Self-propelled micro-/nanomotors based on controlled assembled architectures. *Adv. Mater.* **2016**, *28*, 1060–1072. [CrossRef] [PubMed]
269. Mano, N.; Heller, A. Bioelectrochemical propulsion. *J. Am. Chem. Soc.* **2005**, *127*, 11574–11575. [CrossRef] [PubMed]
270. Ma, X.; Jannasch, A.; Albrecht, U.R.; Hahn, K.; Lopez, A.M.; Schaffer, E.; Sanchez, S. Enzyme-powered hollow mesoporous Janus nanomotors. *Nano Lett.* **2015**, *15*, 7043–7050. [CrossRef] [PubMed]
271. Ma, X.; Hortelao, A.C.; Lopez, A.M.; Sanchez, S. Bubble-free propulsion of ultrasmall tubular nanojets powered by biocatalytic reactions. *J. Am. Chem. Soc.* **2016**, *138*, 13782–13785. [CrossRef] [PubMed]
272. Gao, W.; Dong, R.; Thamphiwatana, S.; Li, J.X.; Gao, W.W.; Zhang, L.F.; Wang, J. Artificial micromotors in the mouse's stomach: A step toward in vivo use of synthetic motors. *ACS Nano* **2015**, *9*, 117–123. [CrossRef] [PubMed]
273. Gregory, D.A.; Zhang, Y.; Smith, P.J.; Zhao, X.; Ebbens, S.J. Reactive inkjet printing of biocompatible enzyme powered silk micro-rockets. *Small* **2016**, *12*, 4048–4055. [CrossRef] [PubMed]
274. Pavlick, R.A.; Sengupta, S.; McFadden, T.; Zhang, H.; Sen, A. A polymerization-powered motor. *Angew. Chem. Int. Ed.* **2011**, *50*, 9374–9397. [CrossRef] [PubMed]
275. Yoshizumi, Y.; Okubo, K.; Yokokawa, M.; Suzuki, H. Programmed transport and release of cells by self-propelled micromotors. *Langmuir* **2016**, *32*, 9381–9388. [CrossRef] [PubMed]
276. Zhou, C.; Yin, J.; Wu, C.; Du, L.; Wang, Y. Efficient target capture and transport by fuel-free micromotors in a multichannel microchip. *Soft Matter* **2017**, *13*, 8064–8069. [CrossRef] [PubMed]
277. Esteban-Fernandez de Avila, B.; Ramirez-Herrera, D.E.; Campuzano, S.; Angsantikul, P.; Zhang, L.; Wang, J. Nanomotor-enabled pH-responsive intracellular delivery of caspase-3: Toward rapid cell apoptosis. *ACS Nano* **2017**, *11*, 5367–5374. [CrossRef] [PubMed]
278. Mou, F.; Chen, C.; Zhong, Q.; Yin, Y.; Ma, H.; Guan, J. Autonomous motion and temperature-controlled drug delivery of Mg/Pt-poly(N-isopropylacrylamide) Janus micromotors driven by simulated body fluid and blood plasma. *ACS Appl. Mater. Interfaces* **2014**, *6*, 9897–9903. [CrossRef] [PubMed]
279. Wang, L.; Chen, J.; Feng, X.; Zeng, W.; Liu, R.; Lin, X.; Ma, Y.; Wang, L. Self-propelled manganese oxide-based catalytic micromotors for drug delivery. *RSC Adv.* **2016**, *6*, 65624–65630. [CrossRef]
280. Orozco, J.; Cortes, A.; Cheng, G.; Sattayasamitsathit, S.; Gao, W.; Feng, X.; Shen, Y.; Wang, J. Molecularly imprinted polymer-based catalytic micromotors for selective protein transport. *J. Am. Chem. Soc.* **2013**, *135*, 5336–5339. [CrossRef] [PubMed]
281. Gao, W.; Wang, J. Synthetic micro/nanomotors in drug delivery. *Nanoscale* **2014**, *6*, 10486–10494. [CrossRef] [PubMed]
282. Patra, D.; Sengupta, S.; Duan, W.; Zhang, H.; Pavlick, R.; Sen, A. Intelligent, self-powered, drug delivery systems. *Nanoscale* **2013**, *5*, 1273–1283. [CrossRef] [PubMed]
283. Chen, A.; Ge, X.-H.; Chen, J.; Zhang, L.; Xu, J.-H. Multi-functional micromotor: Microfluidic fabrication and water treatment application. *Lab Chip* **2017**, *17*, 4220–4224. [CrossRef] [PubMed]

284. Yang, P.P.; Zhai, Y.G.; Qi, G.B.; Lin, Y.X.; Luo, Q.; Yang, Y.; Xu, A.P.; Yang, C.; Li, Y.S.; Wang, L.; et al. NIR light propulsive Janus-like nanohybrids for enhanced photothermal tumor therapy. *Small* **2016**, *12*, 5423–5430. [CrossRef] [PubMed]

285. Bhuyan, T.; Kumar Singh, A.; Dutta, D.; Unal, A.; Sankar Ghosh, S.; Bandyopadhyay, D. Magnetic field guided chemotaxis of imushbots for targeted anticancer therapeutics. *ACS Biomater. Sci. Eng.* **2017**, *3*, 1627–1640. [CrossRef]

286. Chen, F.; Hong, H.; Zhang, Y.; Valdovinos, H.F.; Shi, S.; Kwon, G.S.; Theuer, C.P.; Barnhart, T.E.; Cai, W. In vivo tumor targeting and image-guided drug delivery with antibody-conjugated, radiolabeled mesoporous silica nanoparticles. *ACS Nano* **2013**, *7*, 9027–9039. [CrossRef] [PubMed]

287. Li, J.; Angsantikul, P.; Liu, W.; Esteban-Fernández de Ávila, B.; Thamphiwatana, S.; Xu, M.; Sandraz, E.; Wang, X.; Delezuk, J.; Gao, W.; et al. Micromotors spontaneously neutralize gastric acid for pH-responsive payload release. *Angew. Chem. Int. Ed.* **2017**, *56*, 2156–2161. [CrossRef] [PubMed]

288. De Ávila, B.E.; Angsantikul, P.; Li, J.; Angel Lopez-Ramirez, M.; Ramírez-Herrera, D.E.; Thamphiwatana, S.; Chen, C.; Delezuk, J.; Samakapiruk, R.; Ramez, V.; et al. Micromotor-enabled active drug delivery for in vivo treatment of stomach infection. *Nat. Commun.* **2017**, *8*, 272. [CrossRef] [PubMed]

289. Wu, Z.; Lin, X.; Zou, X.; Sun, J.; He, Q. Biodegradable protein-based rockets for drug transportation and light-triggered release. *ACS Appl. Mater. Interfaces* **2015**, *7*, 250–255. [CrossRef] [PubMed]

290. Tu, Y.; Peng, F.; White, P.B.; Wilson, D.A. Redox-sensitive stomatocyte nanomotors: Destruction and drug release in the presence of glutathione. *Angew. Chem. Int. Ed.* **2017**, *56*, 7620–7624. [CrossRef] [PubMed]

291. Wang, L.; Liu, Y.; He, J.; Hourwitz, M.J.; Yang, Y.; Fourkas, J.T.; Han, X.; Nie, Z. Continuous microfluidic self-assembly of hybrid Janus-like vesicular motors: Autonomous propulsion and controlled release. *Small* **2015**, *11*, 3762–3767. [CrossRef] [PubMed]

292. Solovev, A.A.; Xi, W.; Gracias, D.H.; Harazim, S.M.; Deneke, C.; Sanchez, S.; Schmidt, O.G. Self-propelled nanotools. *Acs Nano* **2012**, *6*, 1751–1756. [CrossRef] [PubMed]

293. Wang, H.; Khezri, B.; Pumera, M. Catalytic DNA-functionalized self-propelled micromachines for environmental remediation. *Chem* **2016**, *1*, 473–481. [CrossRef]

294. Fu, S.; Zhang, X.; Xie, Y.; Wu, J.; Ju, H. An efficient enzyme-powered micromotor device fabricated by cyclic alternate hybridization assembly for DNA detection. *Nanoscale* **2017**, *9*, 9026–9033. [CrossRef] [PubMed]

295. Van Nguyen, K.; Minteer, S.D. DNA-functionalized Pt nanoparticles as catalysts for chemically powered micromotors: Toward signal-on motion-based DNA biosensor. *Chem. Commun.* **2015**, *51*, 4782–4784. [CrossRef] [PubMed]

296. Wu, J.; Balasubramanian, S.; Kagan, D.; Manesh, K.M.; Campuzano, S.; Wang, J. Motion-based DNA detection using catalytic nanomotors. *Nat. Commun.* **2010**, *1*, 36. [CrossRef] [PubMed]

297. Singh, V.V.; Kaufmann, K.; Esteban-Fernández de Ávila, B.; Uygun, M.; Wang, J. Nanomotors responsive to nerve-agent vapor plumes. *Chem. Commun.* **2016**, *52*, 3360–3363. [CrossRef] [PubMed]

298. Li, M.; Zhang, H.; Liu, M.; Dong, B. Motion-based glucose sensing based on a fish-like enzymeless motor. *J. Mater. Chem. C* **2017**, *5*, 4400–4407. [CrossRef]

299. Esteban-Fernández de Ávila, B.; Zhao, M.; Campuzano, S.; Ricci, F.; Pingarrónc, J.M.; Mascinia, M.; Wang, J. Rapid micromotor-based naked-eye immunoassay. *Talanta* **2017**, *167*, 651–657. [CrossRef] [PubMed]

300. Wang, J.; Gao, W. Nano/microscale motors: Biomedical opportunities and challenges. *ACS Nano* **2012**, *6*, 5745–5751. [CrossRef] [PubMed]

301. Weaver, R. *Molecular Biology*, 5th ed.; McGraw-Hill Education—Europe: New York, NY, USA, 2011; ISBN 9789813150508.

302. Adams, L.L.A.; Kodger, T.E.; Kim, S.-H.; Shum, H.C.; Franke, T.; Weitz, D.A. Single step emulsification for the generation of multi-component double emulsions. *Soft Matter* **2012**, *8*, 10719–10724. [CrossRef]

303. Epstein, I.R.; Showalter, K. Nonlinear chemical dynamics: Oscillations, patterns, and chaos. *J. Phys. Chem.* **1996**, *100*, 13132–13147 [CrossRef]

micromachines

MDPI

Review

Tubular Micro/Nanomotors: Propulsion Mechanisms, Fabrication Techniques and Applications

Fengjun Zha [†], Tingwei Wang [†], Ming Luo * and Jianguo Guan *

State Key Laboratory of Advanced Technology for Materials Synthesis and Processing,
International School of Materials Science and Engineering, Wuhan University of Technology,
Wuhan 430070, China; zhafj159@whut.edu.cn (F.Z.); 315387183@whut.edu.cn (T.W.)
* Correspondence: luoming_2016@whut.edu.cn (M.L.); guanjg@whut.edu.cn (J.G.)
† These two authors contributed equally to this paper.

Received: 17 January 2018; Accepted: 11 February 2018; Published: 13 February 2018

Abstract: Micro/nanomotors are self-propelled machines that can convert various energy sources into autonomous movement. With the great advances of nanotechnology, Micro/Nanomotors of various geometries have been designed and fabricated over the past few decades. Among them, the tubular Micro/Nanomotors have a unique morphology of hollow structures, which enable them to possess a strong driving force and easy surface functionalization. They are promising for environmental and biomedical applications, ranging from water remediation, sensing to active drug delivery and precise surgery. This article gives a comprehensive and clear review of tubular Micro/Nanomotors, including propulsion mechanisms, fabrication techniques and applications. In the end, we also put forward some realistic problems and speculate about corresponding methods to improve existing tubular Micro/Nanomotors.

Keywords: tubular Micro/Nanomotors; propulsion mechanisms; fabrication techniques; applications

---

1. Introduction

Motion has been essential for life over millions of years of evolution and is a common phenomenon in every living cell and organism. Microscopic biological motors [1–3] and machines are involved in numerous cellular or organismal activities, such as ATP synthesis, DNA replication and protein synthesis in cells, as well as the free locomotion of living creatures from ants to blue whales. Inspired by the biological motors and machines of nature, considerable efforts have been devoted to developing self-propelled artificial Micro/Nanomotors that can mimic these amazing functions of natural systems [3–5]. Artificial Micro/Nanomotors are machines at a micro/nanometer scale that are able to transduce other forms of energy from their surrounding into autonomous movement [6,7]. They may bring about revolutionary changes in micro/nanoengineering, environment treatment [8–10], biomedicine [11–14], etc.

Thanks to the advances in the nanotechnology over the past decades, Micro/Nanomotors with various geometries have been developed, such as rods [15–19], Janus spheres [20–28], tubes [29–36] and so on. The systematical and in-depth studies demonstrate that the geometry of Micro/Nanomotors has a huge impact on their performance in terms of propulsion mechanisms, motion behaviors and subsequent applications [3]. For example, the Au-Pt bimetallic nanorods synthesized by Sen and co-workers [15] are capable of moving autonomously in a solution of H_2O_2 with a speed of 2–10 body lengths per second. The motion of such bimetallic nanorods follows a self-electrophoresis propulsion mechanism, which is susceptible to the ionic strength of the solution. For Janus spheres Micro/Nanomotors using Pt as a catalyst, the propulsion mechanisms generally vary with their particle sizes [37]. At a small size, they move in a self-phoresis propulsion mechanism, while those with a relative large size may be propelled by bubble recoiling. The latter have a relatively strong driving force and thus move faster than the former. Compared to bimetallic nanorods and Janus spheres, tubular Micro/Nanomotors have a unique hollow structure, providing a confined space for chemical/biochemical reactions or physical transformations. This advantage enables tubular Micro/Nanomotors to be easily propelled by a strong thrust of bubbles [31,32,38]. For instance, the bilayer poly (3,4-ethylenedioxythiophene) (PEDOT)/Pt tubular micromotors fabricated by Wang and co-workers [32] can move at a speed of over 1400 body lengths per second and only require a very low H_2O_2 concentration. The speed of tubular Micro/Nanomotors can be further improved to over 158,000 body lengths per second by using ultrasound as an energy source. These bubble-propelled tubular Micro/Nanomotors can locomote efficiently in high ionic-strength media and relevant biological fluids. By combining the efficient autonomous motion with the large surface areas, tubular Micro/Nanomotors offer great promise for wide important applications, ranging from water remediation [8,9], sensing [39,40] to active drug delivery [41,42] and invasive surgery [43,44].

Tubular Micro/Nanomotors were named as microjets and microrockets for the first time [29]. Up to date, various tubular Micro/Nanomotors have been developed including not only catalytically propelled tubular Micro/Nanomotors but also those propelled by chemical/biochemical reactions, external fields and motile microorganisms. The recent review by Mei and his colleagues, which we noted almost at the same time of our submission, mainly focused on catalytically propelled tubular Micro/Nanomotors [45]. Herein, we attempt to present a comprehensive review on the state-of-the-art of tubular Micro/Nanomotors in balance. After briefly illustrating the dependence of the motion behaviors of Micro/Nanomotors on their morphologies, we introduce in Section 2 the relationship between propulsion mechanisms of tubular Micro/Nanomotors and their hollow structures with typical examples and discuss the corresponding motion behaviors. Then, we summarize in Section 3 the fabrication techniques of tubular Micro/Nanomotors, which mainly include rolled-up and template-assisted methods. In Section 4, we highlight the recent developments of environmental and biological applications based on those tubular Micro/Nanomotors. Finally, we also envision the current challenges and future prospects of tubular Micro/Nanomotors before some concluding remarks. We hope this article will provide readers with useful information and insights and promote the further development of tubular Micro/Nanomotors.

2. Propulsion Mechanisms

The key point to design self-propelled artificial Micro/Nanomotors is to construct an asymmetric field across micro/nanostructures to break the pressure symmetry. Thus far, a variety of asymmetric fields, such as a local electric field, concentration gradients, surface tension gradients, temperature gradients and bubbles, have been employed to drive Micro/Nanomotors. Compared to Micro/Nanomotors with bimetallic nanorods and Janus structures, tubular Micro/Nanomotors possess unique inner hollow structures. This enables chemical/biochemical reactions or physical transformations to be confined in the inner cavity of the tubes. As a result, tubular Micro/Nanomotors commonly possess a relatively strong driving force. In this section, we illustrate the different propulsion mechanisms of tubular Micro/Nanomotors and discuss the corresponding motion behaviors such as driving force or lifetime.

Most tubular Micro/Nanomotors are propelled by (photo)catalytic/chemical reactions [29–32,36,46–55] occurring in the inner hollow structures. For this type of tubular Micro/Nanomotors, a catalyst including noble metals and semiconducting oxides, or active metals should be incorporated in the body of micro/nanotubes to trigger chemical/biochemical reactions in the presence of fuels, such as H_2O_2. As shown in Figure 1A, noble metals, such as Pt, can decompose H_2O_2 into H_2O and O_2 at the inner wall of tubular Micro/Nanomotors [46–49]. The resulting O_2 molecules would nucleate, grow into bubbles, diffuse and finally burst or break away from one open end of tubular Micro/Nanomotors. Once a bubble was ejected from one open end of the tube, the tube started to move in the opposite direction and the symmetry of bubble ejection was broken. In this way, the open end with the first bubble ejection would always act as bubble exhaust nozzle and the other open end of the tube would serve as the feeding inlet of aqueous fuels. For tubes with different structures (cone or cylinder), the situation will be different. When the tubes possess conical structures, the large opening end would always serve as bubble exhaust nozzle and the movement direction would be toward the small opening end [30]. While in the case of cylindrical structures, the two opening ends can both serve as bubble exhaust nozzle. Meanwhile, the influence of the tube length cannot be ignored. When the tube is too long, the first ejected bubble could not move the tube. In this case, the symmetry for the bubble ejection does not break [50]. Our group demonstrated a TiO$_2$-based tubular micromotor [50], as shown in Figure 1B. When irradiated by ultraviolet (UV) light, TiO$_2$ with a band gap of 3.2 eV would excite electron–hole pairs according to Equation (1), these photo-generated electrons and holes could react with H$_2$O$_2$ to produce O$_2$ and H$_2$O described as Equations (2) and (3). When the inner diameter and length of the microtube were tuned, the O$_2$ molecules would preferentially nucleate and grow into bubbles on the inner concave surface rather than on the outer surface, resulting in strong propulsion of the micromotor. The motion behaviors and speed of this micromotor can be reversibly, wirelessly and remotely controlled at will with an ultrafast response rate (less than 0.2 s) by regulating "off-on" switch and intensity of UV irradiation. Using the same propulsion mechanism but alternative fuels of acids, Wang and co-workers [36] reported a polyaniline (PANI)/Zn bilayer tubular micromotor. In a strongly acidic solution, H$_2$ bubbles will be produced in the inner cavity of microtube through a spontaneous redox reaction (Equation (4)), leading to ultrafast propulsion of tubular micromotors (Figure 1C). Such acid-powered tubular micromotors are able to perform in extreme environments. However, owing to the consumption of Zn, these acid-powered tubular micromotors will stop run within a few minutes, which will hinder their practical applications. Sánchez and co-workers [56] reported a ultrasmall tubular SiO$_2$ nanomotor using urea as fuels. This nanomotor was propelled by the turnover of urea substrate triggered by urease according to Equation (5). As shown in Figure 1D, the reaction products, NH$_3$ and CO$_2$, were formed inside the nanotubes, generating internal flows that extent into the external space via the tube opening. Both the inside and outside urease contributed to the motion of the nanomotors. No visible bubbles were observed for this nanomotor, which held great promise for use in biomedical fields. Compared to bubble-propelled micromotors, the driving force of these nanomotors is relatively weak.

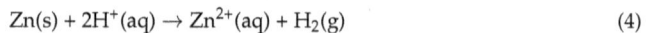

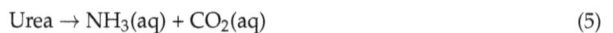

$$TiO_2 + h\nu \rightarrow h^+ + e^- \tag{1}$$

$$H_2O_2 + 2h^+ \rightarrow O_2 + 2H^+ \tag{2}$$

$$H_2O_2 + 2e^- + 2H^+ \rightarrow 2H_2O \tag{3}$$

$$Zn(s) + 2H^+(aq) \rightarrow Zn^{2+}(aq) + H_2(g) \tag{4}$$

$$Urea \rightarrow NH_3(aq) + CO_2(aq) \tag{5}$$

Tubular Micro/Nanomotors can also be propelled by external fields [57–60] or motile microorganisms [61–64]. External field-induced physical transformations can be confined in the inner cavity of the micro-/nanotubes. Wu and co-workers [57] demonstrated a near-infrared (NIR) light-driven polymer multilayer tubular micromotor. Under the irradiation of an NIR laser, the Au shell with a strong plasmon resonance absorption in the NIR region created local temperature gradients around the wall of the microtube, as depicted in Figure 2A. Owing to the asymmetric structure of the microtube, the temperature in the inner surface will be higher than in the outer surface. As a result, two thermophoretic forces ($f_i > f_o$) perpendicular to the inner and outer surface of the microtube were generated. Due to the conical structure of the microtube, the net resultant force (f_r) would face toward the front small-opening along the longitudinal axis of the microtube. This net resultant force pushed the micromotor to move fast toward, in the direction of the front small-opening and roughly maintains a linear trajectory. Wang and co-workers [58] reported a superfast ultrasound-propelled tubular micromotor. As depicted in Figure 2B, the thiolated cysteamine was first modified to the inner Au layer through thiol-Au bond, allowing for electrostatic binding of the anionic perfluorocarbon (PFC) droplets. The PFC droplets can be vaporized by ultrasound to produce ejected gas bubbles, propelling the microtube as other bubble-propelled micromotors. This tubular micromotor can travel at remarkably high average velocities at over 158,000 body lengths per second, providing a strong thrust to deeply penetrate and deform tissues. However, the microtube will stop once it runs out of PFC droplets, suggesting that the propulsion only lasts for a short time. Motile microorganism can be trapped into the inner cavity of a microtube when they have a similar size. Sánchez and co-workers [61] reported a sperm-propelled bio-hybrid micromotor. As shown in Figure 2C, a sperm cell was trapped into the tube cavity of a microtube. The as-prepared bio-hybrid micromotor can be propelled by the sperm cell. By incorporating a magnetic component into the microtube, the direction of the bio-hybrid micromotors can be controlled by external magnetic fields. These sperm-propelled bio-hybrid tubular micromotors promise the biocompatibility, flexibility and untethered operation.

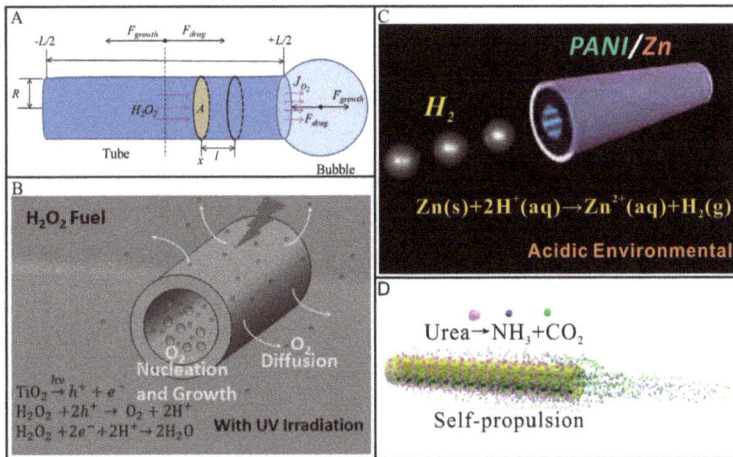

Figure 1. Representative examples of chemically/biochemically propelled tubular Micro/Nanomotors. (**A**) Schematic illustration of tubular catalytic micromotor propelled by O_2-bubble ejection in an aqueous H_2O_2 solution (adapted from Reference [46]); (**B**) Schematic diagram of photocatalytic reaction propelled motion of TiO_2 micromotor in an aqueous H_2O_2 solution under the irradiation of UV light (adapted from [50]); (**C**) Schematic diagram of motion for tubular polyaniline (PANI)/Zn micromotor in an acidic environment (adopted from [36]); (**D**) Schematic illustration of the motion of urease-conjugated SiO_2 tubular micromotor (adapted from [56]).

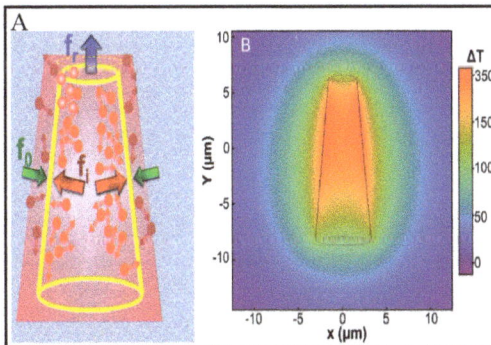

Figure 2. *Cont.*

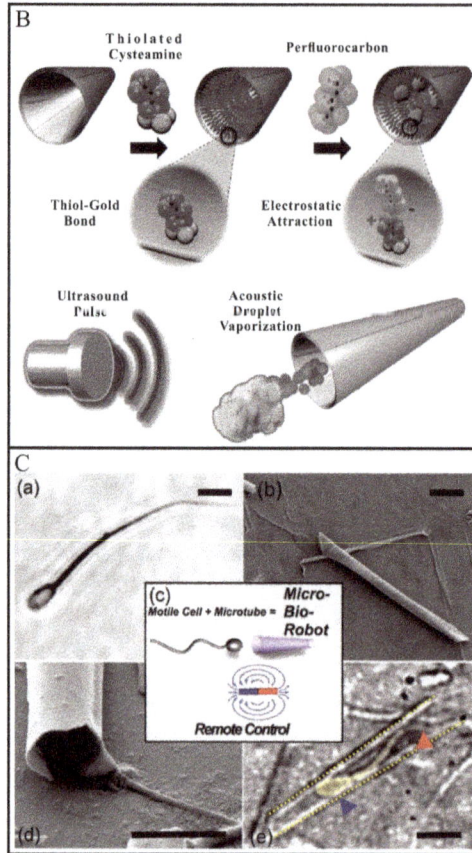

Figure 2. Representative examples of external field or motile microorganism-propelled tubular Micro/Nanomotors. (**A**) Schematic diagram of Au nanoshells powered by self-thermophoresis upon exposure to a NIR laser (adapted from [57]); (**B**) Schematic illustration of acoustic droplet vaporization and propulsion of PFC-loaded micromotors triggered by an ultrasound pulse (adapted from [58]); (**C**) A bio-hybrid tubular micromotor combining a single motile sperm cell with a rolled-up microtube (adapted from [61]).

To date, various energy sources, such as chemical/biochemical reactions, external fields and motile microorganisms, have been employed to propel tubular Micro/Nanomotors. The confinement effect of the hollow structures enables the driving force of tubular Micro/Nanomotors to be much stronger than other Micro/Nanomotors with different geometries.

3. Fabrication Techniques

Fabrication techniques play a vital role in the realization of the practical applications of tubular Micro/Nanomotors, as each application needs special functional units. Over the past decade, a large number of fabrication techniques have been developed to prepare tubular Micro/Nanomotors. Each of these techniques has its own advantages and limitations in terms of scalability, precision, cost, or device size. In this section, we summarize different types of fabrication techniques and divide them into two parts: rolled-up method and template-assisted method.

3.1. Rolled-Up Method

Rolled-up method has been widely used for the design of tubular Micro/Nanomotors, which is generally based on strain engineering [29]. By incorporating an engineered strain gradient in the deposited membranes, the membranes can form into desired structures when released from the substrate. Thus far, a variety of materials or materials combinations have been used for this fabrication method.

Mei and co-workers fabricated several tubular micromotors by using rolled-up method [30,35,65]. As shown in Figure 3A, a photoresist layer was first deposited on the substrate and served as a sacrificial layer [30]. Subsequently, different metals were deposited in order to produce a pre-stressed, multimetallic thin membrane. By selectively removing the photoresist layer with acetone, the thin membrane spontaneously rolled up into a tubular Micro/Nanomotors. The Pt inner layer was served as the catalyst, while the Fe layer was used for magnetic guidance. The diameter and length of the tubular micromotors can be controlled by deposition parameters (e.g., film thickness and strain) and lithographically predefined patterns, respectively. Apart from metals, other materials such as semiconductors, oxides, magnetic materials and polymers can also be used to rolled-up into tubular structures [66,67]. However, the above-mentioned rolled-up methods can only be fit to fabricate few tubular Micro/Nanomotors, when tons of motors are needed, the cost of the harsh preparation conditions (e.g., clean room) will be a huge challenge. Thus, great efforts have been devoted to simplifying the rolled-up process and reducing its cost. Li and co-workers [68] utilized anodic aluminum oxide (AAO) membranes as a sacrificial template to fabricate rolled-up tubular micromotors. As illustrated in Figure 3B, pre-strained Ti/Cr/Pt metallic tri-layers were subsequently deposited on the AAO membranes by e-beam evaporation. After metal deposition, the upper metallic tri-layers were divided into quadrate patterns with a size of 10–30 μm using a fine mechanical scratching process. Finally, potassium hydroxide (KOH) solution was used to selectively dissolve AAO sacrificial templates, leading to the release of the patterned metallic tri-layers. The intrinsic stress in the Ti/Cr/Pt tri-layers made them automatically roll up into microtubular structures. The produced microtubes exhibited higher velocities compared to those with a smooth platinum surface. Yao and co-workers [69] fabricated a tubular micromotor which was scrolled from graphene oxide (GO). In this case, GO has two functions: acting as a support for the metallic layers and providing an easily cleavable interface between the substrate and the metallic layers. Because of material strain and weak bonding between GO layers, microscrolls with GO on the outside and Pt at the inner surface were spontaneously formed upon sonication. The diameter can be modulated by changing the thickness of the deposited metal layers. Pumera and co-workers [48] reported a similar method to fabricate rolled-up tubular micromotors by using fruit tissue cells as the support for the deposited metal layers. The as-prepared tubular micromotors exhibited excellent mobility in the presence of H_2O_2. The same group [70] demonstrated the formation of Pt microtubes by using a transmission electron microscopy grid as templates. In this work, selective removal of the poly (methyl methacrylate) (PMMA) sacrificial layer under the deposited Pt layer or H_2O_2-assisted lift-off of the Pt layer deposited directly on a glass substrate were used to fabricate Pt microtubes. These clean room-free methods are simple and cheap; however, they cannot precisely control the size and morphology of the fabricated tubular micromotors.

To date, various rolled-up methods have been developed to fabricate tubular micromotors with different components. In addition, rolled-up technology has been used to demonstrate the first tubular micromotor with soft reconfiguration and it can also integrate ultra-compact electronic functionality like antennas [71] or integrated circuitry [72] into microtubes. Overall, future efforts should be provided to simplify the fabrication procedure of rolled-up methods.

3.2. Template-Assisted Method

Template assembly is a method to deposit a variety of materials into or onto the templates to form a multilayer tubular structure. Then, get rid of corresponding templates, so the tubular Micro/Nanomotors can be released. This kind of fabrication technique is considered to be pretty easy, low-cost and promises industrial process. Compared with rolled-up method, template-assisted method can prepare Micro/Nanomotors with smaller sizes. According to the position of the template, template-assisted method can be classified into external template-assisted method and internal template-assisted method.

3.2.1. External Template-Assisted Methods

External template-assisted methods prepare tubular Micro/Nanomotors within a template [31,32,36,60,73–75]. A membrane, such as track-etched polycarbonate membranes and porous alumina membranes, is commonly utilized as a template in external template-assisted methods. Each pore of the membrane can be utilized as a separated reactor to synthesize desired tubular micro-/nanomotor with different size and composition. Owing to the monodisperse diameters and large pore densities in the membrane, uniform tubular Micro/Nanomotors can be mass produced. Thus, these methods are powerful and relatively low-cost for preparing tubular Micro/Nanomotors. External template-assisted methods discussed here mainly include membrane template-assisted electrodeposition and layer-by-layer (LbL) assembly.

Membrane template-assisted electrodeposition is the most commonly used methods to fabricate tubular Micro/Nanomotors. For example, Wang and co-workers [31] developed a simplified method for fabricating tubular micromotors within the conically shaped pores of a polycarbonate membrane. As shown in Figure 3C, an Au film was first sputtered on one side of the polycarbonate membrane to serve as the working electrode. Aniline monomers were then deposited, followed by sequential deposition of Pt. Finally, the Au film and the polycarbonate membrane were removed by mechanical polishing and chemical etching, respectively. The resulting bilayer PANI/Pt micromotors were conical in shape with lengths of several micrometers and diameters depending on the pore size of the membrane template. These tubular micromotors exhibit ultrafast speed and require a very low concentration of H_2O_2 fuel. Following this work, the same group [32] investigated the composition and electropolymerization conditions of polymer-based tubular micromotors to realize optimization. PEDOT-based microtubes were found to provide more reproducible yields and consistent morphology. The most favorable condition was the use of a low monomer concentration together with the proper amount of surfactant and appropriate analyte. Instead of taking advantage of Pt-catalyzed decomposition of H_2O_2 as a source of power, Wang's group [36] described tubular PANI/Zn micromotors using acid in the environment as fuel. A layer of Zn was electrodeposited to serve as the inner wall of the microtubes after the electropolymerization of PANI. Except for polymer-based tubular micromotors, complete metallic micro/nanotubes can also be prepared via membrane template-assisted electrodeposition. Pumera and co-workers [73] successfully fabricated bimetallic Cu/Pt microtubes and proposed the use of widely available colloidal graphite ink instead of deposited metal backings in the setup of the electroplating cells to simplify the procedure. A striped metallic nanotube with a diameter of ~300 nm was prepared by electrodeposition employing conductive silver ink on the anodized aluminum oxide (AAO) membrane template together with aluminum foil as the working electrode.

Membrane template-assisted LbL assembly offers an easy and inexpensive method for fabricating multilayer tubular Micro/Nanomotors. The multilayer tubular structures are assembled by oppositely charged materials within the pores of membrane via electrostatic interaction. By incorporating catalysts, such as Pt nanoparticles or catalase, into the wall of multilayer tubular structure, the formed multilayer tubular structure can be propelled in H_2O_2 aqueous solutions. For example, He and co-workers [75] illustrated well-defined polymer multilayer tubular nanomotors by using membrane template-assisted LbL assembly. As depicted in Figure 3D, two oppositely charged biodegradable materials, chitosan (CHI) and sodium alginate (ALG), were alternatively absorbed on the nanopores of polycarbonate membrane and followed by the assembly of Pt nanoparticles into the template pores. Nanotubes with catalytic Pt nanoparticles on the inner wall can be obtained upon dissolution of the membrane. Such multilayer tubular Micro/Nanomotors can encapsulate various materials, such as small organic molecules, inorganic compounds, macromolecules and colloids, thus offer possibilities for directed drug delivery.

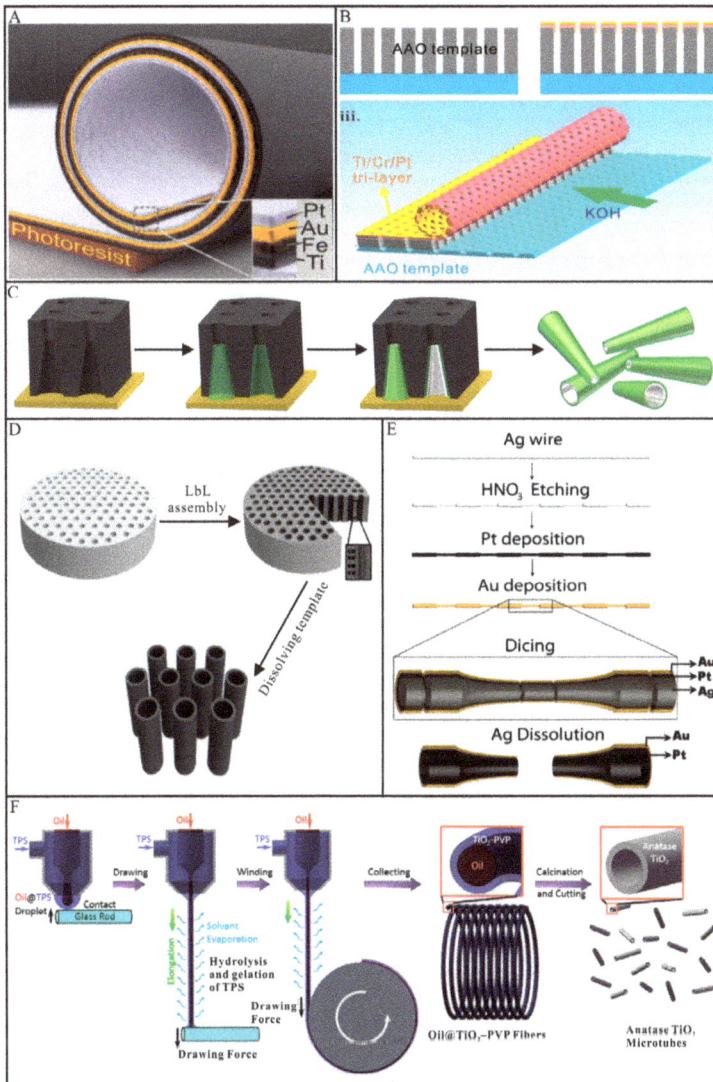

Figure 3. Representative examples of the fabrication techniques of tubular Micro/Nanomotors. (**A**) Schematic diagram of a rolled-up tubular micro/nanomotor consisting of Pt/Au/Fe/Ti multilayers on a photoresist sacrificial layer (adapted from [30]); (**B**) Three-dimensional schematic illustration of the fabrication method for the nanoporous tubular micromotors used anodic aluminum oxide (AAO) as a sacrificial template (adapted from [68]); (**C**) Preparation of bilayer PANI/Pt tubular micromotors using polycarbonate membranes (adapted from [31]); (**D**) Fabrication of polyelectrolyte multilayer tubular nanomotors. Black dots and vertical stripes represent Pt nanoparticles and polyelectrolyte multilayers, respectively (adapted from [75]); (**E**) Silver wire template-assisted layering approach for preparation of tubular micromotors (adapted from [76]); (**F**) Schematics to demonstrate the preparation of the TiO_2 tubular micromotor by a dry spinning method and subsequent calcination (adapted from [50]).

3.2.2. Internal Template-Assisted Methods

In contrast to external template-assisted methods, internal template-assisted methods prepare tubular Micro/Nanomotors on the outside of a template. Wang and co-workers [76] reported an approach to prepare conical tubular micromotors using etched silver wire as a template. As imaged in Figure 3E, after sequential electrodeposition of Pt and Au layers onto the surface, Ag wires were diced into microcones of the desired length. With nitric acid treatment, the Ag wire template was etched and the Pt/Au bilayer on the wire surface became conical microtubes. The diameter of the larger opening is the same as that of the Ag wires (50 µm). However, this approach is not suitable for mass production and the velocity of microtubes fabricated by this method is relatively low. By using Ag nanowires as templates, Sánchez and co-workers [56] fabricated an ultrasmall tubular silica nanomotor. A thin layer of silica was first grown on the surface of Ag nanowires by sol-gel chemistry. The silica coated Ag nanowires were then broken down into shorter segments by sonication. After removal of the Ag nanowires templates by etching in aqua regia, the silica nanotubes were obtained and functionalized with amine group by grafting. Urease was further conjugated onto the surface of the silica nanotubes via covalent conjugation. The as-prepared tubular nanomotors can be propelled by using biofriendly urea as a fuel without visible bubbles. Our group [50] reported a coaxial spinning method to fabricate light-driven tubular TiO_2 micromotors. As shown in Figure 3F, the TiO_2 precursor solution (TPS) and the heavy paraffin oil were added to two separated syringes connected to a coaxial spinneret. The TiO_2 precursor solution was used as an outer water phase and the heavy paraffin oil (Oil) was used as an inner oil phase. Then the Oil@TPS droplet was drawn into a liquid Oil@TPS jet by pulling the glass rod, which was previously contacted with the droplet, away from the spinneret. With the evaporation of solvents as well as the hydrolysis and gelation of TiO_2 precursor, the liquid Oil@TPS jet was solidified into Oil@TiO_2–PVP fiber. By the decomposition of the organic components with calcination, the collected Oil@TiO_2–PVP fibers were finally transformed into the TiO_2 hollow fibers. The TiO_2 hollow fibers were then cut into short microtubes.

Above all, template assembly is an easily accessible method and can be pretty cheap and fast. However, this technique inevitably has some shortcomings. For example, the size and shape of the resulting tubes are limited to presented template pores. In addition, the released tubular Micro/Nanomotors are difficult to be complete and uniform after dissolving the template.

4. Tubular Micro/Nanomotors towards Practical Applications

Tubular Micro/Nanomotors can effectively convert diverse energy sources into fast movement. In addition, the large surface area can be used to functionalize with various functional units. Thus, tubular Micro/Nanomotors have shown great potential applications, such as environmental and biomedical applications. For environmental applications, the bubble-propelled tubular Micro/Nanomotors by using H_2O_2 as fuels are widely explored. For certain biomedical applications, especially in vivo, the biocompatibility of the energy sources should be taken into account. While still in an early stage, attempts to explore environmental and biomedical applications of tubular Micro/Nanomotors are extremely active and encouraging. In this section, we present a detailed description of tubular Micro/Nanomotors towards practical applications in water remediation, sensing, drug delivery and precise surgery, which displays prominent superiorities to traditional methods.

4.1. Water Remediation

Water contaminations, such as organic compounds and heavy metals, are a serious risk to the public health and other life forms on earth. Current advances in nanotechnology have increased the interest of finding useful nanomaterials and nanotools for the fast and efficient removal of pollutants from water. Compared with their static counterparts, tubular micro/nanomotor hold great promise for water remediation. To date, a variety of tubular Micro/Nanomotors have been fabricated to degrade or remove pollutants from waste water.

Sánchez and co-workers [8] demonstrated the first example that utilized the self-propelled tubular micromotors to degrade organic pollutants in aqueous solutions. The tubular Fe/Pt bilayers micromotors were prepared by rolled-up technique. As shown in Figure 4A, these tubular micromotors combined two functions: the inner Pt layer was used to decompose H_2O_2 into O_2 and H_2O for the self-propulsion and the outer Fe layer was used for in situ generation Fe^{2+} ions to trigger the Fenton reaction (Equation (6)). The hydroxyl radical (•OH) generated from Fenton reaction was employed to degrade the organic pollutants. As a result, the degradation efficiency of the model pollutant (Rhodamine 6G) by these self-propelled tubular micromotors was 12 times faster than that of their static counterparts. Almost in the meanwhile, Wang and co-workers [77] demonstrated that the tubular PEDOT/Pt micromotors could accelerate the decontamination of organophosphate nerve agents in waste water (Figure 4B). Similar to the above-mentioned work, H_2O_2 has double functions, acting as a fuel to propel the tubular micromotors that contributed to an efficient fluid mixing and serving as an oxidizing reagent for the in-situ generation OOH^- nucleophiles in the presence of peroxide activator ($NaHCO_3$ or NaOH). Under mild quiescent conditions, the organophosphate nerve agents were oxidized into para-nitrophenol by OOH^- nucleophiles and a decontamination of 100% could be achieved with sufficient micromotors. However, when the same mixture with organophosphate pollutants and H_2O_2 reacted for the same duration, in the absence of the micromotors, the oxidation process was not observed in the aqueous solution. In addition to Pt as a catalyst, Schmidt and co-workers [78] explored a Pd-based catalytically tubular micromotor for efficient water cleaning. This tubular micromotor can decompose the target pollutant ($C_6H_5NO_3$) into nontoxic byproducts (C_6H_7NO) and H_2 bubbles, as described in Equation (7) and thus propel the microtube forward. The improved intermixing ability of the micromotors caused a 10 times faster degradation, as compared to their static counterparts. Moreover, taking the photocatalytic activity of TiO_2 into account, Wang and co-workers [79] reported an internally/externally bubble-propelled photocatalytic tubular nanomotor for the photodecomposition of organic pollutants. Two kinds of tubular nanomotors were fabricated by selective deposition of Pt nanoparticles on the inside or outside surface of TiO_2 nanotubes which can absorb photons. These two tubular nanomotors could efficiently propel at very low concentration of H_2O_2 and without the addition of any surfactant. Under the irradiation of UV light, organic pollutants (e.g., Rhodamine B) can be decomposed by •OH generated during the photocatalytic reaction according to Equation (8). Escarpa and co-workers [80] reported that CdS/PANI/Pt micromotors can also be employed for the photodecomposition of organic pollutants (e.g., bisphenol A). Bisphenol A can be decomposed by •OH generated during the photocatalytic reaction described in Equation (9). The enhanced fluid mixing and transport caused by the fast locomotion of such micromotors, led to a greatly improved degradation yield (~100%) compared to that of their static counterpart structures (~20%).

Besides the degradation of the pollutants, the tubular Micro/Nanomotors were also widely used to remove the pollutants from water. Wang and co-workers [81] demonstrated that ZrO-graphene/Pt tubular micromotors could selectively capture and remove nerve agents from the environmental matrices. In this work, the graphene layer acted as nucleation sites for the growth of nano-sized ZrO and Pt particles. The nano-sized ZrO could selectively interact with the phosphate groups via acid-base Lewis interaction and thus effectively capture nerve agents and enrich phosphopeptides. Compared to their static counterparts, the greatly increased fluid transport caused by the motile micromotors led to a 15-fold faster removal remediation. The same group [34] demonstrated that the tubular micromotors can be utilized for the removal of oil from the polluted water (Figure 4C). For this application, the outer surface of tubular Au/Ni/PEDOT/Pt micromotors was functionalized with a long chain of alkanethiols via covalent conjugation to form a hydrophobic monolayer. Efficient motion of the tubular micromotor obviously facilitated the interaction between oil droplets and the alkanethiol chains. As-prepared tubular micromotors were able to load and transport multiple small olive oil droplets from the solution, while their corresponding static counterparts could not pick up such droplets. Apart from organic pollutants, heavy metals originated from various human industrial activities also do great harm to living systems. Hence, it is essential to develop efficient and inexpensive methods to capture and remove them from waste water [80,82–84]. For this reason, Sánchez and co-workers [82] reported the tubular GO/Ni/Pt micromotors for the efficient removal and recovery of Pb^{2+} ions from waste water (Figure 4D). Owing to the abundant oxygen moieties on GO nanosheets, the Pb^{2+} ions could spontaneously be absorbed into the outer layer of such tubular micromotors. In this case, mobile GO-micromotors could remove Pb^{2+} ten times more efficiently than non-motile GO-micromotors, cleaning water from 1000 ppb down to below 50 ppb in 60 min. After accomplishing decontamination of Pb^{2+} ions, these tubular micromotors could be easily removed from the aqueous solutions by a magnet. Subsequently, by adjusting the pH of solutions to acidity, the adsorbed Pb^{2+} ions would be released from the tubular micromotors, allowing them to be recycled and reused for further decontamination processes. Escarpa and co-workers reported [80] that tubular ZnS/PANI/Pt micromotors could remove Hg^{2+} ions from contaminated solutions. By immersing the tubular ZnS/PANI/Pt micromotors into contaminated water, Hg^{2+} ions were able to replace the Zn^{2+} ions in ZnS nanoparticles through cation exchange (Equation (10)). When ZnS was successfully converted into HgS, the solution color became bright yellow, which can be used to monitor the pollution of Hg^{2+} ions. Similarly, the enhanced fluid mixing and transport caused by the fast locomotion of these micromotors, led to a greatly improved removal yield (~100%) compared to that of their static counterpart structures (~20%).

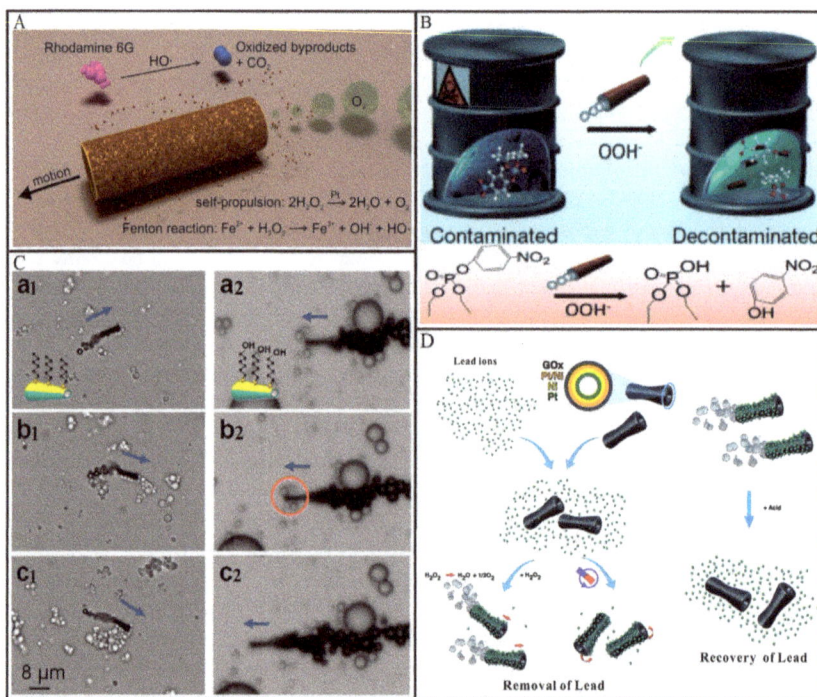

Figure 4. Representative examples of tubular Micro/Nanomotors for water remediation. (**A**) Schematic process for the degradation of polluted water (rhodamine 6G as a model contaminant) into inorganic products by multifunctional micromotors (adapted from [8]); (**B**) Illustration of a micromotor-based accelerated oxidative decontamination of organophospate nerve agents (adapted from [9]); (**C**) C6-SAM-modified micromotors with different head functional groups that can (left) or cannot (right) pick up small olive oil droplets (adapted from [34]); (**D**) Schematic images of GO-micromotors based approach for lead decontamination and recovery (adapted from [82]).

$$Fe^{2+} + H_2O_2 \rightarrow Fe^{3+} + OH^- + \bullet OH \tag{6}$$

$$C_6H_5NO_3 + NaBH_4 + H_2O \rightarrow C_6H_7NO + NaBO_2 + H_2O + H_2 \text{ (Pd as a catalyst)} \tag{7}$$

$$\text{organic pollutant} + \bullet OH \rightarrow CO_2 + H_2O \tag{8}$$

$$BPA + \bullet OH \rightarrow \text{organic acids} \rightarrow CO_2 + H_2O \tag{9}$$

$$ZnS + Hg^{2+} \rightarrow HgS + Zn^{2+} \tag{10}$$

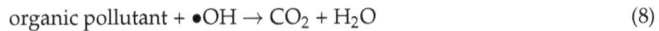

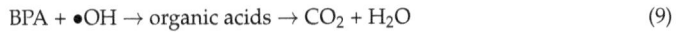

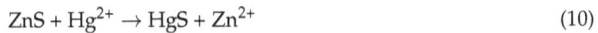

Water remediation conducted by tubular Micro/Nanomotors has been proved to be efficient and environmentally friendly in test tubes. However, there still exist many challenges for future practical applications. For example, various chemical pollutants in the waste water will poison the Pt layer, which greatly affects the motion behaviors of these tubular Micro/Nanomotors. In addition, how to collect these tubular Micro/Nanomotors from the solutions after finishing their missions is still a tricky question.

4.2. Sensing

Owing to the unique propulsion mechanisms, the bubble-propelled tubular Micro/Nanomotors can locomote effectively in high ionic-strength media and relevant biological fluids. In addition, various functional units can be modified on the inner or outer surface of tubular Micro/Nanomotors. Therefore, the tubular Micro/Nanomotors hold considerable promise for various sensing applications in both environmental and biomedical fields.

4.2.1. Environmental Sensing

In recent years, tubular Micro/Nanomotors have been demonstrated to have the capability to sense or monitor the water quality. The speed of the tubular Micro/Nanomotors is commonly used as a signal output for these sensors. Wang and co-workers [39] reported that the motion behavior of tubular PEDOT/Au-catalase micromotor can be used to test the water quality in the presence of aquatic pollutants (Figure 5A). A broad range of contaminants would affect the activity of catalase and thus impair the locomotion and survival time of these tubular micromotors. By monitoring the movement speed of the tubular micromotors, the water quality can be directly assessed. Pumera and co-workers [85] demonstrated that important extracellular thiols as well as basic organic molecules could significantly hamper the motion of Pt catalyzed tubular micromotors, prepared by both rolled-up and electrodeposition approaches, due to the poisoning of the catalytic Pt surface used to decompose H_2O_2 or the quenching of the hydroxyl radicals used to generate O_2 (Figure 5B). The same team [86–88] also demonstrated that electrolytes such as Na^+, K^+, Ca^{2+}, Cl^-, SO_4^{2-} and phosphates, uric acid and blood proteins such as bovine serum albumin, beta-globulin and glucose oxidase enzymes could hamper the mobility of Pt catalyzed tubular micromotors and be detected at small concentrations in the solution by monitoring the speed changes of corresponding micromotors. In order to clarify the influence of real-world environments on the motion of catalytic bubble-propelled micromotors, Pumera et al. [89] exposed the Cu/Pt bimetallic tubular micromotors to various types of water, including tap water, rain water, lake water and sea water, investigating their behaviors under real world conditions. They observed that the viability and mobility of such micromotors were strongly influenced by different water samples. The results were shown that there was a distinct negative correlation between the mobility of the micromotors and the ion content of the water found in real environments. Gao et al. [90] reported that the electrodeposited PEDOT/Pt tubular micromotors displayed remarkably high speeds in fuel-enhanced raw serum, apple juice, sea water, lake water and river water samples. The different results presented by the two research groups have not been found out and may arise from the difference in the material composition of the two micromotors.

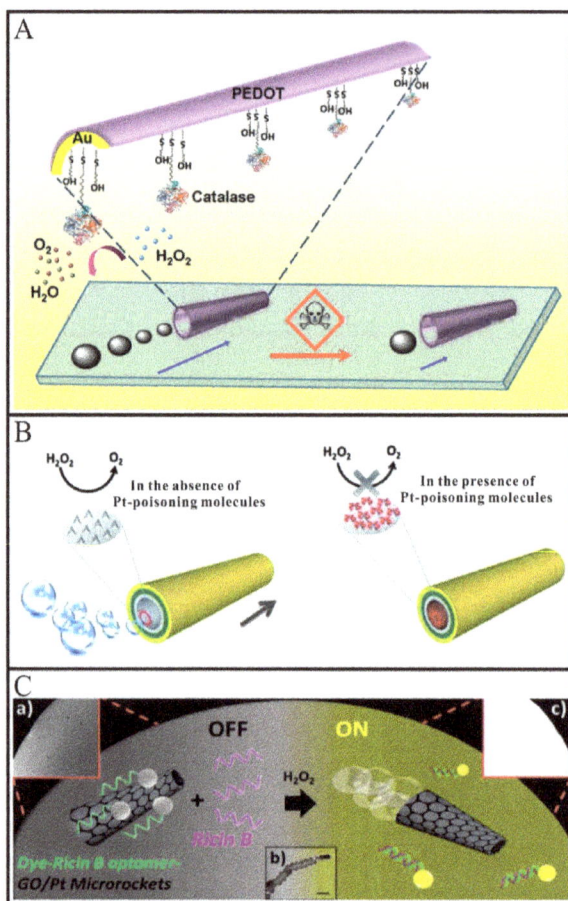

Figure 5. Representative examples of tubular Micro/Nanomotors for environmental sensing. (**A**) Schematic illustration of the pollutant effect on the micromotor locomotion speed through inhibition of the catalase biocatalytic layer (adapted from [39]); (**B**) Poisoning of the Pt-based micromotors with small molecules containing sulphur (adapted from [85]); (**C**) In vitro "off-on" fluorescent detection of ricin-B toxin by FAM-Ricin B aptamer-modified rGO/Pt micromotors (adapted from [91]).

Besides the speed of tubular micromotors, the fluorescence also can be utilized as the signal output [83,91,92]. Wang and co-workers [91] described a tubular micromotor-based sensor for the real-time detection of ricin B toxin in the surrounding environment. As depicted in Figure 5C, when the dye-aptamer adsorbed on the surface of graphene oxide via the π-π interaction, the fluorescence of the dye can be quenched by graphene oxide. While in the presence of ricin B, the dye-aptamer will release from the surface of graphene oxide owing to the specific binding between the ricin B and its aptamer, which resulted in the recovery of the fluorescence of the dye. By monitoring the fluorescence intensities of the solution, the concentration of ricin B can be real-time detected.

Tubular catalytic Micro/Nanomotors-based sensors have shown great potential for environmental monitoring. However, since a variety of species in waste water will affect the motion behaviors of tubular Micro/Nanomotors, the selectivity of those sensors which utilized the speed as the signal output will be not good.

4.2.2. Biosensing

Different from other micro/nanomaterials, the bubble-propelled tubular Micro/Nanomotors are able to autonomously locomote in biological liquids and thus provide an alternative method to isolate and detect biological analytes, such as nucleic acids, proteins, bacteria and living cells in unprocessed samples [40]. For example, Wang and co-workers [93] demonstrated that the tubular micromotor, which was modified with single-strand DNA on its outer surface via Au-S interaction, can selectively capture the target DNA from the raw biological samples and transport them to a clean location for subsequent analysis. Nguyen and Minteer [94] presented a novel DNA biosensor based on a tubular PEDOT/Au micromotor by using the speed as the signal output. As shown in Figure 6A, in the presence of the target DNA, the report DNA-modified Pt nanoparticle was captured to the inner surface of the tubular micromotors through the sandwich DNA hybridization. With the increased concentration of target DNA, the Pt catalyst nanoparticles attached to the inner wall will gradually increase, so the speed of the tubular micromotor will spontaneously increase. Recently, Wu and co-workers [95] demonstrated an efficient catalase-powered micromotor for the detection of target DNA. A multi-layer DNA and catalase were successively assembled on the inner surface of the tubular micromotor with the help of hybridization chain reaction. In the presence of target DNA, the sensing unit would hybrid with target DNA and release the multi-layer DNA as well as the multi-catalase, resulting in a decrease of the motion speed. Therefore, motion speed of such tubular micromotor-based biosensor could be used to roughly show the concentration of target DNA.

Motion-based design is not limited to sense DNA in biomedical field and it can also be extended to isolate and detect other biomacromolecules such as proteins [96–100]. For this purpose, Wang and co-workers [96] demonstrated a molecularly imprinted polymer-based tubular micromotor for the selective capture and transport of avidin. The recognition sites contained in the molecularly imprinted polymers were able to selectively capture the avidin from raw serum and saliva samples and transport them to predetermined destination. The same group [97] also demonstrated an aptamer-modified tubular micromotor for the isolation of thrombin from complex biological samples. This tubular micromotor can selectively capture the thrombin via the specific interaction between thrombin and its aptamer modified on the outer surface of tubular micromotors. The captured thrombin can be released by adding ATP which can bind and displace the immobilized mixed thrombin-ATP aptamer in 20 min. From isolation to detection, Wu and co-workers [98] fabricated a tubular micromotor-based biosensor for the detection of cancer biomarker, such as carcinoembryonic antigen (CEA). As depicted in Figure 6B, in the presence of CEA, the antibody-modified microspheres can be captured to the surface of tubular micromotors, leading to the decrease of the speed of tubular micromotors. The speed of the micromotors or the number of the microspheres conjugated on the micromotor can be used to monitor the concentration of the target proteins. Such microsensors can conveniently distinguish the concentration of CEA in a range of 1–1000 ng/mL and the whole detection procedure for protein target can be completed in 5 min without any washing and separation step. However, the speed of the tubular micromotor is influenced by various factors, thus those biosensors based on the speed as the signal output will be instable. Wang and co-workers [100] demonstrated a molybdenum disulfide (MoS$_2$)-based tubular micromotor for thrombin detection by using the fluorescence as the signal output. The fluorescent dye-modified thrombin aptamer was first immobilized on the surface of MoS$_2$ via π-π interactions. The fluorescent signal of the dye was rapidly quenched through the Förster resonance energy transfer (FRET) between the MoS$_2$ and the fluorescent dye. The continuous movement of the receptor-functionalized tubular micromotors resulted in the self-mixing and enhanced contact with the thrombin. Upon the interaction, the fluorescent dye modified aptamer can specifically recognize the thrombin and lead to the release of the fluorescent dye modified aptamer from the surface of tubular micromotor. Thus, the fluorescent signal will be recovered.

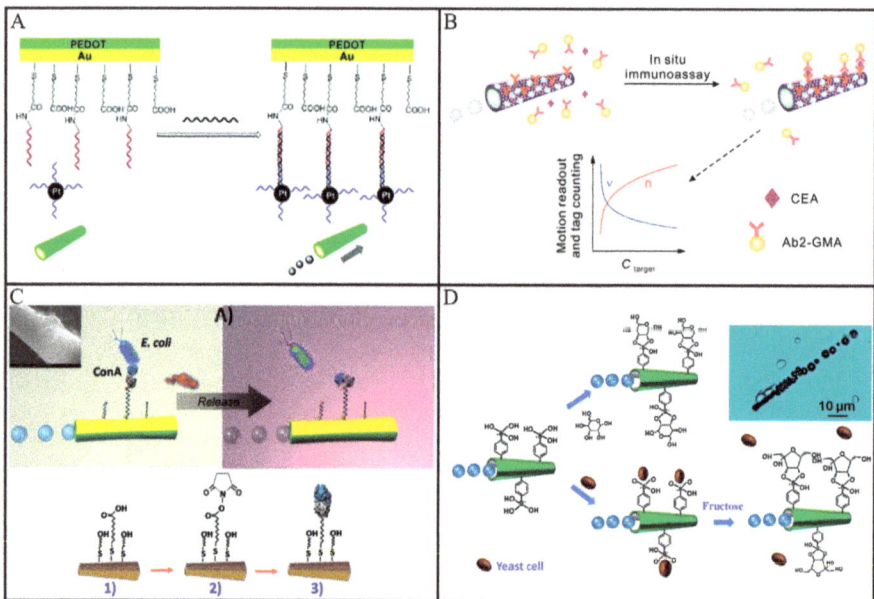

Figure 6. Representative examples of tubular Micro/Nanomotors for biosensing. (**A**) The principle for DNA detection by introducing Pt nanoparticle–DNA conjugate to the microtube via specific DNA hybridization (adapted from [94]); (**B**) Use for in situ immunoassay of protein biomarker via motion readout and tag counting (adapted from [98]); (**C**) A ConA-modified micromotor for selective isolation, transport and release of the target bacteria (adapted from [101]); (**D**) Schematic illustration of the PAPBA/Ni/Pt micromotor and its "on-the-fly" interaction with glucose (top) and yeast cell (bottom), along with triggered (fructose-induced) release of the cell (adapted from [33]).

Apart from biomacromolecules, self-propelled tubular micromotors can also be used to capture and release bacteria and living cells. Wang and co-workers [101] reported a lectin-modified tubular micromotor for the isolation of bacteria (*Escherichia coli*). As shown in Figure 6C, the concanavalin A (ConA) lectin modified on the outer surface of tubular micromotors can be utilized to recognize *Escherichia coli* and thus capture them from the clinical and environmental samples. By moving the tubular micromotors to a low pH-solution, the captured bacteria were released from the tubular micromotors owing to the dissociation of the sugar-lectin complex. An immuno-micromotor-based approach for in vitro isolation of circulating tumor cells was also explored by the same group [102]. By recognizing certain antigenic surface proteins such as CEA expressed on circulating tumor cells, these monoclonal-antibody-functionalized micromotors could selectively bind to target circulating tumor cells and then effectively transport them in phosphate-buffered saline and serum. Afterwards, the same team [33] presented a carbohydrate-sensitive tubular micromotor for the isolation monosaccharides and yeast cells. As imaged in Figure 6D, the outer surface of tubular micromotors was functionalized with boronic acid which can be used to selectively recognize monosaccharide. Thus, these boronic acid-modified tubular micromotors can be employed to bind and transport yeast cells which contain sugar residues on their wall. By the addition of fructose, the captured yeast cells can be release via a competitive sugar binding.

By integrating the sample pretreatment and subsequent detection in the same solution, tubular micro/nanomotor-based biosensors are simple and fast compared to conventional analytical methods [103–105]. The speed or the fluorescence signal of single tubular micro/nanomotor are generally used as the signal output, thus such biosensors can be utilized to the in-situ detection of the target analytes. However, there are still some issues for these biosensors. For example, the speed of Pt-based tubular Micro/Nanomotors is hampered in some biological media (e.g., proteins-rich media) [85]. Thus, the biosensors based on the speed as the signal output are instable. This can be alleviated by covering the Pt layer with a protective layer, such as MnO_2 layer [106] or a polymeric layer [107]. Besides, the sensitivity of these tubular Micro/Nanomotors-based biosensors is very low. Thus, apart from the speed and fluorescence, other sensitive signal output should be explored to improve their sensitivity.

4.3. Active Drug Delivery

Over the past several decades, a large number of nanomaterial-based carriers have been employed for drug delivery. However, because of the lack of the capabilities of self-propulsion and controllable navigation, the drug delivery efficiency of these nanocarriers is still very low (<5%) [108]. Compared to the above-mentioned nanocarriers, Micro/Nanomotors are able to perform controlled navigation to targeted locations under physiological conditions and environments, as well as have the potential to rapidly transport and deliver therapeutic payloads directly to disease sites, thereby improving the therapeutic efficacy and reducing systemic side effects of highly toxic drugs. Thus, they represent a new and attractive class of delivery carriers that could potentially revolutionize drug delivery systems [109,110]. Among them, tubular Micro/Nanomotors have received great attentions owing to their strong thrust in biofluids. In the following subsection, we summarize the recent developments of in vitro and in vivo active drug delivery by tubular Micro/Nanomotors in detail.

Chemically/biochemically propelled tubular Micro/Nanomotors have been widely explored for active drug delivery and tremendous progresses have been made in the past few years. The first example by using a tubular nanomotor for drug delivery was demonstrated by He and co-workers [75]. In this work, the tubular nanomotors were fabricated by a nanoporous template-assisted LbL assembly. Fluorescent anticancer drug doxorubicin (DOX), magnetic Fe_3O_4 nanoparticles and catalytic Pt nanoparticles were encapsulated into the multilayer wall of this tubular polymeric nanomotor, respectively. Under the guidance of magnetic fields, the nanomotors can autonomously transport and steer to the target Hela cells. After the nanomotors were attached to the outer surface of the target HeLa cell, the DOX molecules encapsulated in the nanomotors can be released through the ultrasound irradiation. Wang and co-workers [100] fabricated molybdenum disulfide (MoS_2)/Pt tubular micromotors for active drug delivery. DOX was loaded onto the MoS_2 layer of tubular micromotors via π-π stacking and hydrophobic interactions. At lower pH, DOX will release from MoS_2 surface owing to the protonation of the amino group on DOX. A high concentration of H_2O_2 was considered to be incompatible with living organisms so that scientists have paid much attention to seek more active catalysts than Pt to replace Pt. For example, He and co-workers [41] demonstrated a catalase-based tubular micromotor for drug delivery. As depicted in Figure 7A, these tubular micromotors were fabricated by a template-assisted LbL technique, following the integration of a thermal-sensitive gelatin hydrogel in which catalase, Au nanoparticles, magnetic nanoparticles and DOX were encapsulated. The experimental results demonstrated that this catalase-based tubular micromotor can drive at a very low concentration of H_2O_2. With the magnetic guidance, the DOX-loaded tubular micromotors can transport to the targeted cancer cell. Upon the irradiation of NIR light, Au nanoparticles will absorb energy and thus heat the gelatin hydrogel. After the gelatin hydrogel melted, the DOX will be released from the microtubes and subsequently kill the surrounding cancer cells. In order to totally eliminate the effect of H_2O_2, alternative fuels should be used to drive tubular motors. Wang and co-workers reported [111] an acidic fuel-driven tubular micromotor for combinatorial delivery and release of multiple cargos. The double-conical tubular Zn micromotors were prepared by template-electrodeposition and fully loaded silica and Au nanoparticles through particle-infiltration techniques. When introduced in acidic fuel media, these tubular Zn micromotors were propelled by ejecting H_2 bubbles. When the Zn body was dissolved, the encapsulated cargos were released autonomously from the micromotors. More recently, chemically propelled tubular micromotors have been moved to the living animal models for active drug delivery. For example, Wang and co-workers [112] firstly applied acidic fuel-driven tubular micromotors for active drug delivery in the stomach of living mice. Owing to the harsh acidic environment in stomach, the Zn-based tubular micromotors fully loaded with cargos can display efficient propulsion. As shown in Figure 7B, these Zn-based tubular micromotors led to a dramatically increased retention of cargos in the stomach of mice. Afterwards, the same group [113] demonstrated that the Mg-based tubular micromotors with enteric coating can be utilized for drug delivery in the gastrointestinal tract (Figure 7C). The enteric coating was stable in acidic conditions but soluble in neutral or alkaline intestinal fluid. The enteric coating on the outer surface of tubular micromotors was used to position in the gastrointestinal tract, while the Mg particles loaded in cavity of tubular micromotors were used to propel in intestinal fluid. By tailoring the thickness of the enteric coating, they can tune the time required to dissolve the polymer layer, thereby controlling the distance that the tubular micromotors can travel in the gastrointestinal tract before their propulsion was activated. Upon activation, the tubular micromotors will propel and penetrate into the local tissue and remain there to release payloads.

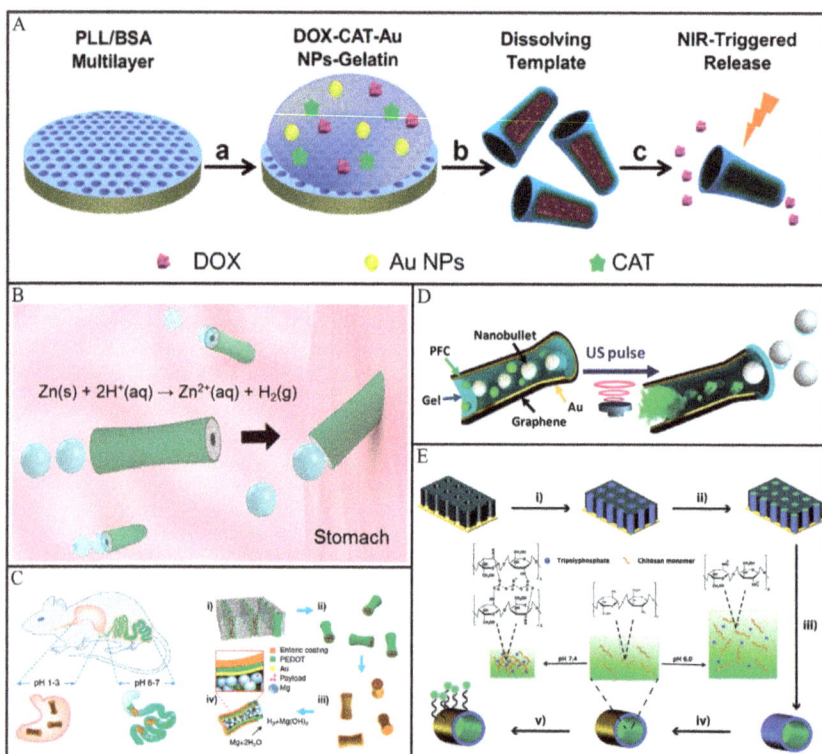

Figure 7. Representative examples of tubular Micro/Nanomotors for active drug delivery. (**A**) Fabrication and light-triggered drug release process of (PLL/BSA)$_{10}$-DOX-CAT-AuNPs-Gelatin micromotors (adapted from [41]); (**B**) Schematic representation of the in vivo propulsion and tissue penetration of zinc-based micromotors (adapted from [112]); (**C**) Schematic illustration of in vivo operation of the enteric Mg micromotors for propulsion and fluorescent images of localized delivery to the gastrointestinal tract (adapted from [113]); (**D**) Schematic illustration of the firing of nanobullets from the microcannon-structured motor by the spontaneous PFC vaporization upon application of US pulse (adapted from [114]); (**E**) modified Ni nanotube motors for active drug delivery (adapted from [60]).

Compared with the so far chemically/biochemically propelled Micro/Nanomotors, external field-propelled Micro/Nanomotors are more suitable for active drug delivery. This is because of the good biocompatibility, long sustainability and robust motion control. Recently, tubular Micro/Nanomotors propelled by ultrasound and magnetic fields have also been explored for active drug delivery. Wang and co-workers [114] demonstrated that the ultrasonically triggered tubular micromotors are able to loading and firing multiple cargos. As imaged in Figure 7D, silica/fluorescent microspheres and PFC emulsions were fully loaded in hollow conically shaped microtubes. Hoop and co-workers [60] fabricated a multifunctional magnetically propelled micromotor for active drug delivery. As depicted in Figure 7E, this micromotor consisted of a magnetic Ni microtube, which can be propelled by means of external magnetic fields. A pH-responsive chitosan hydrogel was functionalized in the inner cavity of Ni microtube, which served as a matrix to load drugs. By altering the pH values to acidity, the loaded drug will be release from micromotors owing to the swelling or dissolution of the hydrogel. A fluorescence tag (FITC labeled thiol-ssDNA) was conjugated on the outer surface of Ni microtube, which was used to trace the tubular micromotors. These magnetically propelled tubular micromotors can precisely navigate to targeted cells and release the encapsulated drugs. More recently, a sperm-biohybrid micromotor has been developed for active drug delivery [115]. This micromotor comprised a motile sperm that served as the propulsion source and drug (DOX) carrier and a 3D printed four-armed microtube, used for the magnetic guidance and mechanical release of the drug-loaded sperm cell in the desired area. The drug delivery occurred when the tubular microstructures bend upon pushing against a tumor spheroid and the sperm squeezed through the cancer cells and fused with cell membrane, thus minimizing toxic effects and unwanted drug accumulation in healthy tissues. These sperm-hybrid micromotors might have the potential application for gynecologic cancer and other diseases in the female reproductive tract treatment.

The proof-of-concept studies of tubular Micro/Nanomotors for active drug delivery have received good performance in vitro and in vivo. To extend these tubular Micro/Nanomotors to practical applications, several key issues, such as the size of tubular motors, the biocompatibility of the energy sources and the controllability of the motion behaviors etc., should be addressed by future research.

4.4. Precise Surgery

Unlike their macroscopical counterparts, Micro/Nanomotors with a strong driving force and deep penetration ability can potentially navigate throughout human body and operate in many hard-to-reach tissue locations [116,117]. Thus, they would have great promise to assist doctors to operate various minimally invasive surgeries in living body with high precision and flexibility. In this subsection, we summarize the recent advances of precise surgery based on tubular Micro/Nanomotors.

Recent advances in tubular Micro/Nanomotors have shown considerable feasibility for applying these tiny devices to precise surgery. Solovev and co-workers [118] utilized a catalytically propelled InGaAs/GaAs/(Cr)Pt tubular micromotor to drill into biomaterials such as those constituting HeLa cells. By tuning the tips of microtubes, the tubular micromotors can perform different movement. An asymmetrically tubular micromotor with a sharp tip could move in a corkscrew-like trajectory in the H_2O_2 solutions, which can drill deeply into the fixed HeLa cells, as imaged in Figure 8A. Owing to the high concentration of H_2O_2 involved, these tubular micromotors are unsuitable for practical applications. Cai and co-workers [59] demonstrated that a magnetically propelled Ni-embedded carbon nanotube can be utilized to penetrate into cell membranes. With the combination of rotating and static gradient magnetic fields, the carbon nanotube can spear and penetrate deeply into cells on a substrate. DNA plasmids loaded on the surface of the nanotubes subsequently released into targeted cells. Srivastava and co-workers [43] reported a magnetically propelled biotube for single-cell surgery and drug release. The biotubes were extracted from the *Dracaena* sp. Plant and coated with a Fe layer via an e-beam deposition process to incorporate magnetic control (Figure 8B). These biotubes possessed two functions that are the creation of cellular incision together with site-directed drug delivery. Apart from operating on single cell levels, the magnetically propelled tubular micromotors can also be operated in tissues or even a living animal. For example, a tubular Ti/Cr/Fe microdriller with a sharp tip was used to operate surgery in porcine liver tissue ex vivo by Sánchez and co-workers [44]. With the propulsion of an external rotational magnetic field, this tubular microdriller can embed inside the porcine liver tissue (Figure 8C). Pané and co-workers prepared a tubular CoNi micromotor for performing surgery in the eye of a living rabbit. As illustrated in Figure 8D, the tubular micromotor can be controlled wirelessly in the central vitreous of the rabbit eye with a rotating magnetic field [119]. These implantable tubular micromotors have the potential for targeting diseases in confined spaces of the human body.

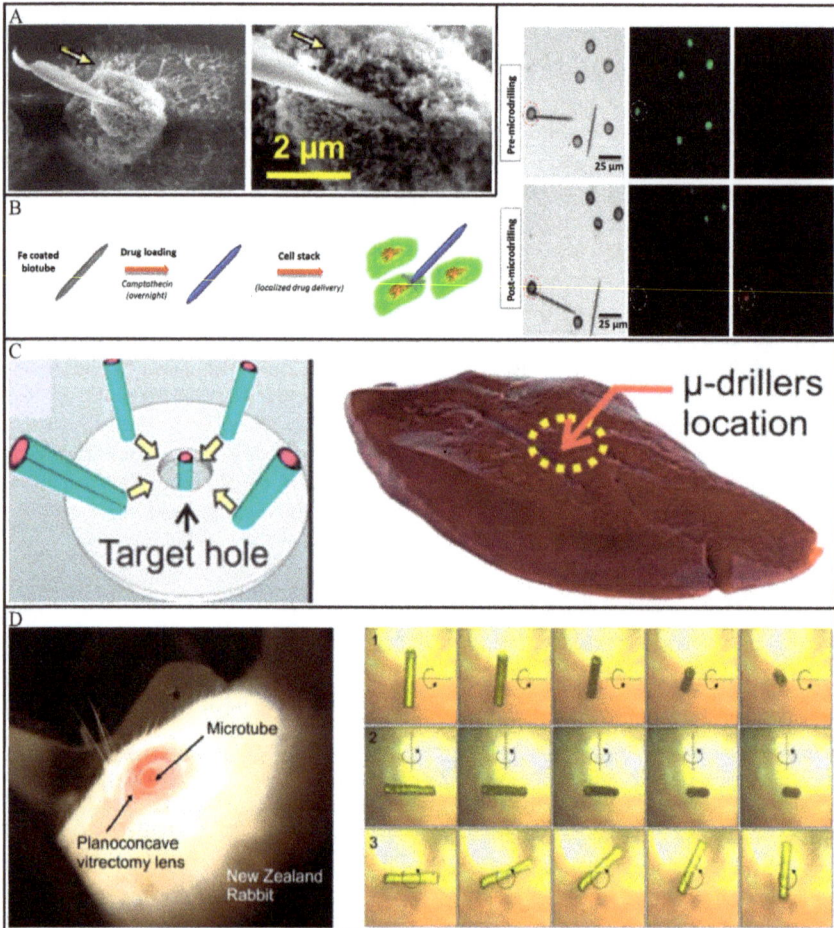

Figure 8. Representative examples of tubular Micro/Nanomotors for precise surgery. (**A**) SEM images of the guided catalytic InGaAs/GaAs/(Cr)Pt micromotors before and after drilling a single cell (adapted from [118]); (**B**) Schematic representation of imparting magnetic and drug delivery properties to the biotube, as well as fluorescent images of live cells (green) and dead cells (red) before and after microdrilling, respectively (adapted from [43]); (**C**) Schematic image showing the fuel-free motion of the micromotors towards the center of magnetic field and the drilling operation on pig liver tissue (adapted from [44]); (**D**) A living New Zealand rabbit eye with a micromotor and rotation of the micromotor around three axes at a rotating magnetic field in the vitreous humor (adapted from [119]).

The above-mentioned studies have demonstrated that tubular Micro/Nanomotors have great potential to operate precision surgery in vitro or even in living animal models. However, this is still an alluring but unmet goal remaining for biomedical researchers. More efforts should be done to extent these surgical tubular Micro/Nanomotors to practical applications.

5. Conclusions and Outlooks

Over the past decades, remarkable progress has been made in both understanding and applying tubular Micro/Nanomotors. Up to date, various energy sources, such as chemical/biochemical reactions, external fields and motile microorganisms, have been used to drive tubular Micro/Nanomotors. With the development of nanotechnology, different fabrication techniques, which mainly include rolled-up and template-assisted methods, have been developed to design and fabricate tubular Micro/Nanomotors with different sizes (the length in tens of nanometers to hundreds of micrometers) and special functionalities. Owing to their unique properties, such as the strong driving force and easy surface functionalization, tubular Micro/Nanomotors have been widely utilized for environmental and biomedical applications in terms of water remediation, sensing, active drug delivery and precise surgery. More recently, one important application of tubular Micro/Nanomotors in food-related fields is also emerging [120,121]. Although great advances have been made for tubular Micro/Nanomotors, several issues are still yet to be prior addressed to realize future practical applications. For example, Schmidt and co-worker [122] pointed out that medical micromotors need better imaging and control. Here, we generally put forward three challenges combining reported literatures and our own thoughts. First, the motion behaviors of tubular Micro/Nanomotors should be further improved. To date, most tubular Micro/Nanomotors are propelled by H_2O_2, which, however, is considered to be incompatible with living organisms at a high concentration. Although bio-friendly fuels, such as acids or urea, can be used as alternative fuels to power tubular Micro/Nanomotors, these tubular Micro/Nanomotors suffer from a weak driving force or limited lifetime. Second, facile and cost-effective fabrication techniques should be developed toward parallel mass production of tubular Micro/Nanomotors. Recently, the rolled-up methods are always dependent on complicated fabrication process and expensive facilities, while the template-assisted methods are difficult to tune the shape or size of the products. The geometry and morphology of the tubular Micro/Nanomotors play a vital role for their motion behaviors. The uniformity of the structures and reproducibility of the methods are of great importance for the practical applications. Thus, quality control cannot be ignored when simplifying the fabrication procedures. The fabrication methods of microtubes developed by our group [50,123] provide a good choice. Third, more efforts should be done to improve the performance of the tubular micro/nanomotor-based tools for practical applications in the future. For environmental applications, tubular Micro/Nanomotors have been demonstrated to be powerful in test tubes. However, how to move these tiny devices from test tubes to real environments is still a big challenge. For tubular Micro/Nanomotors-based sensors, other forms of signal output should be explored to improve their analytical performances (e.g., sensitivity or selectivity) which are inferior to the conventional methods. For active drug delivery and precise surgery, a large number of motors should cooperate and communicate with each other to accomplish tasks. We envision that above challenges can be gradually addressed, eventually expanding the horizon of tubular Micro/Nanomotors in various fields.

Acknowledgments: F.Z. and T.W. contributed equally to this work. This work was financially supported by the National Natural Science Foundation of China (21705123, 21474078 and 51521001), the Natural Science Foundation of Hubei Province (2015CFA003) and Top Talents Lead Cultivation Project of Hubei Province, the Yellow Crane talent plan of Wuhan municipal government and the fundamental Research Funds for the Central University (WUT: 2017 IVA 098, 2017-CL-A1-29).

Author Contributions: Jianguo Guan conceived the idea, and Ming Luo conceived the outline; Fengjun Zha and Tingwei Wang wrote the paper, made the pictures, and got the permission and copyrights. Ming Luo and Jianguo Guan revised the paper. Jianguo Guan supervised the whole project.

Conflicts of Interest: The authors declare no conflict of interest.

References

1. Van den Heuvel, M.G.; Dekker, C. Motor proteins at work for nanotechnology. *Science* **2007**, *317*, 333–336. [CrossRef] [PubMed]
2. Schliwa, M.; Woehlke, G. Molecular motors. *Nature* **2003**, *422*, 759–765. [CrossRef] [PubMed]
3. Guix, M.; Mayorga-Martinez, C.C.; Merkoci, A. Nano/micromotors in (bio)chemical science applications. *Chem. Rev.* **2014**, *114*, 6285–6322. [CrossRef] [PubMed]
4. Ozin, G.A.; Manners, I.; Fournier-Bidoz, S.; Arsenault, A. Dream nanomachines. *Adv. Mater.* **2005**, *17*, 3011–3018. [CrossRef]
5. Wang, J. Can man-made nanomachines compete with nature biomotors. *ACS Nano* **2009**, *3*, 4–9. [CrossRef] [PubMed]
6. Paxton, W.F.; Sundararajan, S.; Mallouk, T.E.; Sen, A. Chemical locomotion. *Angew. Chem. Int. Ed.* **2006**, *45*, 5420–5429. [CrossRef] [PubMed]
7. Xu, T.L.; Gao, W.; Xu, L.P.; Zhang, X.J.; Wang, S.T. Fuel-free synthetic micro-/nanomachines. *Adv. Mater.* **2017**, *29*, 1603250. [CrossRef] [PubMed]
8. Soler, L.; Magdanz, V.; Fomin, V.M.; Sánchez, S.; Schmidt, O.G. Self-propelled micromotors for cleaning polluted water. *ACS Nano* **2013**, *7*, 9611–9620. [CrossRef] [PubMed]
9. Soler, L.; Sánchez, S. Catalytic nanomotors for environmental monitoring and water remediation. *Nanoscale* **2014**, *6*, 7175–7182. [CrossRef] [PubMed]
10. Eskandarloo, H.; Kierulf, A.; Abbaspourrad, A. Nano- and micromotors for cleaning polluted waters: Focused review on pollutant removal mechanisms. *Nanoscale* **2017**, *9*, 13850–13863. [CrossRef] [PubMed]
11. Wang, J.; Gao, W. Nano/Microscale motors: Biomedical opportunities and challenges. *ACS Nano* **2012**, *6*, 5745–5751. [CrossRef] [PubMed]
12. Duan, W.; Wang, W.; Das, S.; Yadav, V.; Mallouk, T.E.; Sen, A. Synthetic nano- and micromachines in analytical chemistry: Sensing, migration, capture, delivery and separation. *Annu. Rev. Anal. Chem.* **2015**, *8*, 311–333. [CrossRef] [PubMed]
13. Li, J.X.; Gao, W.; de Ávila, B.E.; Zhang, L.F.; Wang, J. Micro/nanorobots for biomedicine: Delivery, surgery, sensing and detoxification. *Sci. Robot.* **2017**, *2*, eaam6431. [CrossRef]
14. Campuzano, S.; de Ávila, B.E.; Yáñez-Sedeño, P.; Pingarron, J.M.; Wang, J. Nano/microvehicles for efficient delivery and (bio)sensing at the cellular level. *Chem. Sci.* **2017**, *8*, 6750–6763. [CrossRef] [PubMed]
15. Paxton, W.F.; Kistler, K.C.; Olmeda, C.C.; Sen, A.; Angelo, S.K.; Cao, Y.Y.; Mallouk, T.E.; Lammert, P.E.; Crespi, V.H. Catalytic nanomotors: Autonomous movement of striped nanorods. *J. Am. Chem. Soc.* **2004**, *126*, 13424–13431. [CrossRef] [PubMed]
16. Paxton, W.F.; Baker, P.T.; Kline, T.R.; Wang, Y.; Mallouk, T.E.; Sen, A. Catalytically induced electrokinetics for motors and micropumps. *J. Am. Chem. Soc.* **2006**, *128*, 14881–14888. [CrossRef] [PubMed]
17. Liu, R.; Sen, A. Autonomous nanomotor based on copper-platinum segmented nanobattery. *J. Am. Chem. Soc.* **2011**, *133*, 20064–20067. [CrossRef] [PubMed]
18. De Ávila, B.E.; Martín, A.; Soto, F.; Lopez-Ramirez, M.A.; Campuzano, S.; Vásquez-Machado, G.M.; Gao, W.W.; Zhang, L.F.; Wang, J. Single cell real-time miRNAs sensing based on nanomotors. *ACS Nano* **2015**, *9*, 6756–6764. [CrossRef] [PubMed]
19. Wu, Z.G.; Li, T.L.; Gao, W.; Xu, T.L.; Jurado-Sánchez, B.; Li, J.X.; Gao, W.W.; He, Q.; Zhang, L.F.; Wang, J. Cell-membrane-coated synthetic nanomotors for effective biodetoxification. *Adv. Funct. Mater.* **2015**, *25*, 3881–3887. [CrossRef]
20. Pavlick, R.A.; Sengupta, S.; McFadden, T.; Zhang, H.; Sen, A. A polymerization-powered motor. *Angew. Chem. Int. Ed.* **2011**, *50*, 9374–9377. [CrossRef] [PubMed]
21. Gao, W.; Pei, A.; Feng, X.M.; Hennessy, C.; Wang, J. Organized self-assembly of Janus micromotors with hydrophobic hemispheres. *J. Am. Chem. Soc.* **2013**, *135*, 998–1001. [CrossRef] [PubMed]
22. Simmchen, J.; Baeza, A.; Ruiz, D.; Esplandiu, M.J.; Vallet-Regí, M. Asymmetric hybrid silica nanomotors for capture and cargo transport: Towards a novel motion-based DNA sensor. *Small* **2012**, *8*, 2053–2059. [CrossRef] [PubMed]
23. Mou, F.Z.; Chen, C.R.; Ma, H.R.; Yin, Y.X.; Wu, Q.Z.; Guan, J.G. Self-propelled micromotors driven by the magnesium-water reaction and their hemolytic properties. *Angew. Chem. Int. Ed.* **2013**, *52*, 7208–7212. [CrossRef] [PubMed]

24. Huang, C.X.; Shen, X.T. Janus molecularly imprinted polymer particles. *Chem. Commun.* **2014**, *50*, 2646–2649. [CrossRef] [PubMed]
25. Dong, R.F.; Zhang, Q.L.; Gao, W.; Pei, A.; Ren, B. Highly efficient light-driven TiO$_2$-Au Janus micromotors. *ACS Nano* **2016**, *10*, 839–844. [CrossRef] [PubMed]
26. Mou, F.Z.; Kong, L.; Chen, C.R.; Chen, Z.H.; Xu, L.L.; Guan, J.G. Light-controlled propulsion, aggregation and separation of water-fuelled TiO$_2$/Pt Janus submicromotors and their "on-the-fly" photocatalytic activities. *Nanoscale* **2016**, *8*, 4976–4983. [CrossRef] [PubMed]
27. Ebbens, S.; Gregory, D.A.; Dunderdale, G.; Howse, J.R.; Ibrahim, Y.; Liverpool, T.B.; Golestanian, R. Electrokinetic effects in catalytic platinum-insulator Janus swimmers. *EPL* **2014**, *106*, 58003. [CrossRef]
28. Brown, A.; Poon, W. Ionic effects in self-propelled Pt-coated Janus swimmers. *Soft Matter* **2014**, *10*, 4016–4027. [CrossRef] [PubMed]
29. Mei, Y.F.; Huang, G.S.; Solovev, A.A.; Ureña, E.B.; Mönch, I.; Ding, F.; Reindl, T.; Fu, R.K.; Chu, P.K.; Schmidt, O.G. Versatile approach for integrative and functionalized tubes by strain engineering of nanomembranes on polymers. *Adv. Mater.* **2008**, *20*, 4085–4090. [CrossRef]
30. Solovev, A.A.; Mei, Y.F.; Ureña, E.B.; Huang, G.S.; Schmidt, O.G. Catalytic microtubular jet engines self-propelled by accumulated gas bubbles. *Small* **2009**, *5*, 1688–1692. [CrossRef] [PubMed]
31. Gao, W.; Sattayasamitsathit, S.; Orozco, J.; Wang, J. Highly efficient catalytic microengines: Template electrosynthesis of polyaniline/platinum microtubes. *J. Am. Chem. Soc.* **2011**, *133*, 11862–11864. [CrossRef] [PubMed]
32. Gao, W.; Sattayasamitsathit, S.; Uygun, A.; Pei, A.; Ponedal, A.; Wang, J. Polymer-based tubular microbots: Role of composition and preparation. *Nanoscale* **2012**, *4*, 2447–2453. [CrossRef] [PubMed]
33. Kuralay, F.; Sattayasamitsathit, S.; Gao, W.; Uygun, A.; Katzenberg, A.; Wang, J. Self-propelled carbohydrate-sensitive microtransporters with built-in boronic acid recognition for isolating sugars and cells. *J. Am. Chem. Soc.* **2012**, *134*, 15217–15220. [CrossRef] [PubMed]
34. Guix, M.; Orozco, J.; García, M.; Gao, W.; Sattayasamitsathit, S.; Merkoçi, A.; Escarpa, A.; Wang, J. Superhydrophobic alkanethiol-coated microsubmarines for effective removal of oil. *ACS Nano* **2012**, *6*, 4445–4451. [CrossRef] [PubMed]
35. Mei, Y.F.; Solovev, A.A.; Sánchez, S.; Schmidt, O.G. Rolled-up nanotech on polymers: From basic perception to self-propelled catalytic microengines. *Chem. Soc. Rev.* **2011**, *40*, 2109–2119. [CrossRef] [PubMed]
36. Gao, W.; Uygun, A.; Wang, J. Hydrogen-bubble-propelled zinc-based microrockets in strongly acidic media. *J. Am. Chem. Soc.* **2012**, *134*, 897–900. [CrossRef] [PubMed]
37. Manjare, M.; Yang, B.; Zhao, Y.P. Bubble driven quasioscillatory translational motion of catalytic micromotors. *Phys. Rev. Lett.* **2012**, *109*, 128305. [CrossRef] [PubMed]
38. Sánchez, S.; Ananth, A.N.; Fomin, V.M.; Viehrig, M.; Schmidt, O.G. Superfast motion of catalytic microjet engines at physiological temperature. *J. Am. Chem. Soc.* **2011**, *133*, 14860–14863. [CrossRef] [PubMed]
39. Orozco, J.; García-Gradilla, V.; D'Agostino, M.; Gao, W.; Cortés, A.; Wang, J. Artificial enzyme-powered microfish for water-quality testing. *ACS Nano* **2013**, *7*, 818–824. [CrossRef] [PubMed]
40. Wang, J. Self-propelled affinity biosensors: Moving the receptor around the sample. *Biosens. Bioelectron.* **2016**, *76*, 234–242. [CrossRef] [PubMed]
41. Wu, Z.G.; Lin, X.K.; Zou, X.; Sun, J.M.; He, Q. Biodegradable protein-based rockets for drug transportation and light-triggered release. *ACS Appl. Mater. Interfaces* **2015**, *7*, 250–255. [CrossRef] [PubMed]
42. Solovev, A.A.; Sánchez, S.; Pumera, M.; Mei, Y.F.; Schmidt, O.G. Magnetic control of tubular catalytic microbots for the transport, assembly and delivery of micro-objects. *Adv. Funct. Mater.* **2010**, *20*, 2430–2435. [CrossRef]
43. Srivastava, S.K.; Medina-Sánchez, M.; Koch, B.; Schmidt, O.G. Medibots: Dual-action biogenic microdaggers for single-cell surgery and drug release. *Adv. Mater.* **2016**, *28*, 832–837. [CrossRef] [PubMed]
44. Xi, W.; Solovev, A.A.; Ananth, A.N.; Gracias, D.H.; Sánchez, S.; Schmidt, O.G. Rolled-up magnetic microdrillers: Towards remotely controlled minimally invasive surgery. *Nanoscale* **2013**, *5*, 1294–1297. [CrossRef] [PubMed]
45. Xu, B.; Zhang, B.R.; Wang, L.; Huang, G.S.; Mei, Y.F. Tubular micro/nanomachines: From the basics to recent advances. *Adv. Funct. Mater.* **2018**, *28*, 1705872. [CrossRef]
46. Manjare, M.; Yang, B.; Zhao, Y.P. Bubble-propelled microjets: Model and experiment. *J. Phys. Chem. C* **2013**, *117*, 4657–4665. [CrossRef]

47. Li, J.X.; Liu, W.J.; Wang, J.Y.; Rozen, I.; He, S.; Chen, C.R.; Kim, H.G.; Lee, H.J.; Lee, H.B.; Kwon, S.H.; et al. Nanoconfined atomic layer deposition of TiO$_2$/Pt nanotubes: Toward ultrasmall highly efficient catalytic nanorockets. *Adv. Funct. Mater.* **2017**, *27*, 1700598. [CrossRef]

48. Wang, H.; Moo, J.G.; Pumera, M. Tissue cell assisted fabrication of tubular catalytic platinum microengines. *Nanoscale* **2014**, *6*, 11359–11363. [CrossRef] [PubMed]

49. Jodra, A.; Soto, F.; Lopez-Ramirez, M.A.; Escarpa, A.; Wang, J. Delayed ignition and propulsion of catalytic microrockets based on fuel-induced chemical dealloying of the inner alloy layer. *Chem. Commun.* **2016**, *52*, 11838–11841. [CrossRef] [PubMed]

50. Mou, F.Z.; Li, Y.; Chen, C.R.; Li, W.; Yin, Y.X.; Ma, H.R.; Guan, J.G. Single-component TiO$_2$ tubular microengines with motion controlled by light-induced bubbles. *Small* **2015**, *11*, 2564–2570. [CrossRef] [PubMed]

51. Giudicatti, S.; Marz, S.M.; Soler, L.; Madani, A.; Jorgensen, M.R.; Sanchez, S.; Schmidt, O.G. Photoactive rolled-up TiO$_2$ microtubes: Fabrication, characterization and applications. *J. Mater. Chem. C* **2014**, *2*, 5892–5901. [CrossRef] [PubMed]

52. Safdar, M.; Minh, T.D.; Kinnunen, N.M.; Jänis, J. Manganese oxide based catalytic micromotors: Effect of polymorphism on motion. *ACS Appl. Mater. Interfaces* **2016**, *8*, 32624–32629. [CrossRef] [PubMed]

53. Safdar, M.; Wani, O.M.; Janis, J. Manganese oxide-based chemically powered micromotors. *ACS Appl. Mater. Interfaces* **2015**, *7*, 25580–25585. [CrossRef] [PubMed]

54. Ye, H.; Sun, H.Q.; Wang, S.B. Electrochemical synthesis of graphene/MnO$_2$ in an architecture of bilayer microtubes as micromotors. *Chem. Eng. J.* **2017**, *324*, 251–258. [CrossRef]

55. Sanchez, S.; Solovev, A.A.; Mei, Y.F.; Schmidt, O.G. Dynamics of biocatalytic microengines mediated by variable friction control. *J. Am. Chem. Soc.* **2010**, *132*, 13144–131445. [CrossRef] [PubMed]

56. Ma, X.; Hortelao, A.C.; Miguel-López, A.; Sánchez, S. Bubble-free propulsion of ultrasmall tubular nanojets powered by biocatalytic reactions. *J. Am. Chem. Soc.* **2016**, *138*, 13782–13785. [CrossRef] [PubMed]

57. Wu, Z.G.; Si, T.Y.; Gao, W.; Lin, X.K.; Wang, J.; He, Q. Superfast near-infrared light-driven polymer multilayer rockets. *Small* **2016**, *12*, 577–582. [CrossRef] [PubMed]

58. Kagan, D.; Benchimol, M.J.; Claussen, J.C.; Chuluun-Erdene, E.; Esener, S.; Wang, J. Acoustic droplet vaporization and propulsion of perfluorocarbon-loaded microbullets for targeted tissue penetration and deformation. *Angew. Chem. Int. Ed.* **2012**, *51*, 7519–7522. [CrossRef] [PubMed]

59. Cai, D.; Mataraza, J.M.; Qin, Z.H.; Huang, Z.P.; Huang, J.Y.; Chiles, T.C.; Carnahan, D.; Kempa, K.; Ren, Z.F. Highly efficient molecular delivery into mammalian cells using carbon nanotube spearing. *Nat. Methods* **2005**, *2*, 449–454. [CrossRef] [PubMed]

60. Hoop, M.; Mushtaq, F.; Hurter, C.; Chen, X.Z.; Nelson, B.J.; Pané, S. A smart multifunctional drug delivery nanoplatform for targeting cancer cells. *Nanoscale* **2016**, *8*, 12723–12728. [CrossRef] [PubMed]

61. Magdanz, V.; Sanchez, S.; Schmidt, O.G. Development of a sperm-flagella driven micro-bio-robot. *Adv. Mater.* **2013**, *25*, 6581–6588. [CrossRef] [PubMed]

62. Gaffney, E.A.; Gadêlha, H.; Smith, D.J.; Blake, J.R.; Kirkman-Brown, J.C. Mammalian sperm motility: Observation and theory. *Annu. Rev. Fluid Mech.* **2011**, *43*, 501–528. [CrossRef]

63. Magdanz, V.; Medina-Sánchez, M.; Schwarz, L.; Xu, H.F.; Elgeti, J.; Schmidt, O.G. Spermatozoa as functional components of robotic microswimmers. *Adv. Mater.* **2017**, *29*, 1606301. [CrossRef] [PubMed]

64. Magdanz, V.; Guix, M.; Hebenstreit, F.; Schmidt, O.G. Dynamic polymeric microtubes for the remote-controlled capture, guidance and release of sperm cells. *Adv. Mater.* **2016**, *28*, 4084–4089. [CrossRef] [PubMed]

65. Li, J.X.; Zhang, J.; Gao, W.; Huang, G.S.; Di, Z.F.; Liu, R.; Wang, J.; Mei, Y.F. Dry-released nanotubes and nanoengines by particle-assisted rolling. *Adv. Mater.* **2013**, *25*, 3715–3721. [CrossRef] [PubMed]

66. Harazim, S.M.; Xi, W.; Schmidt, C.K.; Sanchez, S.; Schmidt, O.G. Fabrication and applications of large arrays of multifunctional rolled-up SiO/SiO$_2$ microtubes. *J. Mater. Chem.* **2012**, *22*, 2878–2884. [CrossRef]

67. Magdanz, V.; Stoychev, G.; Ionov, L.; Sanchez, S.; Schmidt, O.G. Stimuli-responsive microjets with reconfigurable shape. *Angew. Chem. Int. Ed.* **2014**, *53*, 2673–2677. [CrossRef] [PubMed]

68. Li, J.X.; Liu, Z.Q.; Huang, G.S.; An, Z.H.; Chen, G.; Zhang, J.; Li, M.L.; Liu, R.; Mei, Y.F. Hierarchical nanoporous microtubes for high-speed catalytic microengines. *NPG Asia Mater.* **2014**, *6*, e94. [CrossRef]

69. Yao, K.; Manjare, M.; Barrett, C.A.; Yang, B.; Salguero, T.T.; Zhao, Y.P. Nanostructured scrolls from graphene oxide for microjet engines. *J. Phys. Chem. Lett.* **2012**, *3*, 2204–2208. [CrossRef] [PubMed]

70. Zhao, G.J.; Ambrosi, A.; Pumera, M. Clean room-free rapid fabrication of roll-up self-powered catalytic microengines. *J. Mater. Chem. A* **2014**, *2*, 1219–1223. [CrossRef]

71. Karnaushenko, D.D.; Karnaushenko, D.; Makarov, D.; Schmidt, O.G. Compact helical antenna for smart implant applications. *NPG Asia Mater.* **2015**, *7*, e188. [CrossRef]

72. Karnaushenko, D.; Münzenrieder, N.; Karnaushenko, D.D.; Koch, B.; Meyer, A.K.; Baunack, S.; Petti, L.; Tröster, G.; Makarov, D.; Schmidt, O.G. Biomimetic microelectronics for regenerative neuronal cuff implants. *Adv. Mater.* **2015**, *27*, 6797–6805. [CrossRef] [PubMed]

73. Zhao, G.J.; Pumera, M. Concentric bimetallic microjets by electrodeposition. *RSC Adv.* **2013**, *3*, 3963–3966. [CrossRef]

74. Zhao, G.J.; Ambrosi, A.; Pumera, M. Self-propelled nanojets via template electrodeposition. *Nanoscale* **2013**, *5*, 1319–1324. [CrossRef] [PubMed]

75. Wu, Z.G.; Wu, Y.J.; He, W.P.; Lin, X.K.; Sun, J.M.; He, Q. Self-propelled polymer-based multilayer nanorockets for transportation and drug release. *Angew. Chem. Int. Ed.* **2013**, *52*, 7000–7003. [CrossRef] [PubMed]

76. Maneshm, K.M.; Cardona, M.; Yuan, R.; Clark, M.; Kagan, D.; Balasubramanian, S.; Wang, J. Template-assisted fabrication of salt-independent catalytic tubular microengines. *ACS Nano* **2010**, *4*, 1799–1804. [CrossRef] [PubMed]

77. Orozco, J.; Cheng, G.Z.; Vilela, D.; Sattayasamitsathit, S.; Vazquez-Duhalt, R.; Valdés-Ramírez, G.; Pak, O.S.; Escarpa, A.; Kan, C.Y.; Wang, J. Micromotor-based high-yielding fast oxidative detoxification of chemical threats. *Angew. Chem. Int. Ed.* **2013**, *52*, 13276–13279. [CrossRef] [PubMed]

78. Srivastava, S.K.; Guix, M.; Schmidt, O.G. Wastewater mediated activation of micromotors for efficient water cleaning. *Nano Lett.* **2016**, *16*, 817–821. [CrossRef] [PubMed]

79. Wang, S.; Jiang, Z.Z.; Ouyang, S.S.; Dai, Z.P.; Wang, T. Internally/externally bubble-propelled photocatalytic tubular nanomotors for efficient water cleaning. *ACS Appl. Mater. Interfaces* **2017**, *9*, 23974–23982. [CrossRef] [PubMed]

80. Jurado-Sánchez, B.; Wang, J.; Escarpa, A. Ultrafast nanocrystals decorated micromotors for on-site dynamic chemical processes. *ACS Appl. Mater. Interfaces* **2016**, *8*, 19618–19625.

81. Singh, V.V.; Martin, A.; Kaufmann, K.; de Oliveira, S.D.; Wang, J. Zirconia/graphene oxide hybrid micromotors for selective capture of nerve agents. *Chem. Mater.* **2015**, *27*, 8162–8169. [CrossRef]

82. Vilela, D.; Parmar, J.; Zeng, Y.F.; Zhao, Y.L.; Sánchez, S. Graphene-based microbots for toxic heavy metal removal and recovery from water. *Nano Lett.* **2016**, *16*, 2860–2866. [CrossRef] [PubMed]

83. Vilela, D.; Hortelao, A.C.; Balderas-Xicohtencatl, R.; Hirscher, M.; Hahn, K.; Ma, X.; Sánchez, S. Facile fabrication of mesoporous silica micro-jets with multi-functionalities. *Nanoscale* **2017**, *9*, 13990–13997. [CrossRef] [PubMed]

84. Wang, H.; Khezri, B.; Pumera, M. Catalytic DNA-functionalized self-propelled micromachines for environmental remediation. *Chem* **2016**, *1*, 473–481. [CrossRef]

85. Zhao, G.J.; Sánchez, S.; Schmidt, O.G.; Pumera, M. Poisoning of bubble propelled catalytic micromotors: The chemical environment matters. *Nanoscale* **2013**, *5*, 2909–2914. [CrossRef] [PubMed]

86. Wang, H.; Zhao, G.J.; Pumera, M. Blood electrolytes exhibit a strong influence on the mobility of artificial catalytic microengines. *Phys. Chem. Chem. Phys.* **2013**, *15*, 17277–17280. [CrossRef] [PubMed]

87. Wang, H.; Zhao, G.J.; Pumera, M. Blood metabolite strongly suppresses motion of electrochemically deposited catalytic self-propelled microjet engines. *Electrochem. Commun.* **2014**, *38*, 128–130. [CrossRef]

88. Wang, H.; Zhao, G.J.; Pumera, M. Blood proteins strongly reduce the mobility of artificial self-propelled micromotors. *Chem. Eur. J.* **2013**, *19*, 16756–16759. [CrossRef] [PubMed]

89. Zhao, G.; Wang, H.; Khezri, B.; Webster, R.D.; Pumera, M. Influence of real-world environments on the motion of catalytic bubble-propelled micromotors. *Lab Chip* **2013**, *13*, 2937–2941. [CrossRef] [PubMed]

90. Gao, W.; Sattayasamitsathit, S.; Orozco, J.; Wang, J. Efficient bubble propulsion of polymer-based microengines in real-life environments. *Nanoscale* **2013**, *5*, 8909–8914. [CrossRef] [PubMed]

91. de Ávila, B.E.; Ramirez, M.A.; Báez, D.F.; Jodra, A.; Singh, V.V.; Kaufmann, K.; Wang, J. Aptamer-modified graphene-based catalytic micromotors: Off–on fluorescent detection of ricin. *ACS Sens.* **2016**, *1*, 217–221. [CrossRef]

92. Jurado-Sánchez, B.; Escarpa, A.; Wang, J. Lighting up micromotors with quantum dots for smart chemical sensing. *Chem. Commun.* **2015**, *51*, 14088–14091. [CrossRef] [PubMed]

93. Kagan, D.; Campuzano, S.; Balasubramanian, S.; Kuralay, F.; Flechsig, G.U.; Wang, J. Functionalized micromachines for selective and rapid isolation of nucleic acid targets from complex samples. *Nano Lett.* **2011**, *11*, 2083–2087. [CrossRef] [PubMed]

94. Nguyen, K.V.; Minteer, S.D. DNA-functionalized Pt nanoparticles as catalysts for chemically powered micromotors: Toward signal-on motion-based DNA biosensor. *Chem. Commun.* **2015**, *51*, 4782–4784. [CrossRef] [PubMed]

95. Fu, S.Z.; Zhang, X.Q.; Xie, Y.Z.; Wu, J.; Ju, H.X. An efficient enzyme-powered micromotor device fabricated by cyclic alternate hybridization assembly for DNA detection. *Nanoscale* **2017**, *9*, 9026–9033. [CrossRef] [PubMed]

96. Orozco, J.; Cortés, A.; Cheng, G.Z.; Sattayasamitsathit, S.; Gao, W.; Feng, X.M.; Shen, Y.F.; Wang, J. Molecularly imprinted polymer-based catalytic micromotors for selective protein transport. *J. Am. Chem. Soc.* **2013**, *135*, 5336–5339. [CrossRef] [PubMed]

97. Orozco, J.; Campuzano, S.; Kagan, D.; Zhou, M.; Gao, W.; Wang, J. Dynamic isolation and unloading of target proteins by aptamer-modified microtransporters. *Anal. Chem.* **2011**, *83*, 7962–7969. [CrossRef] [PubMed]

98. Yu, X.P.; Li, Y.N.; Wu, J.; Ju, H.X. Motor-based autonomous microsensor for motion and counting immunoassay of cancer biomarker. *Anal. Chem.* **2014**, *86*, 4501–4507. [CrossRef] [PubMed]

99. García, M.; Orozco, J.; Guix, M.; Gao, W.; Sattayasamitsathit, S.; Escarpa, A.; Merkoçi, A.; Wang, J. Micromotor-based lab-on-chip immunoassays. *Nanoscale* **2013**, *5*, 1325–1331. [CrossRef] [PubMed]

100. Singh, V.V.; Kaufmann, K.; de Ávila, B.E.; Karshalev, E.; Wang, J. Molybdenum disulfide-based tubular microengines: Toward biomedical applications. *Adv. Funct. Mater.* **2016**, *26*, 6270–6278. [CrossRef]

101. Campuzano, S.; Orozco, J.; Kagan, D.; Guix, M.; Gao, W.; Sattayasamitsathit, S.; Claussen, J.C.; Merkoçi, A.; Wang, J. Bacterial isolation by lectin-modified microengines. *Nano Lett.* **2012**, *12*, 396–401. [CrossRef] [PubMed]

102. Balasubramanian, S.; Kagan, D.; Hu, C.J.; Campuzano, S.; Lobo-Castaoñ, M.J.; Lim, N.; Kang, D.Y.; Zimmerman, M.; Zhang, L.F.; Wang, J. Micromachine-enabled capture and isolation of cancer cells in complex media. *Angew. Chem. Int. Ed.* **2011**, *50*, 4161–4164. [CrossRef] [PubMed]

103. Luo, M.; Chen, X.; Zhou, G.; Xiang, X.; Chen, L.; Ji, X.; He, Z. Chemiluminescence biosensors for DNA detection using graphene oxide and a horseradish peroxidase-mimicking DNAzyme. *Chem. Commun.* **2012**, *48*, 1126–1128. [CrossRef] [PubMed]

104. Luo, M.; Xiang, X.; Xiang, D.; Yang, S.; Ji, X.; He, Z. A universal platform for amplified multiplexed DNA detection based on exonuclease III-coded magnetic microparticle probes. *Chem. Commun.* **2012**, *48*, 7416–7418. [CrossRef] [PubMed]

105. Luo, M.; Li, N.; Liu, Y.; Chen, C.; Xiang, X.; Ji, X.; He, Z. Highly sensitive and multiple DNA biosensor based on isothermal strand-displacement polymerase reaction and functionalized magnetic microparticles. *Biosens. Bioelectron.* **2014**, *55*, 318–323. [CrossRef] [PubMed]

106. Minh, T.D.; Safdar, M.; Jänis, J. Protection of platinum based micromotors from thiol toxicity by using manganese oxide. *Chem. Eur. J.* **2017**, *23*, 8134–8136. [CrossRef] [PubMed]

107. Maria-Hormigos, R.; Jurado-Sánchez, B.; Escarpa, A. Tailored magnetic carbon allotrope catalytic micromotors for 'on- chip' operations. *Nanoscale* **2017**, *9*, 6286–6290. [CrossRef] [PubMed]

108. Bae, Y.H.; Park, K. Targeted drug delivery to tumors: Myths, reality and possibility. *J. Control. Release* **2011**, *153*, 198–205. [CrossRef] [PubMed]

109. Gao, W.; Wang, J. Synthetic micro/nanomotors in drug delivery. *Nanoscale* **2014**, *6*, 10486–10494. [CrossRef] [PubMed]

110. Luo, M.; Feng, Y.Z.; Wang, T.W.; Guan, J.G. Micro/nanorobots at work in active drug delivery. *Adv. Funct. Mater.* **2018**, *28*, 1706100.

111. Sattayasamitsathit, S.; Kou, H.H.; Gao, W.; Thavarajah, W.; Kaufmann, K.; Zhang, L.F.; Wang, J. Fully loaded micromotors for combinatorial delivery and autonomous release of cargoes. *Small* **2014**, *10*, 2830–2833. [CrossRef] [PubMed]

112. Gao, W.; Dong, R.F.; Thamphiwatana, S.; Li, J.X.; Gao, W.W.; Zhang, L.F.; Wang, J. Artificial micromotors in the mouse's stomach: A step toward in vivo use of synthetic motors. *ACS Nano* **2015**, *9*, 117–123. [CrossRef] [PubMed]

113. Li, J.X.; Thamphiwatana, S.; Liu, W.J.; de Ávila, B.E.; Angsantikul, P.; Sandraz, E.; Wang, J.X.; Xu, T.L.; Soto, F.; Ramez, V.; et al. Enteric micromotor can selectively position and spontaneously propel in the gastrointestinal tract. *ACS Nano* **2016**, *10*, 9536–9542. [CrossRef] [PubMed]

114. Soto, F.; Martin, A.; Ibsen, S.; Vaidyanathan, M.; Garcia-Gradilla, V.; Levin, Y.; Escarpa, A.; Esener, S.C.; Wang, J. Acoustic microcannons: Toward advanced microballistics. *ACS Nano* **2016**, *10*, 1522–1528. [CrossRef] [PubMed]

115. Xu, H.F.; Medina-Sánchez, M.; Magdanz, V.; Schwarz, L.; Hebenstreit, F.; Schmidt, O.G. Sperm-hybrid micromotor for targeted drug delivery. *ACS Nano* **2017**. [CrossRef] [PubMed]

116. Mack, M.J. Minimally invasive and robotic surgery. *J. Am. Med. Assoc.* **2001**, *285*, 568–572. [CrossRef]

117. Nelson, B.J.; Kaliakatsos, I.K.; Abbott, J.J. Microrobots for minimally invasive medicine. *Annu. Rev. Biomed. Eng.* **2010**, *12*, 55–85. [CrossRef] [PubMed]

118. Solovev, A.A.; Xi, W.; Gracias, D.H.; Harazim, S.M.; Deneke, C.; Sánchez, S.; Schmid, O.G. Self-propelled nanotools. *ACS Nano* **2012**, *6*, 1751–1756. [CrossRef] [PubMed]

119. Chatzipirpiridis, G.; Ergeneman, O.; Pokki, J.; Ullrich, F.; Fusco, S.; Ortega, J.A.; Sivaraman, K.M.; Nelson, B.J.; Pane, S. Electroforming of implantable tubular magnetic microrobots for wireless ophthalmologic applications. *Adv. Healthcare Mater.* **2015**, *4*, 209–214. [CrossRef] [PubMed]

120. Maria-Hormigos, R.; Jurado-Sánchez, B.; Escarpa, A. Surfactant-free β-galactosidase micromotors for "on-the-move" lactose hydrolysis. *Adv. Funct. Mater.* **2017**, *27*, 1704256. [CrossRef]

121. Molinero-Fernández, A.; Moreno-Guzman, M.; López, M.Á.; Escarpa, A. Biosensing strategy for simultaneous and accurate quantitative analysis of mycotoxins in food samples using unmodified graphene micromotors. *Anal. Chem.* **2017**, *89*, 10850–10857. [CrossRef] [PubMed]

122. Medina-Sánchez, M.; Schmidt, O.G. Medical microbots need better imaging and control. *Nature* **2017**, *545*, 406–408. [CrossRef] [PubMed]

123. Mou, F.Z.; Guan, J.G.; Shi, W.D.; Sun, Z.G.; Wang, S.H. Oriented contraction: A facile nonequilibrium heat-treatment approach for fabrication of maghemite fiber-in-tube and tube-in-tube nanostructures. *Langmuir* **2010**, *26*, 15580–15585. [CrossRef] [PubMed]

micromachines

MDPI

Review
Light-Powered Micro/Nanomotors

Hongxu Chen, Qilong Zhao and Xuemin Du *

Institute of Biomedical & Health Engineering, Shenzhen Institutes of Advanced Technology (SIAT),
Chinese Academy of Sciences (CAS), Shenzhen 518055, China; hx.chen@siat.ac.cn (H.C.);
ql.zhao@siat.ac.cn (Q.Z.)
* Correspondence: xm.du@siat.ac.cn; Tel.: +86-755-8639-2652

Received: 27 December 2017; Accepted: 18 January 2018; Published: 23 January 2018

Abstract: Designed micro/nanomotors are micro/nanoscale machines capable of autonomous motion in fluids, which have been emerging in recent decades owing to their great potential for biomedical and environmental applications. Among them, light-powered micro/nanomotors, in which motion is driven by light, exhibit various advantages in their precise motion manipulation and thereby a superior scope for application. This review summarizes recent advances in the design, manufacture and motion manipulation of different types of light-powered micro/nanomotors. Their structural features and motion performance are reviewed and compared. The challenges and opportunities of light-powered micro/nanomotors are also discussed. With rapidly increasing innovation, advanced, intelligent and multifunctional light-powered micro/nanomachines will certainly bring profound impacts and changes for human life in the future.

Keywords: micro/nanomotors; light-powered; manufacture; motion manipulation

1. Introduction

The homeostasis of biological systems and locomotion of organisms in nature have long been an inspiring topic of research [1]. Inspired by natural microorganisms, considerable efforts have been devoted to achieving artificial self-propelled micro/nanomotors (MNMs) [2–12], which bring about different areas of influential applications, such as environmental remediation [13–18], target drug delivery [19–24], and cell manipulation and isolation [25–31].

MNMs are micro/nanoscale machines capable of converting different energies into mechanical energy that drives machinery movement. The energy sources can be chemical energy, derived from chemical reactions [32–35], or various sources of external stimuli (such as, light, magnetic, ultrasonic or electric field) [31,36–49]. Chemically-powered MNMs can act as reactants and/or catalysts to trigger in situ chemical reactions, subsequently generating chemical gradients or bubbles to autonomously propel themselves in a fluid. Typical ones are propelled by the decomposition of hydrogen peroxide (H_2O_2) [50–60]. The HCl, N_2H_4, I_2 and other fuels have been also reported in succession for MNM propulsion [61,62]. The variety of fuels for propelling chemically-powered MNMs effectively increases their scope of applications, which has been reviewed by Samuel Sánchez et al. [63]. MNMs driven by external physical stimuli have been also widely investigated, of which several reviews have highlighted the advances of different systems, such as light-driven MNMs [64], ultrasound-driven MNMs [65], magnetic-driven MNMs (powered by a rotating magnetic field and an oscillating magnetic field), and electric-driven MNMs (in a direct-current electric field or an alternating-current field) [66]. In addition, recent advances regarding the motion manipulation of these externally-stimulated MNMs by different approaches have also been reviewed [67]. Since these externally stimulated MNMs with preset motion behaviors, long lifetimes, and excellent biocompatibility have shown great promise in various fields of technology, with the design and development of MNMs enabling quick responses to stimuli and precise motion manipulation, which are of great significance and attracting broad interest in research.

Light is one of the most versatile power sources that is renewable and easy to control. With these unique features, light is an excellent candidate for the energy source to drive MNM movement, since the motion of elaborately-designed light-powered MNMs can be non-invasively controlled at highly precise spatial and temporal resolutions. Light-powered MNMs are propelled through converting light energy into mechanical energy, which is initiated from the development of molecular motors on the basis of light-responsive molecules. Significant milestones for the development of light-powered MNMs are summarized in Figure 1. Azobenzene-based artificial molecular machines were reported in the 1980s, propelled by the photoisomerization of chemical structures of azobenzene, were the first prototype of the initial light-powered MNMs. Inspired by kinesin, researchers have developed artificial walkers from DNA, and the molecules can take a step forward based on DNA cleavage and ligation. By using light as an energy source, DNA walkers can mimic the function of biological motors in cargo transport and biosynthesis [68–71]. Since then, various types of light-powered molecular motors were developed gradually. In the 1990s, researchers found that light can propel liquid droplet motion, which engendered a new research field named optofluidics, where the motion of a liquid is driven by optical forces, light-induced capillary forces or a combination of optical and electrical effects [72,73]. In 2004, an Au-Pt bimetal nanomotor reported by the Sen and Mallouk research group differentiated a new branch, namely solid state motors, which were inspired by the self-propelling plates reported by Whitesides et al. in 2002 [74,75]. The premise of the motion of solid state motors is the formation of an asymmetrical gradient field around the motors initiated by light. As the gradient field is unstable, a certain force around the motors is required to stabilize the gradient field to maintain its steady state in a fluid. In the end, this force drives the motors movement. Solid state light-powered MNMs offer the possibility to develop novel light-powered micro/nano robots with advanced properties and functions, making them an emerging topic in both the academic and industrial fields.

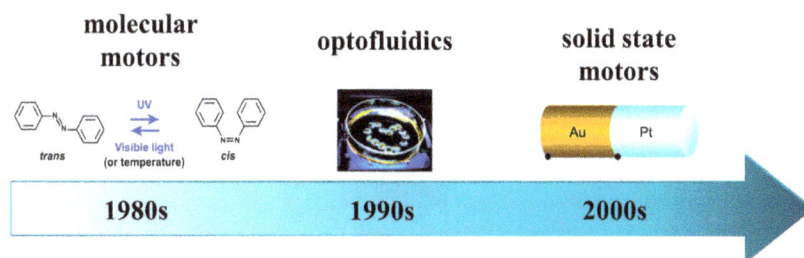

Figure 1. The milestones for the development process of light-powered MNMs: Molecular motors. Reproduced with permission [72]. Copyright 2012, Royal Society of Chemistry. Optofluidics. Reproduced with permission [73]. Copyright 2009, WILEY-VCH. Solid state motors. Reproduced with permission [74]. Copyright 2004, American Chemical Society.

The objective of this review is to highlight various light-powered strategies to drive MNMs. In order to provide the reader with a general overview of the light-powered MNMs discussed in this review, we summarize some typical geometries, light sources, driving mechanisms and motion behaviors of light-driven MNMs, as shown in Table 1. By focusing on the fabrication of light-powered MNMs based on photoactive materials and structural design, we intend to discuss the importance of motion manipulation with regard to different light sources (e.g., ultraviolet (UV), visible and Near-Infrared (NIR) light) and motion behaviors. After briefly introducing the potential applications, we finally review the opportunities and challenges of the field.

Table 1. Typical geometries, light sources, driving mechanisms and motion behaviors of light-powered MNMs.

	Geometries of MNMs	Light Source	Driving Mechanism	Motion Behavior	References
1	Hydrogel ribbon	Near-Infrared light	Photothermal effect	Translational motion	[76]
2	Wheel and spring-like ribbon	Ultraviolet light	Photoisomerization of azobenzene and strain energy	Controlled direction and speed	[77]
3	Tubular liquid crystal polymer	Blue light	Capillary forces arising from photodeformation	Controllable velocity and direction	[78]
4	TiO$_2$-Au Janus micromotor	Ultraviolet light	Self-electrophoresis	25 body length/s	[79]
5	BiOI-metal Janus motor	Visible light	Self-electrophoresis	1.62 μm/s in pure water	[80]
6	Polymer multilayer rockets	Near-Infrared light	Thermophoretic force	High speed of 160 μm/s	[81]
7	Au/B-TiO$_2$ Janus micromotor	Multiple light wavelengths	Self-electrophoresis	Maximus speed in H$_2$O$_2$: 30.1 μm/s	[82]
8	Nanotree	Ultraviolet light	Self-electrophoresis	Positive and negative phototaxis behaviors	[83]
9	Peanut-shaped colloid	Blue light	Diffusion-osmotic flow	Phototactic behavior	[84]

2. Fabrication of Light-Powered MNMs

Light-powered MNMs are micro/nanodevices that can convert light energy into mechanical energy. The key to propelling the motion of light-powered MNMs is the formation of an asymmetrical gradient field around the motors initiated by light. To serve this purpose, light-powered MNMs are normally built either by employing photoactive materials (e.g., photothermal materials and photoisomerized materials) or by constructing asymmetrical structures/geometries (e.g., nanowires, Janus spheres, micro/nanotubes, microcapsules, etc.). In the following section, we will introduce the fabrication of light-powered MNMs based on photoactive materials and structural design, respectively.

2.1. Fabrication of Light-Powered MNMs Based on Photoactive Materials

Photoactive materials can absorb energy from the incident light and convert it into mechanical energy. Among them, photothermal materials, those that generate thermal effects under light irradiation, have been widely studied, by which many light-responsive actuators have been fabricated [85]. Furthermore, a series of tunable photo responsive actuators consisting of photothermal materials was demonstrated by Peng et al., which achieved an integration of complex movements triggered by light [86]. By adjusting the pre-programmed nanostructures, a light-manipulated mechanical arm was assembled and an energy harvesting system was used to execute complex but well-controlled motions. This mechanical arm was able to conduct movements of grasping/releasing and elongation/contraction manipulated by light illuminated areas. The four-step movements of the mechanical arm are shown in Figure 2A. The real-time response, remote controllability and light sensitivity of the mechanical arm offer high competency, as the arm can be adapted to perform different functions and be involved in different activities. Recently, the Martin Möller research group designed a new actuation mechanism for morphing a microswimmer with fast cyclic sequences of shape configurations, subsequently leading to translational motion, as shown in Figure 2B [76]. Light irradiation effectuated a thermal response for a purposefully designed hydrogel ribbon. Then the out-of-equilibrium response yielded precise and fast shape deformation with a rigorous and versatile control of complex motility modes, as needed for mobile microscale robots. They demonstrated the simple hydrogel ribbon motion in water. The ribbon not only followed a purposeful spatial configuration, but also underwent cyclic variations in its spatial configuration that followed a different forward and backward path in space and thus created a thrust to propel the hydrogel ribbon in water.

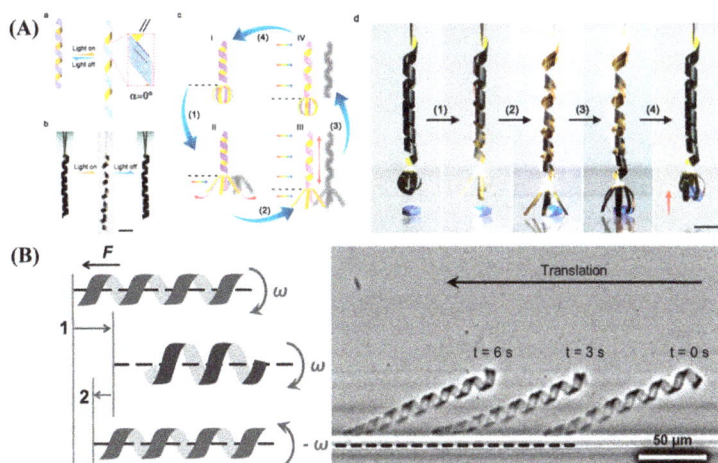

Figure 2. (A) Schematic illustration of helical Strip B (a). Photographs of helical Strip B before and after light irradiation (b). Schematic illustration of a mechanical arm completing a catching (releasing) movement (c). (d) Photographs of an object being lifted up by the mechanical arm. Scale bars, 1 cm in (b,d). Reproduced with permission [86]. Copyright 2016, American Chemical Society; (B) Illustration of the locomotion generated by non-reciprocal deformations of the helix (left); Directing the rotational motion to a linear translocation when the oscillating helix is confined close to a flat wall that impedes the rotation around the axis normal to the helix direction (right). Reproduced with permission [76]. Copyright 2016, WILEY-VCH.

As an alternative photoactive material, liquid crystalline elastomer (LCE) has been attracting broad and growing interest in recent years because of their versatility in creating moving devices. Liquid-crystalline networks are smart materials that combine the anisotropic properties of liquid crystals with the good mechanical behavior of polymeric networks. They exhibit a shape change depending on the local alignment of the liquid-crystal director field inside the network by light illumination, inducing the mobility of LCEs. For example, Zhao et al. demonstrated the tunable photo-controlled motions of malleable azobenzene liquid crystalline polymer actuators [77], of which motion was driven by the UV light-triggered transformation of energy from stored mechanical strain energy in the polymer into mechanical force. This results in a variety of robust, tunable, and continuous motions at the macroscopic scale, as shown in Figure 3A,B. In another example, a photonic liquid-crystalline network microhand was reported by Wiersma et al., which was able to be remotely controlled by optical illumination, act autonomously and grab small particles resulting from their optical properties [87]. As shown in Figure 3C, the elastic reshaping properties of liquid-crystalline networks played a finger-like grasping action under light irradiation. Different deformations and motions could be also achieved by programming the alignment of liquid crystalline, which allow the polymer to perform a wider range of humanized actions in order to complete more delicate tasks.

Early studies of light-powered MNMs mainly focused on molecular machines based on molecular photoisomerization. Apart from the above-mentioned photo-controlled motions of liquid crystalline polymers, optofluidics enables more complex photo-powered motions. For example, Yu et al. reported a strategy to manipulate fluid slugs by photo-induced asymmetric deformation of tubular liquid crystal polymer microactuators, which induces capillary forces for liquid propulsion [78]. These microactuators are able to control a wide diversity of liquids over a long distance with controllable velocities and directions by light, as shown in Figure 4. The development of sophisticated light-powered MNMs by optofluidic approaches will be of great significance in the future.

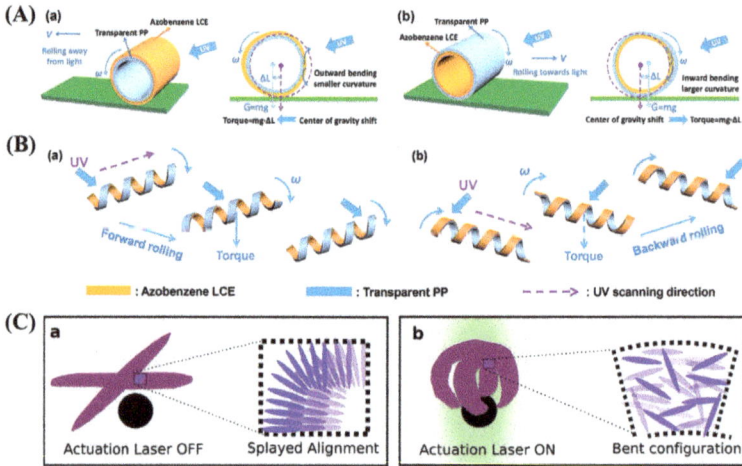

Figure 3. Light-controlled motion of liquid crystalline polymer. (**A**) Schematic showing (a) leftward and (b) rightward shift of the center of gravity in the wheel due to the UV-light-induced asymmetric deformation. Reproduced with permission [77]. Copyright 2017, WILEY-VCH; (**B**) Schematic of light-pushing forward rolling (a) and light-pulling backward rolling (b) of the helical ribbons due to UV-light-induced torque. Reproduced with permission [77]. Copyright 2017, WILEY-VCH; (**C**) Schematic of photonic microhand design (a) Illustration of a microhand and related mesogen alignment. (b) Illustration of the closed microfingers in response to an optical stimulus and the related change in molecular alignment. Reproduced with permission [87]. Copyright 2017, WILEY-VCH.

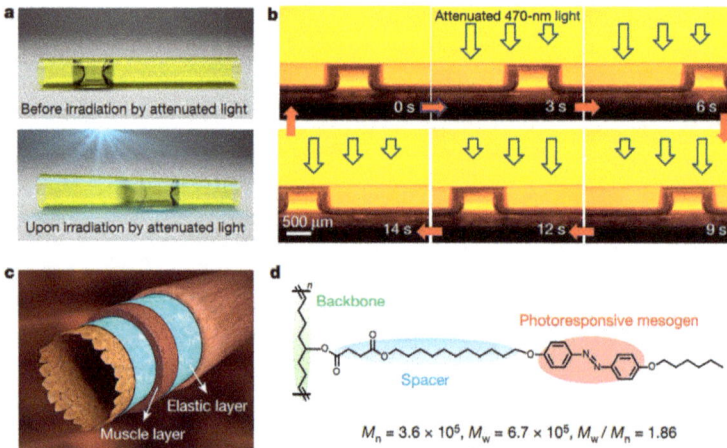

Figure 4. Design of tubular microactuators. (**a**) Schematics showing the motion of a slug of fully wetting liquid confined in a tubular microactuator (TMA) driven by photodeformation; (**b**) Lateral photographs of the light-induced motion of a silicone oil slug in a TMA fixed on a substrate; (**c**) Schematic illustration of the structure of artery walls; (**d**) Molecular structure of a novel linear liquid crystal polymer (LLCP). Reproduced with permission [78]. Copyright 2016, Nature Publishing Group.

2.2. Fabrication of Light-Powered MNMs with Different Geometries

Popular geometries of light-powered MNMs include nanowires, Janus spheres, micro/nanotubes, microcapsules, etc. The template method is a common method for fabricating light-powered MNMs with asymmetrical structures/geometries. Anodic alumina (AAO) membranes and polycarbonate (PC) containing cylindrical or conical pores have been used as preferred templates for the growth of nanowires or nanorockets by electrodeposition. The structure of nanowires or nanorockets could be controlled by the diameter of the membrane pores, deposition time and charges passed during its plating process. Different metals were used to form metallic nanowires or striped nanostructures with heterogenous composition and asymmetrical geometries by sequential deposition. Monodispersed metallic nanowires or nanorockets could be obtained by subsequently dissolving the membrane in the solvent (Figure 5A,B) [58,88]. Apart from nanowires, Janus spheres with distinct properties in the two faces of particles also favor the generation of gradient fields, thereby becoming interesting structures for fabricating light-powered MNMs. In order to obtain the half-coated particles, Janus spherical light-powered MNMs were fabricated by using monodisperse polystyrene (PS) or silica (SiO$_2$) microspheres as the templates, followed by the deposition of metallic thin films on the microspheres (Figure 5C) [81]. To be specific, a suspension of PS or SiO$_2$ microspheres was dropped onto a cleaned substrate to form a monolayer of the microspheres. The density of the monolayer covering the substrate could be controlled by varying the concentration of microsphere suspension, and the size of the microspheres could be adjusted as needed. Metal layers were then coated onto the monolayer of the microspheres by electron-beam evaporation or sputtering to obtain multi-metallic half-coated particles, which were finally released from the substrate to form Janus MNMs.

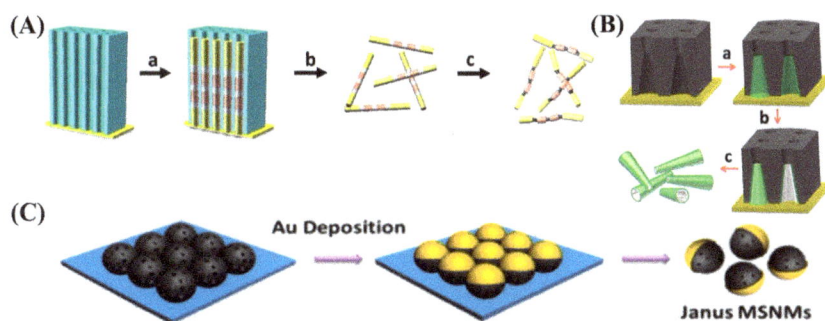

Figure 5. Fabrication schemes of template-assisted MNMs. (**A**) Schematic illustration of AAO template-assisted fabrication of the metal nanowires. Reproduced with permission [88]. Copyright 2011, American Chemical Society; (**B**) The fabrication process of PC template-assisted electrodeposition of micro/nanorockets. Reproduced with permission [58]. Copyright 2016, WILEY-VCH; (**C**) Fabrication scheme of spherical Janus MNMs. Reproduced with permission [81]. Copyright 2016, American Chemical Society.

The layer-by-layer (LbL) assembly technique, involving alternate deposition of positively and negatively charged polyelectrolytes, has been proven to be a versatile and convenient way to construct micro-/nanodevices with a precise structure and composition. He et al. have presented recent progress on the fabrication of MNMs [12]. By LbL assembly, polymers, nanoparticles, proteins and even anonymous assemblies can be conveniently integrated into or onto the LbL-assembled capsules or nanotubes through multiple weak interactions, including electrostatic interactions, hydrogen-bonds, coordination bonds, charge-transfer interactions, biologically specific interactions, and the combined interaction of the above forces, etc. He et al. fabricated the MNMs by LbL assembly of polyelectrolytes [89,90]. The negatively charged poly (styrenesulfonic acid) (PSS) and positively charged poly (allylamine hyhrochloride) (PAH) polyelectrolyte multilayers were adsorbed on the microspheres. Metal was subsequently deposited onto the (PSS/PAH)$_5$-coated microspheres. The hollow Janus capsules partially covered by the Au layer could be obtained by removing the silica templates, as shown in Figure 6A [90]. Apart from Janus spheres, polymeric multilayer tubular rockets could be also prepared by the LbL method [91–93], as shown in Figure 6B. Briefly, the framework of the rockets was prepared by alternatively assembling PSS and PAH onto the inner walls of nanoporous polycarbonate membranes by LbL technique. Then negatively charged gold nanoparticles (AuNPs) were assembled into the (PSS/PAH)$_{20}$-modified porous membranes via electrostatic interactions. The gold nanoshells (AuNSs) inside the rockets were formed through a seeding-growth procedure and finally the tubular rockets were released by dissolving the templates. The resulting rockets could perform NIR-triggered "on/off" motions in a remotely-controlled manner. However, the movement behavior of such a rocket is not stable. To solve this problem, He's group fabricated a near-infrared-light-powered torpedo micromotor by the layer-by-layer sol-gel method, which performs stable movement in a straight line in various media [94]. In consideration of the manufacturing of light-powered MNMs, the LbL assembly method has the advantages of mass production and a convenient operation process. More ingeniously, some new methods such as colloidal lithography have recently emerged for fabricating light-powered MNMs with heterogeneous compositions and/or asymmetrical structures/geometries, which has proven to be a simple, inexpensive and versatile technique enabling rapid and large area patterning, as well as the formation of different conic materials with ordered structures [95,96]. The exploration of new manufacturing methods creates more possibilities for better control over the structures and geometries of light-powered MNMs, widening the functions of light-powered MNMs with more advanced properties.

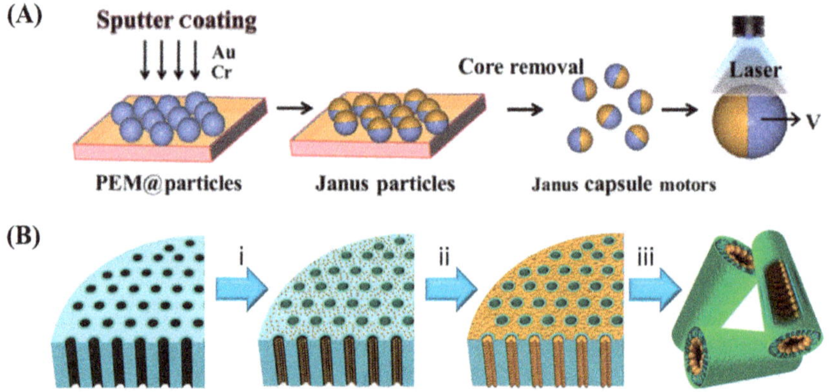

Figure 6. Schematic illustration of MNMs fabricated by the LbL method. (**A**) Scheme of light-triggered Janus capsule motors. Reproduced with permission [90]. Copyright 2014, American Chemical Society; (**B**) Illustration of the fabrication of tubular rockets: (i) LbL assembly of (PAH/PSS)$_{20}$ films, and subsequent deposition of AuNPs into the pores of templates; (ii) Formation of AuNSs though surface-seeding growth method; (iii) Removal of the templates to release the rockets. Reproduced with permission [92]. Copyright 2015, Wiley-VCH.

3. Motion Manipulation of Light-Powered MNMs

For all types of MNMs, motion mode and motion manipulation are critical topics. Recently, many researchers reported the motion behavior of state-of-the-art light-powered MNMs together with their major propulsion mechanisms, including light-induced phoresis propulsion, bubble recoil, interfacial tension gradient, deformation propulsion, self-thermophoresis, and combination force [64,97–99].

For practical applications of light-powered MNMs, the stimulating light at a specific wavelength range (i.e., UV light, visible light and infrared light) is usually required to be in accordance with the nature of the MNMs. Hence, the effect of different light sources on the motion of light-powered MNMs is of high significance and will be reviewed in the following section. To achieve a higher level of motion manipulation for more sophisticated tasks, the controllability of the motion direction of the MNMs is another key aspect. Recent progress has demonstrated some novel light-powered MNMs enabling directional motion in remotely-controlled manners, which will also be summarized in this section.

3.1. Motion of MNMs Manipulated by Different Light Sources

3.1.1. UV Light

For the first time, Guan et al. demonstrated a bubble-propelled photo-activated single component metal oxide tubular microengine by utilizing the photocatalytic H_2O_2 decomposition over TiO_2 under UV irradiation [100]. Upon UV light irradiation, the photogenerated O_2 molecules on the inner surface nucleate and grow into bubbles. Then the generated O_2 bubbles are ejected from a one-end large opening to propel the TiO_2 tubular microengine (Figure 7A). More importantly, the motion state and speed of the microengines can be reversibly, wirelessly, and remotely controlled by turning the "on/off" switch and regulating the intensity of the UV source. Figure 7B shows a highly efficient UV light-driven photocatalytic TiO_2-Au Janus micromotor with wireless steering and velocity control. This Janus micromotor can be powered in pure water under an extremely low UV light intensity (2.5×10^{-3} W/cm^2), and can reach a high speed of 25 body length/s at UV light intensity of 40×10^{-3} W/cm^2 [79]. The propulsion of the TiO_2-Au micromotors dominantly originates from the light-induced self-electrophoresis. Upon UV irradiation, charge separation occurs within the TiO_2 and electrons are injected from the TiO_2 conduction band into the Au hemisphere. Protons are produced from the oxidation of water at TiO_2 and the resultant electrons are consumed during the reduction of protons at Au. The flux of H^+ generates a fluid flow toward the Au hemisphere, generating a slip velocity and propelling the micromotors with the TiO_2 hemisphere forward. In addition, Guan and Zhang et al. demonstrated a disruptive strategy to design micromotors by using isotropic structures. As shown in Figure 7C, the micromotors can continuously move, which induces a net concentration gradient of photocatalyzed products, independent of the random rotation of themselves. Both motion direction and speed were precisely controlled by UV irradiation. In this work, by taking advantage of the limited penetration depth of light in semiconductor materials, the asymmetrical surface chemical reactions on the isotropic semiconductor particles can take place, which induces concentration gradients of photocatalytic products to propel the micro/nanomotors. Due to their isotropic structures, the motion directionality of the as-developed micro/nanomotors is not interfered by their rotational Brownian diffusion or local flows, but always along the irradiated light direction [101].

3.1.2. Visible Light

To the best of our knowledge, most of the existing light-powered MNMs are propelled autonomously by either UV or NIR light. In comparison with that, visible light may serve as an ideal external stimulus for propelling MNMs, as it is more easily available and convenient for operation. Recently, Li et al. reported visible-light-powered Si-Au micromotors, which could move in either deionized water or organic solvents without the addition of chemical fuels [102]. As shown in Figure 8a, the propulsion mechanism is the self-electrophoresis modulated by the photoconductivity of the amorphous silicon segment. Cai et al. presented visible-light-powered Janus micromotors based on BiOI microspheres with one hemisphere coated with a metal layer [80], propelled by the self-electrophoresis mechanism (Figure 8b). Although visible-light-powered MNMs possess many valuable properties for future biomedical and environmental applications, they still confront many challenges and further investigation may be required.

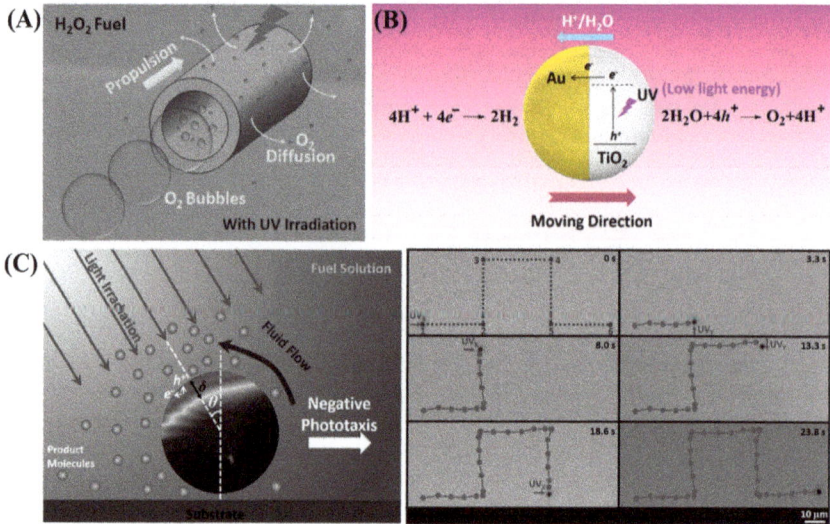

Figure 7. UV light-powered MNMs. (**A**) The UV-induced bubble propulsion mechanism of the TiO$_2$ tubular microengine in H$_2$O$_2$ fuel, the generated O$_2$ bubbles are ejected from a one-end large opening to propel the TiO$_2$ tubular microengine. Reproduced with permission [100]. Copyright 2015, WILEY-VCH; (**B**) The mechanism schematic of TiO$_2$-Au Janus micromotors powered by UV light in water. Reproduced with permission [79]. Copyright 2016, American Chemical Society; (**C**) The mechanism illustration of the phototaxis of a spherical TiO$_2$ micromotor based on the limited penetration depth of light (graph on the left). Time-lapse images and the motion trajectory of a TiO$_2$ micromotor in an aqueous solution containing 0.001 wt % H$_2$O$_2$ as fuel. The predesigned pathway for the micromotor is represented as dashed–dotted lines (graph on the right). Reproduced with permission [101]. Copyright 2017, WILEY-VCH.

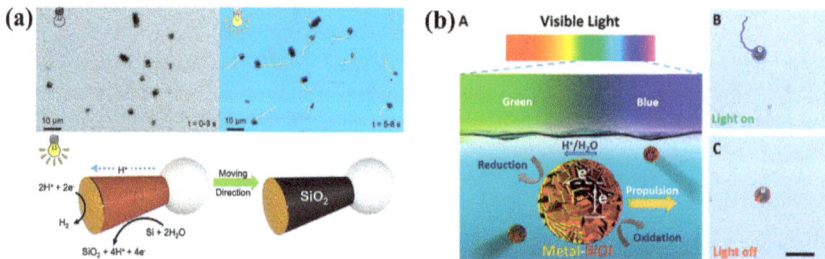

Figure 8. Visible-light-powered MNMs. (**a**) Trajectories of the Si-Au micromotors in water, from left to right, without illumination and with illumination at a light intensity of 13.6 mW mm^{-2} (top figures). Propulsion mechanism of the Si-Au micromotors activated by visible light in deionized (DI) water (down figure). Reproduced with permission [102]. Copyright 2017, Royal Society of Chemistry; (**b**) Mechanism illustration of visible-light-driven BiOI-metal Janus micromotors (A) and the movement trajectories of BiOI-metal Janus micromotors with and without light irradiation (B,C). Reproduced with permission [80]. Copyright 2017, American Chemical Society.

3.1.3. NIR Light

In comparison with UV and visible lights, NIR light is of special interest in the consideration of biomedical applications since light absorption by biological tissues is minimal in this region and NIR light is safe for living organisms. He's research group has focused on the construction of the NIR-driven MNMs, which pave the way to apply self-propelled synthetic MNMs in biomedical fields. In 2014, they demonstrated a new strategy for photothermally triggering the "on-demand" launch of gold-shell-functionalized polymer multilayer micromotors using a NIR laser at the critical concentration of peroxide fuel (Figure 9A). The process was based on the fact that the NIR illumination of the micromotors caused a spontaneous photothermal effect and thus a localized sharp increase in temperature around the micromotors. Accordingly, the increase in temperature induced the accelerated kinetics of the catalytic decomposition, the increased rates of mass transport, and the enhanced release frequency of oxygen bubbles [91]. Further, they presented a polymeric tubular rocket functionalized with AuNSs, which can move at a speed of up to 160 μm s^{-1} [92]. The strong plasma resonance absorption of AuNSs in the NIR region created localized temperature gradients on the inner and outer surfaces of asymmetric AuNSs (Figure 9B). The higher thermal gradient on the inner surface and the asymmetric structure of the rockets resulted in the difference in thermophoretic forces along the elongated axis of rockets, which in turn drove the rockets to move toward the direction of the front small-opening. Similarly, they constructed fuel-free, NIR-driven Janus mesoporous silica nanoparticle motors in 2016 [81]. As shown in Figure 9C, a localized photothermal effect on the Au half-shells resulted in the formation of thermal gradients across the Janus mesoporous silica nanoparticle motors. Thus, the generated self-thermophoresis could actively drive the nanomotors to move at an ultrafast speed upon exposure to an NIR laser. These NIR-powered MNMs demonstrate a novel strategy for overcoming the necessity of chemical fuels and exhibit significant improvement in the maneuverability of MNMs, which provide competent candidates for loading transportation in an eco-friendly manner.

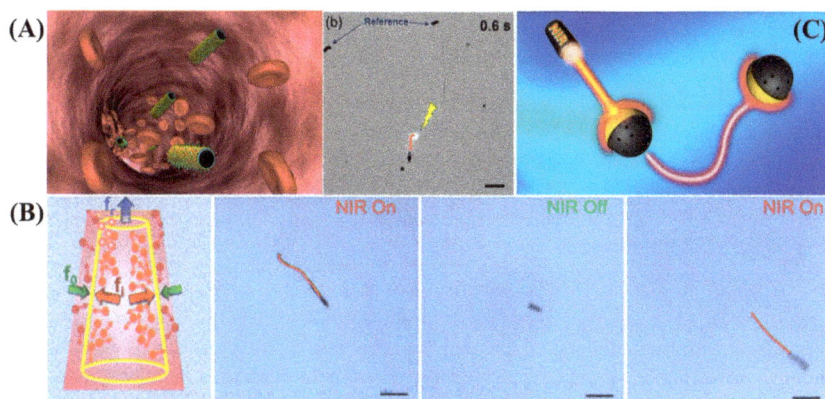

Figure 9. NIR light-driven MNMs. (**A**) NIR-induced launch of a microengine in 0.1% (*v/v*) H$_2$O$_2$ solution. Reproduced with permission [91]. Copyright 2014, American Chemical Society; (**B**) Schematic mechanism of NIR-driven rockets (Small arrows represent the inner and outer thermophoretic forces, and the large arrow indicates the direction of the resultant force) and time-lapse images of NIR controllable launch, stop, and restarted movement of the rocket. Reproduced with permission [92]. Copyright 2015, Wiley-VCH; (**C**) Schematic of NIR-driven Janus mesoporous silica nanoparticle motors. Reproduced with permission [81]. Copyright 2016, American Chemical Society.

3.1.4. Multi-Wavelength Light

The above-mentioned light-powered MNMs are limited to the use of light at a specific wavelength. The narrow light absorption spectrum limits narrow ranges of wavelengths for locomotion. Multi-wavelength-light-powered MNMs were therefore gradually developed. For example, Nelson and Pane et al. reported multiwavelength light-responsive Au/B-TiO_2 Janus micromotors [82]. As shown in Figure 10, the Janus micromotors showed directional motion under multiple light wavelengths including UV, blue, cyan, green, and red light not only in H_2O_2 solution but also in pure water. Because of their good photocatalytic activity at the entire spectrum of UV and visible light, their applications are considerably broad. In addition, Tang et al. have also successfully demonstrated a light-powered silicon nanowire-based nanomotor enabling response to multi-wavelength light (i.e., ultralow-intensity visible light and NIR light) [103]. The research into the multi-wavelength-light-powered MNMs is now just at the primary stage. With advances in materials and manufacturing technologies, novel smart light-powered MNMs, for which the motion behavior (e.g., motion speed, motion direction, etc.) can be manipulated and tuned by the light with different wavelengths, will be developed to perform complicated and multiple tasks intelligently in a controllable way.

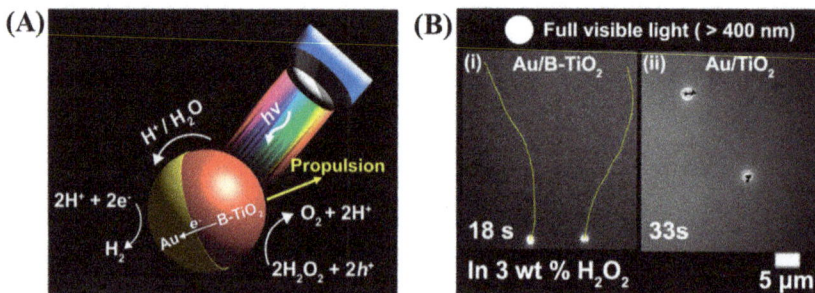

Figure 10. Full visible light (>400 nm) driven Au/B-TiO_2 Janus micromotors. (**A**) Schematic of the propulsion mechanism of Au/B-TiO_2 Janus micromotors; (**B**) Trajectories of (i) Au/B-TiO_2 Janus micromotors over 18 s and (ii) Au/TiO_2 Janus micromotors (control sample) over 33 s. Reproduced with permission [82]. Copyright 2017, American Chemical Society.

3.2. Manipulation of Motion Behaviors

One important goal of MNMs is to manipulate their motion behaviors, mimicking those of live organisms. Organisms in nature can create highly complex collective behaviors through local interactions. The collective behaviors of flocking and schooling make organisms perform cooperative tasks. For example, motile bacteria exhibit organizational behaviors ranging from simple pairwise alignment and aggregation into swarms, to complex transport of other nonmotile species by symbiosis to detoxify their environment. Therefore, it is of great importance to control the collective motion behaviors of MNMs to biomimetic modes. Currently, most MNMs can only make curves or random movements unless an external magnetic field is applied [104,105]. The most challenging point is to control the movement speed and the movement direction of MNMs, which limits their applications. Compared to magnetic navigation, light navigation is an emerging method to manipulate MNMs.

Tang et al. presented a light-controlled programmable artificial phototactic microswimmer [83]. This microswimmer was Janus-nanotree-structured, containing a nanostructured photocathode and a photoanode at the opposite ends where cations and anions were released, respectively, subsequently propelling the microswimmer by self-electrophoresis. These microswimmers self-aligned at the direction of light propagation and mimicked the collective phototactic behavior of green algae in a solution (Figure 11A). By controlling the head and overall surface charges independently via chemical modification, the positive and negative phototaxis behaviors of microswimmers could be successfully programmed. Recently, the self-organization of a self-propelled peanut-shaped hematite colloid triggered by blue light was investigated by Qiang He [84]. Figure 11B shows the dynamic self-assembly of active colloid ribbons perpendicular to their long axis and the positive phototactic behavior of motile colloid ribbons in a solution of hydrogen peroxide fuel. The motion of colloid motors is ascribed to the diffusion-osmotic flow in a chemical gradient by the photocatalytic decomposition of hydrogen peroxide fuel. The phototactic behavior of colloid ribbons stems from the fact that the Gaussian beam distribution of light intensity leads to a higher rate of photocatalytic reaction in the center of the light spot and causes a larger hydrogen peroxide concentration gradient accordingly.

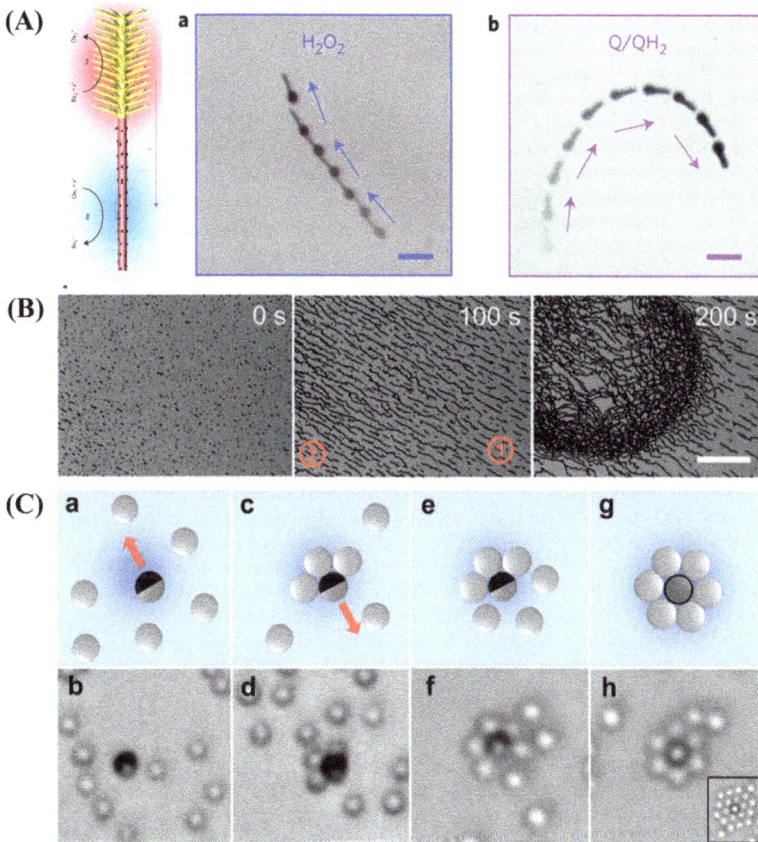

Figure 11. (**A**) Schematic of a Janus artificial microswimmer. Superimposed images of sequential frames show the migration of individual Janus nanotrees under global illumination in 0.1% H_2O_2 (a) and a mixture solution of 1,4-benzoquinone and hydroquinone (b). Reproduced with permission [83]. Copyright 2016, Nature Publishing Group; (**B**) Time-lapse optical images of collective behavior of peanut-shaped colloid motors under illumination of blue light. Reproduced with permission [84]. Copyright 2017, Wiley-VCH; (**C**) (a,b) Under UV illumination, an active particle adopts a tilted orientation, and moves with its TiO_2 face leading. (c,d) Once trapped, passive particles preferentially attach to the TiO_2 half (black region), and the active particle's direction of propulsion reverses so that it moves toward its SiO_2 face. (e–h) When more passive particles attach, the active particle usually reorients into a symmetric configuration with the active TiO_2 surface facing up or down. Reproduced with permission [106]. Copyright 2017, WILEY-VCH.

In addition to the aforementioned methods for achieving the programmed assemblies and collective motion manipulation of individual MNMs, a new method was developed and studied for the precise collective motion manipulation of light-powered MNMs [106]. It was found that self-propelled active colloids could induce the crystallization of passive silica colloids into well-controlled 2D assemblies when illuminated by UV light. The strength of the attractive interaction between the active colloids and the passive colloids, as well as the extent of the assembled clusters are modulated by the diffusiophoretic effects arising from a local chemical gradient activated by UV illumination, as shown in Figure 11C. Using this method, the collective motion of individual MNMs could be controlled, resulting in different assembly modes, e.g., isolated square, pentagonal, hexagonal, heptagonal clusters and some large assemblies with ordered and disordered translating symmetries, which offered a novel platform technology for making rationally designed colloidal clusters and crystals with controllable sizes, shapes, and symmetries.

4. Application Prospects

With their various excellent properties, light-powered MNMs have shown superior application prospects in environmental remediation and biomedicine. In terms of environmental remediation, light-powered MNMs have many advantages including remote operation, adjustable velocity and reutilization. For instance, specifically-designed light-powered MNMs modified by different active layers are able to adsorb certain metallic ions or remove specific oil pollutants, and are thereby promising for wastewater purification [13,14,16]. As for biomedical applications, soft infrared-powered MNMs are supposed to have significant potentials and merits. On the one hand, these soft infrared-powered MNMs can be fabricated by using polymers with proper mechanical properties and specific biological properties (e.g., biostability, biocompatibility, biodegradability and bioactivity), making them reliable and excellent candidates for implantation and clinical applications [31]. On the other hand, because of the deep penetration of infrared across live tissues, soft NIR-powered MNMs can be traced and triggered upon implantation in a noninvasive and remote way [21,64,89]. These advantages lay solid foundations for the preparation of advanced light-powered MNMs with novel properties and functions for diagnostic and therapeutic applications.

5. Conclusions and Outlook

In conclusion, research into light-powered MNMs has facilitated great progress in design and manufacture, showing different possibilities in various application fields. However, there are still many challenges for light-powered MNMs, including the limits of the operation environment, the adaptability of light wavelength, and the difficulty of motion direction control. Specifically, at present, the majority of light-powered MNMs are propelled in the fluids of H_2O_2 and H_2O [80,82]. However, the toxicity of H_2O_2 severely restricts their scope of application and the speed of light-powered MNMs in pure water still needs to improve even though H_2O is an ideal environment [80]. Besides, existing light-powered MNMs are mostly driven by UV and NIR irradiation [79,81], which may respectively cause damage to live organisms or unexpected thermal effects to the motor. The exploration of visible-light-powered MNMs, particularly enabling responses to visible light with different colors/wavelengths and desired driving efficiency, which remains a major challenge, is necessary but now relatively insufficient in the current investigations. Moreover, the precise control and manipulation of the directional motion of light-powered MNMs is a significant and challenging goal. The use of an external magnetic field is by far the most common and visible approach reported to realize the directional motion of MNMs to an intended direction or location. For light-powered MNMs, extensive efforts will still be made to achieve the precise regulation of their movement direction in innovative ways.

In the future, we believe that the exploration of highly efficient light response materials and the design of well-defined micro/nanostructures shall be of great importance, particularly for the development of advanced light-powered MNMs with excellent performance with low-cost, environmentally friendly and facile approaches. Furthermore, the research and development (R&D) of intelligent light-powered MNMs is an important direction via a combination of bio-inspired design and bio-inspired smart materials [64,107–109]. With intelligent abilities, these novel light-powered MNMs can perform complex tasks autonomously, and/or perform specific tasks in special environments, therefore hugely improving their scope of application.

Acknowledgments: The authors gratefully acknowledge the financial support from the Special Support Project for Outstanding Young Scholars of Guangdong Province (Grant No. 2015TQ01R292), Guangdong-Hong Kong Technology Cooperation Funding (Grant No. 2017A050506040), Shenzhen Science and Technology Innovation Committee (Grant Nos. JCYJ20170307164610282, JCYJ20170413152640731), Guangdong Innovative and Entrepreneurial Research Team Program (No. 2013S046), Shenzhen Peacock Plan, and China Postdoctoral Science Foundation (Grant No. 2017M612780).

Author Contributions: Hongxu Chen wrote the whole manuscript. Qilong Zhao reviewed the manuscript. Xuemin Du supervised the work.

Conflicts of Interest: The authors declare no conflict of interest.

References

1. Hess, H.; Bachand, G.D.; Vogel, V. Powering nanodevices with biomolecular motors. *Chem. Eur. J.* **2004**, *10*, 2110–2116. [CrossRef] [PubMed]
2. Ozin, G.A.; Manners, I.; Fournier-Bidoz, S.; Arsenault, A. Dream nanomachines. *Adv. Mater.* **2005**, *17*, 3011–3018. [CrossRef]
3. Wang, J. Can man-made nanomachines compete with nature biomotors? *ACS Nano* **2009**, *3*, 4–9. [CrossRef] [PubMed]
4. Aranson, I.S. Microrobotics: Swimmers by design. *Nature* **2016**, *531*, 312–313. [CrossRef] [PubMed]
5. Heuvel, M.G.; Dekker, C. Motor proteins at work for nanotechnology. *Science* **2007**, *317*, 333–336. [CrossRef] [PubMed]
6. Palagi, S.; Mark, A.G.; Reigh, S.Y.; Melde, K.; Qiu, T.; Zeng, H.; Parmeggiani, C.; Martella, D.; Sanchez-Castillo, A.; Kapernaum, N.; et al. Structured light enables biomimetic swimming and versatile locomotion of photoresponsive soft microrobots. *Nat. Mater.* **2016**, *15*, 647–653. [CrossRef] [PubMed]
7. Guix, M.; Mayorga-Martinez, C.C.; Merkoci, A. Nano/micromotors in (bio)chemical science applications. *Chem. Rev.* **2014**, *114*, 6285–6322. [CrossRef] [PubMed]
8. Mei, Y.; Solovev, A.A.; Sanchez, S.; Schmidt, O.G. Rolled-up nanotech on polymers: From basic perception to self-propelled catalytic microengines. *Chem. Soc. Rev.* **2011**, *40*, 2109–2119. [CrossRef] [PubMed]
9. Wang, W.; Duan, W.T.; Ahmed, S.; Mallouk, T.E.; Sen, A. Small power: Autonomous nano- and micromotors propelled by self-generated gradients. *Nano Today* **2013**, *8*, 531–554. [CrossRef]
10. Kim, K.; Guo, J.; Xu, X.; Fan, D.L. Recent progress on man-made inorganic nanomachines. *Small* **2015**, *11*, 4037–4057. [CrossRef] [PubMed]
11. Lin, X.; Wu, Z.; Wu, Y.; Xuan, M.; He, Q. Self-propelled micro/nanomotors based on controlled assembled architectures. *Adv. Mater.* **2016**, *28*, 1060–1072. [CrossRef] [PubMed]
12. Wu, Z.; Lin, X.; Si, T.; He, Q. Recent progress on bioinspired self-propelled micro/nanomotors via controlled molecular self-assembly. *Small* **2016**, *12*, 3080–3093. [CrossRef] [PubMed]
13. Gao, W.; Wang, J. The environmental impact of micro/nanomachines: A review. *ACS Nano* **2014**, *8*, 3170–3180. [CrossRef] [PubMed]
14. Guix, M.; Orozco, J.; Garcia, M.; Gao, W.; Sattayasamitsathit, S.; Merkoci, A.; Escarpa, A.; Wang, J. Superhydrophobic alkanethiol-coated microsubmarines for effective removal of oil. *ACS Nano* **2012**, *6*, 4445–4451. [CrossRef] [PubMed]
15. Soler, L.; Sanchez, S. Catalytic nanomotors for environmental monitoring and water remediation. *Nanoscale* **2014**, *6*, 7175–7182. [CrossRef] [PubMed]
16. Soler, L.; Magdanz, V.; Fomin, V.M.; Sanchez, S.; Schmidt, O.G. Self-propelled micromotors for cleaning polluted water. *ACS Nano* **2013**, *7*, 9611–9620. [CrossRef] [PubMed]
17. Moo, J.G.; Pumera, M. Chemical energy powered nano/micro/macromotors and the environment. *Chem. Eur. J.* **2015**, *21*, 58–72. [CrossRef] [PubMed]

18. Orozco, J.; Vilela, D.; Valdes-Ramirez, G.; Fedorak, Y.; Escarpa, A.; Vazquez-Duhalt, R.; Wang, J. Efficient biocatalytic degradation of pollutants by enzyme-releasing self-propelled motors. *Chem. Eur. J.* **2014**, *20*, 2866–2871. [CrossRef] [PubMed]

19. Kagan, D.; Laocharoensuk, R.; Zimmerman, M.; Clawson, C.; Balasubramanian, S.; Kang, D.; Bishop, D.; Sattayasamitsathit, S.; Zhang, L.; Wang, J. Rapid delivery of drug carriers propelled and navigated by catalytic nanoshuttles. *Small* **2010**, *6*, 2741–2747. [CrossRef] [PubMed]

20. Gao, W.; Kagan, D.; Pak, O.S.; Clawson, C.; Campuzano, S.; Chuluun-Erdene, E.; Shipton, E.; Fullerton, E.E.; Zhang, L.; Lauga, E.; et al. Cargo-towing fuel-free magnetic nanoswimmers for targeted drug delivery. *Small* **2012**, *8*, 460–467. [CrossRef] [PubMed]

21. Wu, Z.; Wu, Y.; He, W.; Lin, X.; Sun, J.; He, Q. Self-propelled polymer-based multilayer nanorockets for transportation and drug release. *Angew. Chem. Int. Ed.* **2013**, *52*, 7000–7003. [CrossRef] [PubMed]

22. Gao, W.; Wang, J. Synthetic micro/nanomotors in drug delivery. *Nanoscale* **2014**, *6*, 10486–10494. [CrossRef] [PubMed]

23. Mhanna, R.; Qiu, F.; Zhang, L.; Ding, Y.; Sugihara, K.; Zenobi-Wong, M.; Nelson, B.J. Artificial bacterial flagella for remote-controlled targeted single-cell drug delivery. *Small* **2014**, *10*, 1953–1957. [CrossRef] [PubMed]

24. Patra, D.; Sengupta, S.; Duan, W.; Zhang, H.; Pavlick, R.; Sen, A. Intelligent, self-powered, drug delivery systems. *Nanoscale* **2013**, *5*, 1273–1283. [CrossRef] [PubMed]

25. Laurell, T.; Petersson, F.; Nilsson, A. Chip integrated strategies for acoustic separation and manipulation of cells and particles. *Chem. Soc. Rev.* **2007**, *36*, 492–506. [CrossRef] [PubMed]

26. Kuralay, F.; Sattayasamitsathit, S.; Gao, W.; Uygun, A.; Katzenberg, A.; Wang, J. Self-propelled carbohydrate-sensitive microtransporters with builtIn boronic acid recognition for isolating sugars and cells. *J. Am. Chem. Soc.* **2012**, *134*, 15217–15220. [CrossRef] [PubMed]

27. Kim, S.; Qiu, F.; Kim, S.; Ghanbari, A.; Moon, C.; Zhang, L.; Nelson, B.J.; Choi, H. Fabrication and characterization of magnetic microrobots for three-dimensional cell culture and targeted transportation. *Adv. Mater.* **2013**, *25*, 5863–5868. [CrossRef] [PubMed]

28. Marx, V. Biophysics: Using sound to move cells. *Nat. Methods* **2014**, *12*, 41–44. [CrossRef] [PubMed]

29. Guo, F.; Li, P.; French, J.B.; Mao, Z.; Zhao, H.; Li, S.; Nama, N.; Fick, J.R.; Benkovic, S.J.; Huang, T.J. Controlling cell–cell interactions using surface acoustic waves. *Proc. Natl. Acad. Sci. USA* **2015**, *112*, 43–48. [CrossRef] [PubMed]

30. Balasubramanian, S.; Kagan, D.; Hu, C.M.; Campuzano, S.; Lobo-Castanon, M.J.; Lim, N.; Kang, D.Y.; Zimmerman, M.; Zhang, L.; Wang, J. Micromachine-enabled capture and isolation of cancer cells in complex media. *Angew. Chem. Int. Ed.* **2011**, *50*, 4161–4164. [CrossRef] [PubMed]

31. Wang, J.; Gao, W. Nano/microscale motors: Biomedical opportunities and challenges. *ACS Nano* **2012**, *6*, 5745–5751. [CrossRef] [PubMed]

32. Ma, X.; Jannasch, A.; Albrecht, U.-R.; Hahn, K.; MiguelLópez, A.; Schäffer, E.; Sánchez, S. Enzyme-powered hollow mesoporous Janus nanomotors. *Nano Lett.* **2015**, *15*, 7043–7050. [CrossRef] [PubMed]

33. Uygun, M.; Singh, V.V.; Kaufmann, K.; Uygun, D.A.; De Oliveira, S.D.; Wang, J. Micromotor-based biomimetic carbon dioxide sequestration: Towards mobile microscrubbers. *Angew. Chem. Int. Ed.* **2015**, *54*, 12900–12904. [CrossRef] [PubMed]

34. Singh, V.V.; Soto, F.; Kaufmann, K.; Wang, J. Micromotor-based energy generation. *Angew. Chem. Int. Ed.* **2015**, *54*, 6896–6899. [CrossRef] [PubMed]

35. Wu, Z.; Li, J.; Ávila, B.E.-F.; Li, T.; Gao, W.; He, Q.; Zhang, L.; Wang, J. Water-powered cell-mimicking Janus micromotor. *Adv. Funct. Mater.* **2015**, *26*, 7497–7501. [CrossRef]

36. Wezenberg, S.J.; Chen, K.-Y.; Feringa, B.L. Visible-light-driven photoisomerization and increased rotation speed of a molecular motor acting as a ligand in a ruthenium(II) complex. *Angew. Chem. Int. Ed.* **2015**, *54*, 11457–11461. [CrossRef] [PubMed]

37. Rikken, R.S.; Nolte, R.J.; Maan, J.C.; Hest, J.C.; Wilson, D.A.; Christianen, P.C. Manipulation of micro- and nanostructure motion with magnetic fields. *Soft Matter* **2014**, *10*, 1295–1308. [CrossRef] [PubMed]

38. Li, J.; Li, T.; Xu, T.; Kiristi, M.; Liu, W.; Wu, Z.; Wang, J. Magneto–acoustic hybrid nanomotor. *Nano Lett.* **2015**, *15*, 4814–4821. [CrossRef] [PubMed]

39. Peyer, K.E.; Tottori, S.; Qiu, F.; Zhang, L.; Nelson, B.J. Magnetic helical micromachines. *Chem. Eur. J.* **2013**, *19*, 28–38. [CrossRef] [PubMed]

40. Fischer, P.; Ghosh, A. Magnetically actuated propulsion at low Reynolds numbers: Towards nanoscale control. *Nanoscale* **2011**, *3*, 557–563. [CrossRef] [PubMed]

41. Chaturvedi, N.; Hong, Y.; Sen, A.; Velegol, D. Magnetic enhancement of phototaxing catalytic motors. *Langmuir* **2010**, *26*, 6308–6313. [CrossRef] [PubMed]

42. Campuzano, S.; Kagan, D.; Orozco, J.; Wang, J. Motion-driven sensing and biosensing using electrochemically propelled nanomotors. *Analyst* **2011**, *136*, 4621–4630. [CrossRef] [PubMed]

43. Mou, F.; Kong, L.; Chen, C.; Chen, Z.; Xu, L.; Guan, J. Light-controlled propulsion, aggregation and separation of water-fuelled TiO_2/Pt Janus submicromotors and their "on-the-fly" photocatalytic activities. *Nanoscale* **2016**, *8*, 4976–4983. [CrossRef] [PubMed]

44. Xu, T.; Soto, F.; Gao, W.; Garcia-Gradilla, V.; Li, J.; Zhang, X.; Wang, J. Ultrasound-modulated bubble propulsion of chemically powered microengines. *J. Am. Chem. Soc.* **2014**, *136*, 8552–8555. [CrossRef] [PubMed]

45. Wang, W.; Li, S.; Mair, L.; Ahmed, S.; Huang, T.J.; Mallouk, T.E. Acoustic propulsion of nanorod motors inside living cells. *Angew. Chem. Int. Ed.* **2014**, *53*, 3201–3204. [CrossRef] [PubMed]

46. Xu, T.; Soto, F.; Gao, W.; Dong, R.; Garcia-Gradilla, V.; Magana, E.; Zhang, X.; Wang, J. Reversible swarming and separation of self-propelled chemically powered nanomotors under acoustic fields. *J. Am. Chem. Soc.* **2015**, *137*, 2163–2166. [CrossRef] [PubMed]

47. Balk, A.L.; Mair, L.O.; Mathai, P.P.; Patrone, P.N.; Wang, W.; Ahmed, S.; Mallouk, T.E.; Liddle, J.A.; Stavis, S.M. Kilohertz rotation of nanorods propelled by ultrasound, traced by microvortex advection of nanoparticles. *ACS Nano* **2014**, *8*, 8300–8309. [CrossRef] [PubMed]

48. Ahmed, S.; Gentekos, D.T.; Fink, C.A.; Mallouk, T.E. Self-assembly of nanorod motors into geometrically regular multimers and their propulsion by ultrasound. *ACS Nano* **2014**, *8*, 11053–11060. [CrossRef] [PubMed]

49. Garcia-Gradilla, V.; Sattayasamitsathit, S.; Soto, F.; Kuralay, F.; Yardimci, C.; Wiitala, D.; Galarnyk, M.; Wang, J. Ultrasound-propelled nanoporous gold wire for efficient drug loading and release. *Small* **2014**, *10*, 4154–4159. [CrossRef] [PubMed]

50. Wang, W.; Chiang, T.-Y.; Velegol, D.; Mallouk, T.E. Understanding the efficiency of autonomous nano- and microscale motors. *J. Am. Chem. Soc.* **2013**, *135*, 10557–10565. [CrossRef] [PubMed]

51. Baraban, L.; Harazim, S.M.; Sanchez, S.; Schmidt, O.G. Chemotactic behavior of catalytic motors in microfluidic channels. *Angew. Chem. Int. Ed.* **2013**, *52*, 5552–5556. [CrossRef] [PubMed]

52. Gao, W.; Pei, A.; Feng, X.; Hennessy, C.; Wang, J. Organized self-assembly of Janus micromotors with hydrophobic hemispheres. *J. Am. Chem. Soc.* **2013**, *135*, 998–1001. [CrossRef] [PubMed]

53. Lee, T.-C.; Alarcón-Correa, M.; Miksch, C.; Hahn, K.; Gibbs, J.G.; Fischer, P. Self-propelling nanomotors in the presence of strong brownian forces. *Nano Lett.* **2014**, *14*, 2407–2412. [CrossRef] [PubMed]

54. Gao, W.; Pei, A.; Wang, J. Water-driven micromotors. *ACS Nano* **2012**, *6*, 8432–8438. [CrossRef] [PubMed]

55. Mei, Y.; Huang, G.; Solovev, A.A.; Ureña, E.B.; Mönch, I.; Ding, F.; Reindl, T.; Fu, R.K.Y.; Chu, P.K.; Schmidt, O.G. Versatile approach for integrative and functionalized tubes by strain engineering of nanomembranes on polymers. *Adv. Mater.* **2008**, *20*, 4085–4090. [CrossRef]

56. Sanchez, S.; Solovev, A.A.; Mei, Y.; Schmidt, O.G. Dynamics of biocatalytic microengines mediated by variable friction control. *J. Am. Chem. Soc.* **2010**, *132*, 13144–13145. [CrossRef] [PubMed]

57. Orozco, J.; Garcia-Gradilla, V.; D'Agostino, M.; Gao, W.; Cortes, A.; Wang, J. Artificial enzyme-powered microfish for water quality testing. *ACS Nano* **2013**, *7*, 818–824. [CrossRef] [PubMed]

58. Gao, W.; Sattayasamitsathit, S.; Orozco, J.; Wang, J. Highly efficient catalytic microengines: Template electrosynthesis of polyaniline/platinum microtubes. *J. Am. Chem. Soc.* **2011**, *133*, 11862–11864. [CrossRef] [PubMed]

59. Gao, W.; Uygun, A.; Wang, J. Hydrogen-bubble-propelled zinc-based microrockets in strongly acidic media. *J. Am. Chem. Soc.* **2012**, *134*, 897–900. [CrossRef] [PubMed]

60. Campuzano, S.; Orozco, J.; Kagan, D.; Guix, M.; Gao, W.; Sattayasamitsathit, S.; Claussen, J.C.; Merkoci, A.; Wang, J. Bacterial isolation by lectin-modified microengines. *Nano Lett.* **2012**, *12*, 396–401. [CrossRef] [PubMed]

61. Gao, W.; Pei, A.; Dong, R.F. Catalytic iridium-based Janus micromotors powered by ultralow levels of chemical fuels. *J. Am. Chem. Soc.* **2014**, *136*, 2276–2279. [CrossRef] [PubMed]

62. Wong, F.; Sen, A. Progress toward light-harvesting self-electrophoretic motors: Highly efficient bimetallic nanomotors and micropumps in halogen media. *ACS Nano* **2016**, *10*, 7172–7179. [CrossRef] [PubMed]

63. Sanchez, S.; Soler, L.; Katuri, J. Chemically powered micro- and nanomotors. *Angew. Chem. Int. Ed.* **2015**, *54*, 1414–1444. [CrossRef] [PubMed]

64. Xu, L.; Mou, F.; Gong, H.; Luo, M.; Guan, J. Light-driven micro/nanomotors: From fundamentals to applications. *Chem. Soc. Rev.* **2017**, *46*, 6905–6926. [CrossRef] [PubMed]

65. Xu, T.; Xu, L.; Zhang, X. Ultrasound propulsion of micro/nanomotors. *Appl. Mater. Today* **2017**, *9*, 493–503. [CrossRef]

66. Xu, T.; Gao, W.; Xu, L.; Zhang, X.; Wang, S. Fuel-free synthetic micro-/nanomachines. *Adv. Mater.* **2017**, *29*, 1603250. [CrossRef] [PubMed]

67. Tu, Y.; Peng, F.; Wilson, D.A. Motion manipulation of micro- and nanomotors. *Adv. Mater.* **2017**, *29*, 1701970. [CrossRef] [PubMed]

68. You, M.; Huang, F.; Chen, Z.; Wang, R.-W.; Tan, W. Building a nanostructure with reversible motions using photonic energy. *ACS Nano* **2012**, *6*, 7935–7941. [CrossRef] [PubMed]

69. Cheng, J.; Sreelatha, S.; Hou, R.; Efremov, A.; Liu, R.; Maarel, J.R.C.; Wang, Z. Bipedal nanowalker by pure physical mechanisms. *Phys. Rev. Lett.* **2012**, *109*, 238104. [CrossRef] [PubMed]

70. Loh, I.Y.; Cheng, J.; Tee, S.R.; Efremov, A.; Wang, Z. From bistate molecular switches to self-directed track-walking nanomotors. *ACS Nano* **2014**, *8*, 10293–10304. [CrossRef] [PubMed]

71. Yeo, Q.Y.; Loh, I.Y.; Tee, S.R.; Chiang, Y.H.; Cheng, J.; Liu, M.H.; Wang, Z.S. A DNA bipedal nanowalker with a piston-like expulsion stroke. *Nanoscale* **2017**, *9*, 12142–12149. [CrossRef] [PubMed]

72. Baigl, D. Photo-actuation of liquids for light-driven microfluidics: State of the art and perspectives. *Lab Chip* **2012**, *12*, 3637–3653. [CrossRef] [PubMed]

73. Diguet, A.; Guillermic, R.-M.; Magome, N.; Saint-Jalmes, A.; Chen, Y.; Yoshikawa, K.; Baigl, D. Photomanipulation of a droplet by the chromocapillary effect. *Angew. Chem. Int. Ed.* **2009**, *48*, 9281–9284. [CrossRef] [PubMed]

74. Paxton, W.F.; Kistler, K.C.; Olmeda, C.C.; Sen, A.; Angelo, S.K.; Cao, Y.; Mallouk, T.E.; Lammert, P.E.; Cresp, V.H. Catalytic nanomotors: Autonomous movement of striped nanorods. *J. Am. Chem. Soc.* **2004**, *126*, 13424–13431. [CrossRef] [PubMed]

75. Ismagilov, R.F.; Schwartz, A.; Bowden, N.; Whitesides, G.M. Autonomous movement and self-assembly. *Angew. Chem. Int. Ed.* **2002**, *41*, 652–654. [CrossRef]

76. Mourran, A.; Zhang, H.; Vinokur, R.; Möller, M. Soft microrobots employing nonequilibrium actuation via plasmonic heating. *Adv. Mater.* **2017**, *29*, 1604825. [CrossRef] [PubMed]

77. Lu, X.; Guo, S.; Tong, X.; Xia, H.; Zhao, Y. Tunable photocontrolled motions using stored strain energy in malleable azobenzene liquid crystalline polymer actuators. *Adv. Mater.* **2017**, *29*, 1606467. [CrossRef] [PubMed]

78. Lv, J.; Liu, Y.; Wei, J.; Chen, E.; Qin, L.; Yu, Y. Photocontrol of fluid slugs in liquid crystal polymer microactuators. *Nature* **2016**, *537*, 179–184. [CrossRef] [PubMed]

79. Dong, R.; Zhang, Q.; Gao, W.; Pei, A.; Ren, B. Highly efficient light-driven TiO_2-Au Janus micromotors. *ACS Nano* **2016**, *10*, 839–844. [CrossRef] [PubMed]

80. Dong, R.; Hu, Y.; Wu, Y.; Gao, W.; Ren, B.; Wang, Q.; Cai, Y. Visible-light-driven BiOI-based Janus micromotor in pure water. *J. Am. Chem. Soc.* **2017**, *139*, 1722–1725. [CrossRef] [PubMed]

81. Xuan, M.; Wu, Z.; Shao, J.; Dai, L.; Si, T.; He, Q. Near infrared light-powered Janus mesoporous silica nanoparticle motors. *J. Am. Chem. Soc.* **2016**, *138*, 6492–6497. [CrossRef] [PubMed]

82. Jang, B.; Hong, A.; Kang, H.E.; Alcantara, C.; Charreyron, S.; Mushtaq, F.; Pellicer, E.; Büchel, R.; Sort, J.; Lee, S.S.; et al. Multiwavelength light-responsive $Au/B-TiO_2$ Janus micromotors. *ACS Nano* **2017**, *11*, 6146–6154. [CrossRef] [PubMed]

83. Dai, B.; Wang, J.; Xiong, Z.; Zhan, X.; Dai, W.; Li, C.-C.; Feng, S.-P.; Tang, J. Programmable artificial phototactic microswimmer. *Nat. Nanotechnol.* **2016**, *11*, 1087–1092. [CrossRef] [PubMed]

84. Lin, Z.; Si, T.; Wu, Z.; Gao, C.; Lin, X.; He, Q. Light-activated active colloid ribbons. *Angew. Chem. Int. Ed.* **2017**, *56*, 13517–13520. [CrossRef] [PubMed]

85. Du, X.; Cui, H.; Sun, B.; Wang, J.; Zhao, Q.; Xia, K.; Wu, T.; Humayun, M.S. Photothermally triggered shape-adaptable 3D flexible electronics. *Adv. Mater. Technol.* **2017**, *2*, 1700120. [CrossRef]

86. Deng, J.; Li, J.; Chen, P.; Fang, X.; Sun, X.; Jiang, Y.; Weng, W.; Wang, B.; Peng, H. Tunable photothermal actuators based on a pre-programmed aligned nanostructure. *J. Am. Chem. Soc.* **2016**, *138*, 225–230. [CrossRef] [PubMed]

87. Martella, D.; Nocentini, S.; Nuzhdin, D.; Parmeggiani, C.; Wiersma, D.S. Photonic microhand with autonomous action. *Adv. Mater.* **2017**, *29*, 1704047. [CrossRef] [PubMed]

88. Li, T.; Li, J.; Zhang, H.; Chang, X.; Song, W.; Hu, Y.; Shao, G.; Sandraz, E.; Zhang, G.; Li, L.; et al. Magnetically propelled fish-like nanoswimmers. *Small* **2016**, *12*, 6098–6105. [CrossRef] [PubMed]

89. Wu, Y.J.; Lin, X.K.; Wu, Z.G.; Mohwald, H.; He, Q. Self-propelled polymer multilayer Janus capsules for effective drug delivery and light-triggered release. *ACS Appl. Mater. Interfaces* **2014**, *6*, 10476–10481. [CrossRef] [PubMed]

90. Wu, Y.; Si, T.; Shao, J.; Wu, Z.; He, Q. Near-infrared light-driven Janus capsule motors: Fabrication, propulsion, and simulation. *Nano Res.* **2016**, *9*, 3747–3756. [CrossRef]

91. Wu, Z.; Lin, X.; Wu, Y.; Si, T.; Sun, J.; He, Q. Near-infrared light-triggered"on/off" motion of polymer multilayer rockets. *ACS Nano* **2014**, *8*, 6097–6105. [CrossRef] [PubMed]

92. Wu, Z.; Si, T.; Gao, W.; Lin, X.; Wang, J.; He, Q. Superfast near-infrared light-driven polymer multilayer rockets. *Small* **2016**, *12*, 577–582. [CrossRef] [PubMed]

93. Wu, Z.G.; Lin, X.K.; Zou, X.; Sun, J.M.; He, Q. Biodegradable protein-based rockets for drug transportation and light-triggered release. *ACS Appl. Mater. Interfaces* **2015**, *7*, 250–255. [CrossRef] [PubMed]

94. Rao, Q.; Si, T.; Wu, Z.; Xuan, M.; He, Q. A light-activated explosive micropropeller. *Sci. Rep.* **2017**, *7*, 4621. [CrossRef] [PubMed]

95. Chen, H.; Wang, T.; Shen, H.; Liu, W.; Wang, S.; Liu, K.; Zhang, J.; Yang, B. Ag nanoparticle/polymer composite barcode nanorods. *Nano Res.* **2015**, *8*, 2871–2880. [CrossRef]

96. Chen, H.; Mu, S.; Fang, L.; Shen, H.; Zhang, J.; Yang, B. Polymer-assisted fabrication of gold nanoring arrays. *Nano Res.* **2017**, *10*, 3346–3357. [CrossRef]

97. Gomez-Solano, J.R.; Blokhuis, A.; Bechinger, C. Dynamics of self-propelled Janus particles in viscoelastic fluids. *Phys. Rev. Lett.* **2016**, *116*, 138301. [CrossRef] [PubMed]

98. Jiang, H.-R.; Yoshinaga, N.; Sano, M. Active motion of a Janus particle by self-thermophoresis in a defocused laser beam. *Phys. Rev.Lett.* **2010**, *105*, 268302. [CrossRef] [PubMed]

99. Qian, B.; Montiel, D.; Bregulla, A.; Cichosb, F.; Yang, H. Harnessing thermal fluctuations for purposeful activities: The manipulation of single micro-swimmers by adaptive photon nudging. *Chem. Sci.* **2013**, *4*, 1420–1429. [CrossRef]

100. Mou, F.; Li, Y.; Chen, C.; Li, W.; Yin, Y.; Ma, H.; Guan, J. Single-component TiO$_2$ tubular microengines with motion controlled by light-induced bubbles. *Small* **2015**, *11*, 2564–2570. [CrossRef] [PubMed]

101. Chen, C.; Mou, F.; Xu, L.; Wang, S.; Guan, J.; Feng, Z.; Wang, Q.; Kong, L.; Li, W.; Wang, J.; et al. Light-steered isotropic semiconductor micromotors. *Adv. Mater.* **2017**, *29*, 1603374. [CrossRef] [PubMed]

102. Zhou, D.; Li, Y.C.; Xu, P.; Ren, L.; Zhang, G.; Mallouk, T.E.; Li, L. Visible-light driven Si–Au micromotors in water and organic solvents. *Nanoscale* **2017**, *9*, 11434–11438. [CrossRef] [PubMed]

103. Wang, J.; Xiong, Z.; Zhan, X.; Dai, B.; Zheng, J.; Liu, J.; Tang, J. A silicon nanowire as a spectrally tunable light-driven nanomotor. *Adv. Mater.* **2017**, *29*, 1701451. [CrossRef] [PubMed]

104. Li, T.; Li, J.; Morozov, K.I.; Wu, Z.; Xu, T.; Rozen, I.; Leshansky, A.M.; Li, L.; Wang, J. Highly efficient freestyle magnetic nanoswimmer. *Nano Lett.* **2017**, *17*, 5092–5098. [CrossRef] [PubMed]

105. Zhu, W.; Li, J.; Leong, Y.J.; Rozen, I.; Qu, X.; Dong, R.; Wu, Z.; Gao, W.; Chung, P.H.; Wang, J.; et al. 3D-printed artificial microfish. *Adv. Mater.* **2015**, *27*, 4411–4417. [CrossRef] [PubMed]

106. Singh, D.P.; Choudhury, U.; Fischer, P.; Mark, A.G. Non-equilibrium assembly of light-activated colloidal mixtures. *Adv. Mater.* **2017**, *29*, 1701328. [CrossRef] [PubMed]

107. Du, X.; Wang, J.; Cui, H.; Zhao, Q.; Chen, H.; He, L.; Wang, Y. Breath-taking patterns: Discontinuous hydrophilic regions for photonic crystal beads assembly and patterns revisualization. *ACS Appl. Mater. Interfaces* **2017**, *9*, 38117–38124. [CrossRef] [PubMed]

108. Du, X.; Lei, N.-Y.; Hu, P.; Zhang, L.; Ong, D.H.-C.; Ge, X.; Zhang, Z.; Lam, M.H.-W. In vivo imaging of the morphology and changes in pH along the gastrointestinal tract of Japanese medaka by photonic band-gap hydrogel microspheres. *Anal. Chim. Acta* **2013**, *787*, 193–202. [CrossRef] [PubMed]
109. Zhang, L.; Naumov, P.; Du, X.; Hu, Z.; Wang, J. Vapomechanically responsive motion of microchannel programmed actuators. *Adv. Mater.* **2017**, *29*, 1702231. [CrossRef] [PubMed]

micromachines

MDPI

Communication

Mini-EmulsionFabricated Magnetic and Fluorescent Hybrid Janus Micro-Motors

Jiapu Jiao [1,2], Dandan Xu [1,2], Yuhuan Liu [2], Weiwei Zhao [1,2], Jiaheng Zhang [1,2], Tingting Zheng [3], Huanhuan Feng [2,*] and Xing Ma [1,2,4,*]

[1] State Key Lab of Advanced Welding and Joining, Harbin Institute of Technology (Shenzhen), Shenzhen 518055, China; j15602980509@163.com (J.J.); xudandan@stu.hit.edu.cn (D.X.); wzhao@hit.edu.cn (W.Z.); zhangjiaheng@hit.edu.cn (J.Z.)
[2] School of Materials Science and Engineering, Harbin Institute of Technology (Shenzhen), Shenzhen 518055, China; hitlyh182371@163.com
[3] Department of Ultrasound, Peking University Shenzhen Hospital, Shenzhen-PKU-HKUST Medical Center, Shenzhen 518036, China; kyzs_018@126.com
[4] Key Laboratory of Micro-systems and Micro-structures Manufacturing of Ministry of Education, Harbin Institute of Technology, Harbin 150001, China
* Correspondence: Fenghuanhuan@hit.edu.cn (H.F.); maxing@hit.edu.cn (X.M.); Tel.: +86-186-8236-8507 (H.F.); +86-137-1466-4507 (X.M.)

Received: 15 January 2018; Accepted: 11 February 2018; Published: 15 February 2018

Abstract: Self-propelling micro/nano-motors have attracted great attention due to their controllable active motion and various functional attributes. To date, a variety of technologies have been reported for the fabrication of micro/nano-motors. However, there are still several challenges that need to be addressed. One of them is to endow micro/nano-motors with multi-functionalities by a facile fabrication process. Here, we present a universal approach, adopted from the emulsion templating method, for the fabrication of Janus micro-motors. With a one-step process, magnetic nanoparticles and fluorescent dyes are simultaneously embedded into the microparticles. The self-propelled motors can be used as an active label or fluorescent tracer through manipulation of their motion using magnetic guidance.

Keywords: Janus particles; micro-motors; multiple functionalization; self-propelling; motion manipulation

1. Introduction

Self-propelling motors are small autonomous devices which are capable of harvesting energy from their environment and mechanically driving themselves in fluids [1–4]. They have attracted great attention during the last decade due to their ability to drive and build intelligent micro/nano-robots that can accomplish designed tasks for different applications [5]. Although considerable challenges remain in order to move these micro/nano-motors towards practical real-world use [6], the promising future of this field continues to attract more and more scientists. Therefore, there have been a large number of researchers contributing great effort to fabricating various types of micro/nano-motors, and studying their motion behavior and potential applications during the last decade [3,4]. From the view of motor structure, researchers have reported on bimetallic nanorods [7,8], typical spherical Janus particles [9,10], bubble propelled tubular micro-jets [11–14], and magnetically driven micro/nano-helix, etc. [15–17]. Based on their special application or required functionality, different fabrication technologies, including electron beam deposition [7,8,11], supramolecular self-assembly [18], template-assisted electrochemical deposition [12,13], sol–gel chemistry [19–22], layer-by-layer assembly [14,23], and rolled-up method, etc. [24], have been utilized to produce micro/nano-motors. The demonstrated proof-of-concept applications include targeted cargo (drug) delivery [25,26] and imaging for biomedical purposes [27–29], pollutant degradation or remediation in environmental sedimentation [30–32], precise particles or cells' manipulation [33–38], and biochemical sensing [39–43]. Aiming at those potential applications, the self-propelled micro/nano-motors should be controllable with regards to movement direction, as well as apparent labeling for motion tracking or potential in-situ imaging applications.

For the manipulation of the motors' movement, the most direct approach is to incorporate a magnetic element into the motors' building structure and use an external magnetic field to guide the motors' moving direction. Several examples were reported. First of all, the motors can be directly constructed with a magnetic component, such as the magnetic powered Helix, to achieve both magnetically driven self-propulsion and control of the motion direction [17,44–47]. Then, a common method is the use of magnetic motors via the deposition of a magnetic layer such as a ferromagnetic Ni, on the motors' surface [48,49]. Baraban et al. reported progress on the magnetic guidance of spherical Janus particles [50]. They manage to achieve an alignment of the magnetic moment along the main symmetry axis of the spherical Janus micromotors by using a special magnetic cap structure based on ultrathin multilayers of $[Co./Pt(Pd)]_N$. This specific composition exhibits a perpendicular magnetic anisotropy (magnetic moment points perpendicular to the sample surface), even when deposited on arrays of spherical particles with sizes ranging from 50 nm to 5 mm [51,52]. Samuel et. al. used a similar strategy to control the movement direction of mesoporous silica micro-motors for target cargo delivery as well [53]. For tubular micro/nano-jets, a common method is to electrochemically deposit a magnetic layer of Ni between the motors' substrate and catalytic layers, so that the movement direction can be aligned by an external magnetic field [54]. Another strategy is to incorporate magnetic nanoparticles onto the motor's body, by simple physical adsorption, which can not only provide magnetic properties to the motors but also make the motors capable of acting as a magnetic resonance imaging (MRI) tracer in biomedical use [55].

Additionally, the ability to fluorescently label or load micro/nano-motors molecules demonstrates both research and practical applications. For instance, fluorescent-labeled spheres were used to investigate the rotational behavior of active colloids [56]. Samuel et al. developed a series of mesoporous silica micro/nano-motors that could load and encapsulate fluorescent dyes or drug molecules for delivery and release [20,53,57]. Furthermore, the long-term scientific goal of the micro/nano-motors community is to utilize these motors for biomedical purposes, of which, the field of fluorescent imaging is one of the most common. Fluorescent imaging techniques are used for both in vitro and in vivo theranostic applications. Recently, Li Zhang et al. produced a biodegradable, fluorescent and magnetic helix micro-motor for in vivo imaging-guided drug delivery, which demonstrated the utility of producing motors with multiple capabilities [55]. Currently, a post-synthesis process of magnetic component incorporation, dye staining or fluorescent labeling is required to confer multiple functions on the motors. Thus, there are still challenges that require further development towards a new and facile strategy to fabricate micro/nano-motors with multiple functionalities, such as magnetism and fluorescence.

Here we present a universal approach adopted from the emulsion templating method for Janus micro-motors fabrication (see Scheme 1). By high speed stirring in a toluene and water mixture containing polystyrene (PS) polymers, we can obtain an oil/water emulsion containing large amounts of oil (toluene) micro-droplets in which the PS polymers were loaded. After evaporation of the toluene, the solid PS microparticles can be produced in a facile way. By adding hydrophobic magnetic nanoparticles and fluorescent dyes into the emulsion system, the PS microparticles or motors can be functionalized with both magnetism and fluorescence properties during one single emulsion process. This means that the motors can be used as an active biomedical label with motion manipulation via a magnetic field. By depositing a platinum (Pt) layer onto one side of the particles, we obtained Janus motors whose motion can be driven by asymmetric catalytic deposition of hydrogen peroxide (H_2O_2). We believe the multiple functionalized micro-motors hold great potential for varied applications, such as targeted cargo transportation, micro-manipulation, drug delivery, or future bio-imaging-guided micro–mini-scaled surgery.

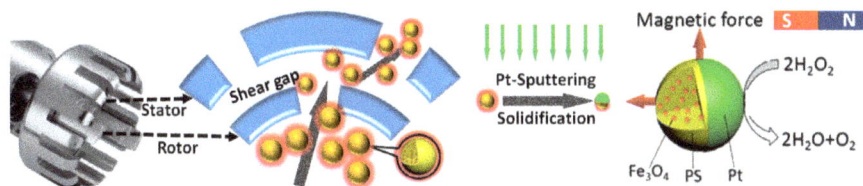

Scheme 1. Scheme of magnetic/fluorescent hybrid micro-motors fabricated by mini-emulsion.

2. Materials and Methods

All materials were purchased from Sigma Aldrich (St. Louis, MI, USA) and used without any further purification.

2.1. Synthesis of Magnetic Particles Fe_3O_4

Magnetic nanoparticles (Fe_3O_4) were synthesized using the hydrothermal method. We used a modified method outlined in Dongyuan Zhao [58]. The procedure is described as follows. An amount of 0.675 g $FeCl_3 \cdot 6H_2O$ was dissolved into 35 mL glycol with constant sonication. An amount of 1.925 g CH_3COONH_4 was added into the solution and stirred for 30 min. The solution was loaded into a reactor and heated at 200 °C for 12 h to complete the reaction. Then, the samples were collected, washed by centrifugation, and air dried for future use.

2.2. Synthesis of Magnetic Particles $Fe_3O_4@SiO_2$

An amount of 40 mg of the previously produced magnetic nanoparticles are dispersed into a mixture solution containing 40 mL deionized (DI) water and 200 mL isopropanol and treated with sonication for 30 min. An amount of 7 mL ammonia (25 wt %) and 0.6 mL tetraethyl orthosilicate (TEOS) were added into the solution which was mechanically stirred for 4 h to complete silication. The samples were collected by a magnet and rinsed 3 times with ethanol.

2.3. Fabrication of Magnetic Particles $Fe_3O_4@SiO_2$ with Fluorosilane Surface Modification

An amount of 50 mg of previously prepared magnetic $Fe_3O_4@SiO_2$ nanoparticles were dispersed into 50 mL ethanol containing 0.5 mL fluorosilane. The solution was refluxed at 80 °C for 24 h. The samples were centrifuged and air dried for future use. The collected nanoparticles were deposited onto a glass slide for contact angle measurement.

2.4. Fabrication of Magnetic and Fluorescent Polystyrene (PS) Microparticles

The micromotors were fabricated via a typical emulsion templating approach [59,60]. The protocol we used is as follows. An amount of 70 mg magnetic nanoparticles $Fe_3O_4@SiO_2$ with fluorosilane surface modification and 0.1 g PS were dispersed into 10 mL toluene containing Nile Red dye (0.07 wt %) with sonication treatment for 30 min. The magnetic PS toluene solution was mixed with 70 mL sodium dodecyl sulfate (SDS) aqueous solution (10 mM concentration). The mixed solution was emulsified by Ultra Turrax T18 for 10 min with rotation speed 6000 rpm. The emulsion was mechanically stirred for 2 days to evaporate the toluene to produce solid magnetic PS microparticles.

2.5. Platinum Janus Polystyrene (PS) Microparticles

Magnetic PS microparticles were collected and placed on a glass slide to form a monolayer by drop coating. Then, the microparticles were coated with a thin layer of Pt by sputtering for 110 s at

10 mA current. Then, the obtained Janus motors were collected by sonication and suspended in DI water for further investigation.

3. Results and Discussion

In order to have magnetic PS microparticles, we first synthesized magnetic nanoparticles (Fe_3O_4) and modified the surface properties, switching from hydrophilic, to hydrophobic as seen in Figure 1a. This conversion of surface property is necessary because during the emulsion process, instead of being suspended in the aqueous solution, the hydrophobic magnetic nanoparticles and water-insoluble fluorescent dyes prefer to be embedded in the hydrophobic toluene/PS spheres formed by high-speed shearing-induced emulsion. In other words, the magnetic nanoparticles need to stay in the oil phase (droplets) during the whole process. During further evaporation of toluene the solid PS microparticles loaded with both magnetic nanoparticles and fluorescent dyes will spontaneously form in the solution.

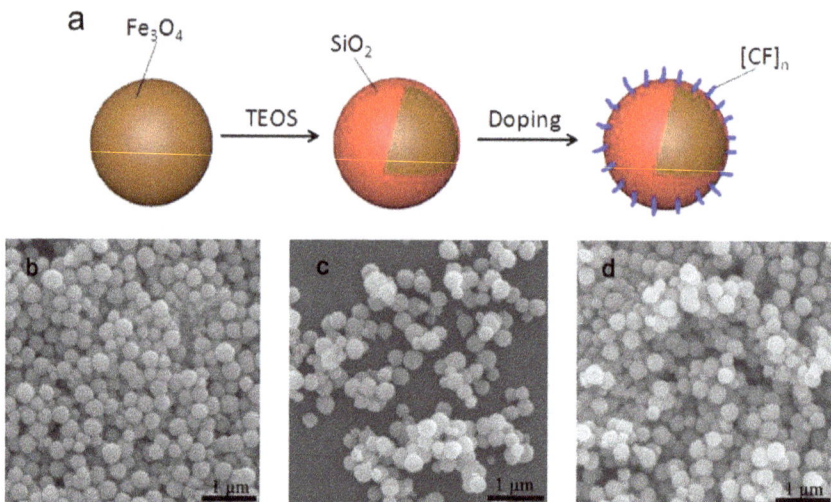

Figure 1. (**a**) Schematic illustration of the preparation hydrophobic magnetic nanoparticles and scanning electron microscopy (SEM) images of (**b**) the original magnetic nanoparticles; (**c**) after silica growth and (**d**) further modification with fluorosilane.

3.1. Magnetic Particles Synthesis and Modification

As shown in Figure 1a, based on the magnetic nanoparticles we synthesized, a silica shell was grown on their surface through a modified Stöber reaction. Then, we continued surface modification via refluxing with perfluorodecyltrichlorosilane (fluorosilane) in ethanol solution to form a hydrophobic surface. The samples were characterized by scanning electron microscopy (SEM) as shown in Figure 1b–d.

The original size of the magnetic nanoparticles was around 200 nm. The size increased to 300 nm due to the growth of the silica shell. The particles' size did not change significantly after perfluorodecyltrichlorosilane modification (see Figure 1c) since the grafting polymer is quite short. The magnetic nanoparticles formed numerous aggregates after surface modification since their surface was modified to switch from hydrophilic to hydrophobic. The silica growth and surface modification were both verified by Energy-dispersive X-ray spectroscopy (EDS) analysis (see Figure S1). During every surface modification step, silicon element and fluoride appeared as expected. The surface modification is quite successful based on the contact angle measurement. The contact angle between water and the hydrophobic surface of the modified magnetic nanoparticles increased from 10 degrees to 165 degrees (the detailed measurements and result information are listed in Figure S2). The size of the Fe_3O_4 nanoparticles was quantified by dynamic light scattering (DLS) in origin, after silica growth, and after fluorosilane surface modification (results are presented in Figure S3). The size increased after silica shell growth due to actual size increasing, as well as slight aggregation as seen in the SEM image (Figure 1c). After surface grafting of fluorosilane, the size increased and was confirmed by DLS measurement, which may be due to aggregation of particles by hydrophobic interaction in aqueous solution. It is understandable that the particles' surface shifting from hydrophilic to hydrophobic would lead to their aggregation in aqueous condition.

3.2. Magnetic/Fluorescent PS Microparticles Fabrication

Magnetic and fluorescent PS microparticles were fabricated through the emulsion templating method as described above. By using this method, the particles' size can be tuned easily by adjusting the rotation speed of Ultra Turrax. We investigated the relationship between the particles' size and rotation speed (results can be seen in Figure S4). In general, the particle size decreases exponentially with the increasing of the rotation speed. We have chosen an optimized condition at 6000 rpm for fabrication of micro beads, which results in an approximate particle size of 30 μm. Toluene evaporation during air drying resulted in smaller particles due to shrinkage attributed to volume loss. Then, we obtained the magnetic and fluorescent PS particles. The results of the fabrication process were characterized by SEM and EDX analysis (see Figure S5). We found iron in the formed magnetic microparticles (see Figure S5b).

The Janus motors were readily acquired by depositing a thin layer of Pt (10–20 nm) onto the PS particles by sputtering machine. Then, the finally obtained Janus magnetic and fluorescent micro/motors half coated with Pt were characterized by SEM and EDS (see Figure 2). SEM image analysis revealed a Pt shell covering the upper half of the PS particles (see figure 2a). EDS mapping also confirms the same element distribution of Janus shape (see Figure 2b,c. Green represents the presence of carbon and red represents the presence of Pt). We used fluorescent microscopy to observe the Janus motors and clearly observed red florescence emitted by Nile Red inside the motors (see Figure 3a). Then, we further measured the fluorescence spectrum of the micro-motors in aqueous solution (see Figure 3b). The spectrum confirmed the presence of Nile Red inside the micro-motors. Furthermore, although Nile Red possesses excellent optical properties, its hydrophobic characteristics and fluorescence emission would be quenched (see Figure 3b), which limits its biomedical application. However, in our strategy, we incorporated such water-insoluble dye into microparticles, thereby preserving their optical properties in aqueous environments. Therefore, we have successfully achieved the production of multifunctional, hybrid micro-motors which were equipped with both magnetic and fluorescent properties. Moreover, based on different applications or requirements, we can easily not only tune the concentration and thus emission strength of fluorescent dyes inside the motors, but also load different kinds of dyes (e.g., near infrared (NIR) for in vitro or in vivo imaging) based on the current strategy, which can greatly extend the utility of the fluorescent motors.

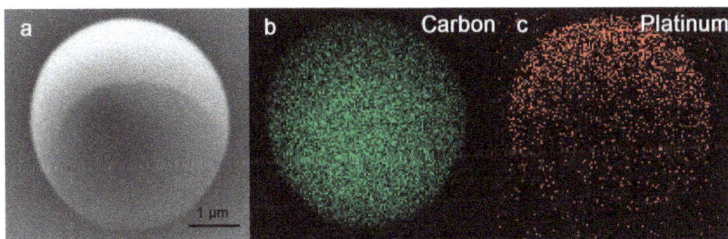

Figure 2. SEM photos of (**a**) magnetic polystyrene (PS) micro-particle, element mapping of (**b**) Carbon and (**c**) Platinum.

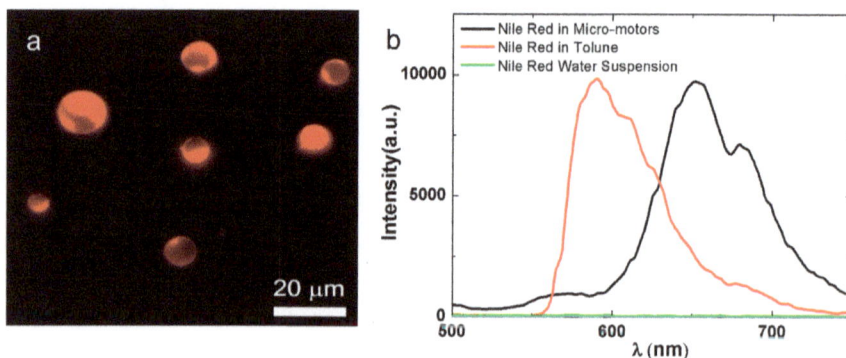

Figure 3. (a) Fluorescent microscopy image and (b) fluorescent emission spectrum of the magnetic/fluorescent hybrid micro-motors in aqueous solution.

3.3. Motion and Manipulation of the Micro-Motors

After obtaining the micro-motors, we studied their motion behavior by placing them into an aqueous solution fueled with H_2O_2 (3 wt %). A typical movement result is shown in Figure 4 together with a scheme illustrating the motion mechanism. The active motor was apparently self-propelled and moved from right to left as indicated by a red star mark. The green star marked passive spheres, which served as a position reference to clearly show the active motion of the Janus motors. The tracking trajectory was shown by a red line. The average velocity of this specific Janus motor was found to be 11.4 µm/s (RSD = 22.7%, $n = 15$), which agrees with other reported phoretic motors powered by Pt/H_2O_2 reaction. [3] The supporting video (Video S1) can be found in Section Supplementary Materials.

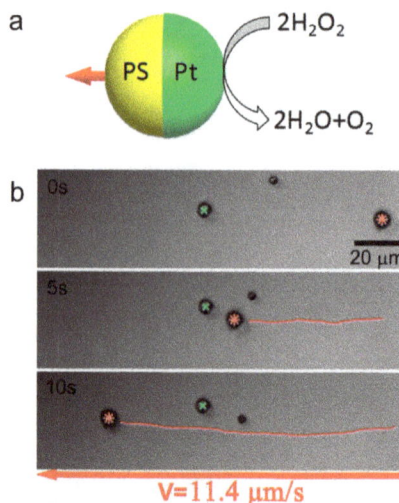

Figure 4. (a) Schematic illustration of the self-propelled Janus motors and (b) video snapshots of the motion of a micro-motor (red) with the addition of peroxide (3 wt %), without any surfactant. (The green particle stands as a reference particle.)

The manipulation of the magnetic micro-motors is shown in Figure 5. The magnetic micro-motor was moving from right to left at the beginning without any magnetic force manipulation. It is clear that the motor was moving almost straightforward. Then, we applied a magnetic field by using a permanent magnetic bar. The magnetic field at the motors' position was about 40 mT. After we applied a magnetic field from the perpendicular direction to give external guidance to the motors, we found an extra movement originating from the bottom side upwards which can be seen in the video (see Figure 5b). The motors showed corresponding motion after the magnetic force was applied. Due to the spherical shape and random distribution of magnetic nanoparticles inside the motors, no exact orientation was attributable to the magnetic moment. However, in our case, both magnetic attraction and orientation were working on the motors to change the moving direction, which works very well from our observation. The micro-motors' moving direction changed from horizontal to vertical after application of the magnetic field. The manipulated motion can be projected onto X and Y—two directions to simplify the analysis. The speed of self-propulsion, as measured on the X axis, was 3.2 μm/s. The speed of magnetic force, as measured on the Y axis, was 2.3 μm/s (Video S2 can be seen in Section Supplementary Materials).

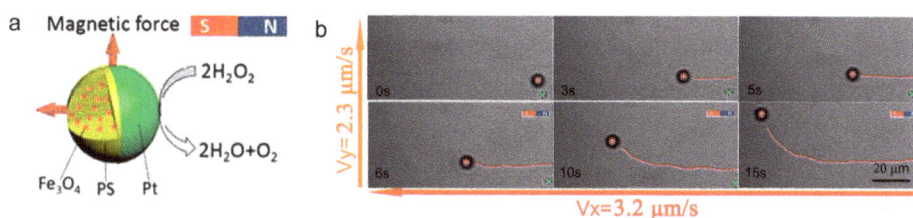

Figure 5. (a) Scheme and (b) video snapshots of a manipulated magnetic micro-motor with 3 wt % H_2O_2, no surfactant. The green particle stands as a reference particle.

4. Conclusions

We have presented a universal approach, adopted from the emulsion templating method, for the fabrication of Janus micro motors. The motors can be functionalized with both magnetism and fluorescence within a one-step process. Such self-propelled systems can move inside aqueous solutions when fueled with 3 wt% H_2O_2, and exhibit strong fluorescent emission in aqueous solution for potential applications in bio-imaging and particle-tracking, etc. Furthermore, due to the presence of magnetic nanoparticles inside the motors, their directionality can be easily manipulated by applying an external magnetic field, leading to controllable and multi-functional motors. Therefore, our motors not only provide a new strategy to fabricate magnetic/fluorescent motors in a facile way but also prove the potential of these motors for future applications, such as targeted cargo delivery and micro-manipulation.

Supplementary Materials: The following are available online at http://www.mdpi.com/2072-666X/9/2/83/s1. Supplemental Figure S1: Energy-dispersive X-ray spectroscopy (EDS) of Fe_3O_4 (S1a), Fe_3O_4@SiO_2 (S1b) and fluorosilane surface modification Fe_3O_4@SiO_2 (S1c); Figure S2: Contact angle measurement of Fe_3O_4, Fe_3O_4@SiO_2 and fluorosilane surface modification Fe_3O_4@SiO_2 in photos and tables; Figure S3: Size distribution of of Fe_3O_4, Fe_3O_4@SiO_2 and fluorosilane surface modification Fe_3O_4@SiO_2 by dynamic light scattering measurement; Figure S5: Energy-dispersive X-ray spectroscopy (EDS) of Polystyrene micro particles (S5a), magnetic PS micro particle (S5b) and platinum coated magnetic PS micro particles (S5c) Supplemental Video S1. Self-propulsion of the Janus motor in H_2O_2; Video S2: Magnetic guidance on the motor's movement; Video S3: Magnetic control on the orientation of the motors.

Acknowledgments: The authors acknowledge the support from JCYJ20170413102735544, Shenzhen Peacock Plan (KQTD201507161644225), the Natural Science Foundation of Guangdong Province, China (2015A030313889), the Science and Technology Planning Project of Shenzhen Municipality, China (ZDSYS201504301045406).

Author Contributions: X.M. and H.F. conceived and designed the experiments; J.J., D.X. and Y.L. performed the experiments; X.M., H.F., W.Z., J.Z., and T.Z. analyzed the data and contributed to the discussions on the article; T.Z. contributed analysis tools; J.J., H.F. and X.M. wrote the paper.

Conflicts of Interest: The authors declare no conflict of interest.

References

1. Fischer, P.; Ghosh, A. Magnetically actuated propulsion at low Reynolds numbers: Towards nanoscale control. *Nanoscale* **2011**, *3*, 557–563. [CrossRef] [PubMed]
2. Wang, J. *Nanomachines: Fundamentals and Applications*; John Wiley & Sons: Hoboken, NJ, USA, 2013.
3. Sánchez, S.; Soler, L.; Katuri, J. Chemically powered micro- and nanomotors. *Angew. Chem. Int. Ed.* **2015**, *54*, 1414–1444. [CrossRef] [PubMed]
4. Wang, H.; Pumera, M. Fabrication of micro/nanoscale motors. *Chem. Rev.* **2015**, *115*, 8704–8735. [CrossRef] [PubMed]
5. Katuri, J.; Ma, X.; Stanton, M.M.; Sánchez, S. Designing micro- and nanoswimmers for specific applications. *Acc. Chem. Res.* **2016**, *50*, 2–11. [CrossRef] [PubMed]
6. Ma, X.; Sánchez, S. Self-propelling micro-nanorobots: Challenges and future perspectives in nanomedicine. *Nanomedicine* **2017**, *12*. [CrossRef] [PubMed]
7. Paxton, W.F.; Kistler, K.C.; Olmeda, C.C.; Sen, A.; St. Angelo, S.K.; Cao, Y.; Mallouk, T.E.; Lammert, P.E.; Crespi, V.H. Catalytic nanomotors: Autonomous movement of striped nanorods. *J. Am. Chem. Soc.* **2004**, *126*, 13424–13431. [CrossRef] [PubMed]
8. Fournier-Bidoz, S.; Arsenault, A.C.; Manners, I.; Ozin, G.A. Synthetic self-propelled nanorotors. *Chem. Commun.* **2005**, *28*, 441–443. [CrossRef] [PubMed]
9. Gao, W.; Pei, A.; Wang, J. Water-driven micromotors. *ACS Nano* **2012**, *6*, 8432–8438. [CrossRef] [PubMed]
10. Wheat, P.M.; Marine, N.A.; Moran, J.L.; Posner, J.D. Rapid fabrication of bimetallic spherical motors. *Langmuir* **2010**, *26*, 13052–13055. [CrossRef] [PubMed]
11. Solovev, A.A.; Mei, Y.; Ureña, E.B.; Huang, G.; Schmidt, O.G. Catalytic microtubular jet engines self-propelled by accumulated gas bubbles. *Small* **2009**, *5*, 1688–1692. [CrossRef] [PubMed]
12. Gao, W.; Sattayasamitsathit, S.; Orozco, J.; Wang, J. Highly efficient catalytic microengines: Template electrosynthesis of polyaniline/platinum microtubes. *J. Am. Chem. Soc.* **2011**, *133*, 11862–11864. [CrossRef] [PubMed]
13. Zhao, G.; Ambrosi, A.; Pumera, M. Self-propelled nanojets via template electrodeposition. *Nanoscale* **2013**, *5*, 1319–1324. [CrossRef] [PubMed]
14. Wu, Z.; Wu, Y.; He, W.; Lin, X.; Sun, J.; He, Q. Self-propelled polymer-based multilayer nanorockets for transportation and drug release. *Angew. Chem. Int. Ed.* **2013**, *52*, 7000–7003. [CrossRef] [PubMed]
15. Zhang, L.; Abbott, J.J.; Dong, L.; Kratochvil, B.E.; Bell, D.; Nelson, B.J. Artificial bacterial flagella: Fabrication and magnetic control. *Appl. Phys. Lett.* **2009**, *94*, 064107. [CrossRef]
16. Ghosh, A.; Fischer, P. Controlled propulsion of artificial magnetic nanostructured propellers. *Nano Lett.* **2009**, *9*, 2243–2245. [CrossRef] [PubMed]
17. Schuerle, S.; Pané, S.; Pellicer, E.; Sort, J.; Baró, M.D.; Nelson, B.J. Helical and tubular lipid microstructures that are electroless-coated with CoNiReP for wireless magnetic manipulation. *Small* **2012**, *8*, 1498–1502. [CrossRef] [PubMed]
18. Tu, Y.; Peng, F.; Sui, X.; Men, Y.; White, P.B.; van Hest, J.C.M.; Wilson, D.A. Self-propelled supramolecular nanomotors with temperature-responsive speed regulation. *Nat. Chem.* **2016**, *9*, 480–486. [CrossRef] [PubMed]
19. Ma, X.; Hahn, K.; Sanchez, S. Catalytic mesoporous janus nanomotors for active cargo delivery. *J. Am. Chem. Soc.* **2015**, *137*, 4976–4979. [CrossRef] [PubMed]
20. Ma, X.; Hortelao, A.C.; Miguel-López, A.; Sánchez, S. Bubble-free propulsion of ultrasmall tubular nanojets powered by biocatalytic reactions. *J. Am. Chem. Soc.* **2016**, *138*, 13782–13785. [CrossRef] [PubMed]
21. Vilela, D.; Hortelao, A.C.; Balderas-Xicohtencatl, R.; Hirscher, M.; Hahn, K.; Ma, X.; Sanchez, S. Facile fabrication of mesoporous silica micro-jets with multi-functionalities. *Nanoscale* **2017**, *9*, 13990–13997. [CrossRef] [PubMed]
22. Wang, Y.-S.; Xia, H.; Lv, C.; Wang, L.; Dong, W.-F.; Feng, J.; Sun, H.-B. Self-propelled micromotors based on Au-mesoporous silica nanorods. *Nanoscale* **2015**, *7*, 11951–11955. [CrossRef] [PubMed]

23. Wu, Y.; Wu, Z.; Lin, X.; He, Q.; Li, J. Autonomous movement of controllable assembled Janus capsule motors. *ACS Nano* **2012**, *6*, 10910–10916. [CrossRef] [PubMed]

24. Li, J.; Zhang, J.; Gao, W.; Huang, G.; Di, Z.; Liu, R.; Wang, J.; Mei, Y. Dry-released nanotubes and nanoengines by particle-assisted rolling. *Adv. Mater.* **2013**, *25*, 3715–3721. [CrossRef] [PubMed]

25. Gao, W.; Kagan, D.; Pak, O.S.; Clawson, C.; Campuzano, S.; Chuluun-Erdene, E.; Shipton, E.; Fullerton, E.E.; Zhang, L.; Lauga, E. Cargo-towing fuel-free magnetic nanoswimmers for targeted drug delivery. *Small* **2012**, *8*, 460–467. [CrossRef] [PubMed]

26. Gao, W.; Wang, J. Synthetic micro/nanomotors in drug delivery. *Nanoscale* **2014**, *6*, 10486–10494. [CrossRef] [PubMed]

27. Ceylan, H.; Giltinan, J.; Kozielski, K.; Sitti, M. Mobile microrobots for bioengineering applications. *Lab Chip* **2017**, *17*, 1705–1724. [CrossRef] [PubMed]

28. Peng, F.; Tu, Y.; Wilson, D.A. Micro/nanomotors towards in vivo application: Cell, tissue and biofluid. *Chem. Soc. Rev.* **2017**, *46*, 5289–5310. [CrossRef] [PubMed]

29. Li, J.; de Ávila, B.E.; Gao, W.; Zhang, L.; Wang, J. Micro/nanorobots for biomedicine: Delivery, surgery, sensing, and detoxification. *Sci. Robot.* **2017**, *2*. [CrossRef]

30. Guix, M.; Orozco, J.; García, M.; Gao, W.; Sattayasamitsathit, S.; Merkoçi, A.; Escarpa, A.; Wang, J. Superhydrophobic alkanethiol-coated microsubmarines for effective removal of oil. *ACS Nano* **2012**, *6*, 4445–4451. [CrossRef] [PubMed]

31. Orozco, J.; Cheng, G.; Vilela, D.; Sattayasamitsathit, S.; Vazquez-Duhalt, R.; Valdés-Ramírez, G.; Pak, O.S.; Escarpa, A.; Kan, C.; Wang, J. Micromotor-based high-yielding fast oxidative detoxification of chemical threats. *Angew. Chem. Int. Ed.* **2013**, *52*, 13276–13279. [CrossRef] [PubMed]

32. Soler, L.; Magdanz, V.; Fomin, V.M.; Sanchez, S.; Schmidt, O.G. Self-propelled micromotors for cleaning polluted water. *ACS Nano* **2013**, *7*, 9611–9620. [CrossRef] [PubMed]

33. Campuzano, S.; Orozco, J.; Kagan, D.; Guix, M.; Gao, W.; Sattayasamitsathit, S.; Claussen, J.C.; Merkoçi, A.; Wang, J. Bacterial isolation by lectin-modified microengines. *Nano Lett.* **2011**, *12*, 396–401. [CrossRef] [PubMed]

34. Kuralay, F.; Sattayasamitsathit, S.; Gao, W.; Uygun, A.; Katzenberg, A.; Wang, J. Self-propelled carbohydrate-sensitive microtransporters with built-in boronic acid recognition for isolating sugars and cells. *J. Am. Chem. Soc.* **2012**, *134*, 15217–15220. [CrossRef] [PubMed]

35. Balasubramanian, S.; Kagan, D.; Hu, C.M.J.; Campuzano, S.; Lobo-Castañon, M.J.; Lim, N.; Kang, D.Y.; Zimmerman, M.; Zhang, L.; Wang, J. Micromachine-enabled capture and isolation of cancer cells in complex media. *Angew. Chem. Int. Ed.* **2011**, *50*, 4161–4164. [CrossRef] [PubMed]

36. Sanchez, S.; Solovev, A.A.; Schulze, S.; Schmidt, O.G. Controlled manipulation of multiple cells using catalytic microbots. *Chem. Commun.* **2011**, *47*, 698–700. [CrossRef] [PubMed]

37. Kagan, D.; Campuzano, S.; Balasubramanian, S.; Kuralay, F.; Flechsig, G.-U.; Wang, J. Functionalized micromachines for selective and rapid isolation of nucleic acid targets from complex samples. *Nano Lett.* **2011**, *11*, 2083–2087. [CrossRef] [PubMed]

38. Orozco, J.; Campuzano, S.; Kagan, D.; Zhou, M.; Gao, W.; Wang, J. Dynamic isolation and unloading of target proteins by aptamer-modified microtransporters. *Anal. Chem.* **2011**, *83*, 7962–7969. [CrossRef] [PubMed]

39. Ergeneman, O.; Chatzipirpiridis, G.; Pokki, J.; Marin-Suárez, M.; Sotiriou, G.A.; Medina-Rodriguez, S.; Sanchez, J.F.F.; Fernandez-Gutiérrez, A.; Pané, S.; Nelson, B.J. In vitro oxygen sensing using intraocular microrobots. *IEEE Trans. Biomed. Eng.* **2012**, *59*, 3104–3109. [CrossRef] [PubMed]

40. Campuzano, S.; Kagan, D.; Orozco, J.; Wang, J. Motion-driven sensing and biosensing using electrochemically propelled nanomotors. *Analyst* **2011**, *136*, 4621–4630. [CrossRef] [PubMed]

41. Rojas, D.; Jurado-Sánchez, B.; Escarpa, A. "Shoot and sense" Janus micromotors-based strategy for the simultaneous degradation and detection of persistent organic pollutants in food and biological samples. *Anal. Chem.* **2016**, *88*, 4153–4160. [CrossRef] [PubMed]

42. Jurado-Sánchez, B.; Pacheco, M.; Rojo, J.; Escarpa, A. Magnetocatalytic graphene quantum dots Janus micromotors for bacterial endotoxin detection. *Angew. Chem. Int. Ed.* **2017**, *56*, 6957–6961. [CrossRef] [PubMed]

43. Vilela, D.; Orozco, J.; Cheng, G.; Sattayasamitsathit, S.; Galarnyk, M.; Kan, C.; Wang, J.; Escarpa, A. Multiplexed immunoassay based on micromotors and microscale tags. *Lab Chip* **2014**, *14*, 3505–3509. [CrossRef] [PubMed]

44. Zeeshan, M.A.; Grisch, R.; Pellicer, E.; Sivaraman, K.M.; Peyer, K.E.; Sort, J.; Özkale, B.; Sakar, M.S.; Nelson, B.J.; Pané, S. Hybrid helical magnetic microrobots obtained by 3D template-assisted electrodeposition. *Small* **2014**, *10*, 1284–1288. [CrossRef] [PubMed]

45. Tottori, S.; Zhang, L.; Qiu, F.; Krawczyk, K.K.; Franco-Obregón, A.; Nelson, B.J. Magnetic helical micromachines: Fabrication, controlled swimming, and cargo transport. *Adv. Mater.* **2012**, *24*, 811–816. [CrossRef] [PubMed]

46. Solovev, A.A.; Sanchez, S.; Pumera, M.; Mei, Y.F.; Schmidt, O.G. Magnetic control of tubular catalytic microbots for the transport, assembly, and delivery of micro-objects. *Adv. Funct. Mater.* **2010**, *20*, 2430–2435. [CrossRef]

47. Manesh, K.M.; Cardona, M.; Yuan, R.; Clark, M.; Kagan, D.; Balasubramanian, S.; Wang, J. Template-assisted fabrication of salt-independent catalytic tubular microengines. *ACS Nano* **2010**, *4*, 1799–1804. [CrossRef] [PubMed]

48. Gao, W.; Pei, A.; Dong, R.; Wang, J. Catalytic iridium-based Janus micromotors powered by ultralow levels of chemical fuels. *J. Am. Chem. Soc.* **2014**, *136*, 2276–2279. [CrossRef] [PubMed]

49. Gao, W.; Feng, X.; Pei, A.; Gu, Y.; Li, J.; Wang, J. Seawater-driven magnesium based Janus micromotors for environmental remediation. *Nanoscale* **2013**, *5*, 4696–4700. [CrossRef] [PubMed]

50. Baraban, L.; Makarov, D.; Streubel, R.; Mönch, I.; Grimm, D.; Sanchez, S.; Schmidt, O.G. Catalytic Janus motors on microfluidic chip: Deterministic motion for targeted cargo delivery. *ACS Nano* **2012**, *6*, 3383–3389. [CrossRef] [PubMed]

51. Albrecht, M.; Hu, G.; Guhr, I.L.; Ulbrich, T.C.; Boneberg, J.; Leiderer, P.; Schatz, G. Magnetic multilayers on nanospheres. *Nat. Mater.* **2005**, *4*, 203–206. [CrossRef] [PubMed]

52. Ulbrich, T.; Bran, C.; Makarov, D.; Hellwig, O.; Risner-Jamtgaard, J.; Yaney, D.; Rohrmann, H.; Neu, V.; Albrecht, M. Effect of magnetic coupling on the magnetization reversal in arrays of magnetic nanocaps. *Phys. Rev. B* **2010**, *81*, 054421. [CrossRef]

53. Ma, X.; Sanchez, S. A bio-catalytically driven Janus mesoporous silica cluster motor with magnetic guidance. *Chem. Commun.* **2015**, *51*, 5467–5470. [CrossRef] [PubMed]

54. Vilela, D.; Parmar, J.; Zeng, Y.; Zhao, Y.; Sánchez, S. Graphene-based microbots for toxic heavy metal removal and recovery from water. *Nano Lett.* **2016**, *16*, 2860–2866. [CrossRef] [PubMed]

55. Yan, X.; Zhou, Q.; Vincent, M.; Deng, Y.; Yu, J.; Xu, J.; Xu, T.; Tang, T.; Bian, L.; Wang, Y.X.-J. Multifunctional biohybrid magnetite microrobots for imaging-guided therapy. *Sci. Robot.* **2017**, *2*, eaaq1155. [CrossRef]

56. Liu, B.; Boker, A. Measuring rotational diffusion of colloidal spheres with confocal microscopy. *Soft Matter* **2016**, *12*, 6033–6037. [CrossRef] [PubMed]

57. Ma, X.; Jang, S.; Popescu, M.N.; Uspal, W.E.; Miguel-López, A.; Hahn, K.; Kim, D.-P.; Sánchez, S. Reversed Janus Micro/Nanomotors with Internal Chemical Engine. *ACS Nano* **2016**, *10*, 8751–8759. [CrossRef] [PubMed]

58. Liu, J.; Sun, Z.; Deng, Y.; Zou, Y.; Li, C.; Guo, X.; Xiong, L.; Gao, Y.; Li, F.; Zhao, D. Highly water-dispersible biocompatible magnetite particles with low cytotoxicity stabilized by citrate groups. *Angew. Chem.* **2009**, *121*, 5989–5993. [CrossRef]

59. Zhang, H.; Cooper, A.I. Synthesis of Monodisperse emulsion-templated polymer beads by oil-in-water-in-oil (O/W/O) sedimentation polymerization. *Chem. Mater.* **2002**, *14*, 4017–4020. [CrossRef]

60. Zhang, H.; Hardy, G.C.; Khimyak, Y.Z.; Rosseinsky, M.J.; Cooper, A.I. Synthesis of hierarchically porous silica and metal oxide beads using emulsion-templated polymer scaffolds. *Chem. Mater.* **2004**, *16*, 4245–4256. [CrossRef]

micromachines

MDPI

Article

Multiple Electrohydrodynamic Effects on the Morphology and Running Behavior of Tiny Liquid Metal Motors

Yue Sun [1,2], Shuo Xu [1,2], Sicong Tan [1,2] and Jing Liu [1,2,3,*]

1   Beijing Key Lab of Cryo-Biomedical Engineering and Key Lab of Cryogenics,
     Technical Institute of Physics and Chemistry, Chinese Academy of Sciences, Beijing 100190, China;
     sunyue161@mails.ucas.ac.cn (Y.S.); xushuo15@mails.ucas.ac.cn (S.X.); tansicong@mail.ipc.ac.cn (S.T.)
2   School of Future Technology, University of Chinese Academy of Sciences, Beijing 100039, China
3   Department of Biomedical Engineering, School of Medicine, Tsinghua University, Beijing 100084, China
*   Correspondence: jliu@mail.ipc.ac.cn; Tel.: +86-10-8254-3765

Received: 20 March 2018; Accepted: 13 April 2018; Published: 18 April 2018

Abstract: Minimized motors can harvest different types of energy and transfer them into kinetic power to carry out complex operations, such as targeted drug delivery, health care, sensing and so on. In recent years, the liquid metal motor is emerging as a very promising tiny machine. This work is dedicated to investigate the motion characteristics of self-powered liquid metal droplet machines under external electric field, after engulfing a small amount of aluminum. Two new non-dimensional parameters, named Ä and Ö, are put forward for the first time to evaluate the ratio of the forces resulting from the electric field to the fluidic viscous force and the ratio of the friction force to the fluidic viscous force. Forces exerted on liquid metal droplets, the viscosity between the droplet and the surrounding fluid, the pressure difference on both ends, the friction between the bottom of the droplet and the sink base, and bubble propulsion force are evaluated and estimated regarding whether they are impetus or resistance. Effects of electric field intensity, droplet size, solution concentration and surface roughness etc. on the morphology and running behavior of such tiny liquid metal motors are clarified in detail. This work sheds light on the moving mechanism of the liquid metal droplet in aqueous solutions, preparing for more precise and complicated control of liquid metal soft machines.

Keywords: liquid metal; electrical field; self-propulsion; droplet motor

1. Introduction

Minimized motors retain the property of their macroscale counterparts on harvesting different types of energy namely thermal [1,2], electrical [3,4], magnetic [5–7], chemical [8–10], ultrasound [11,12], light [13,14] power and so on, transferring into kinetic energy to carry out complex operations. Liquid metal, typically Gallium-based alloys, inherited with favorable fluidity, high thermal and electrical conductivity, large surface tension, low toxicity, and other excellent physical and chemical properties, has attracted huge interest across various fields, like chip cooling [15], nerve connection [16], drug delivery [17], flexible electronics [18], soft-bodied robot [19,20], micro pump [21], etc. Tiny liquid metal motors can be driven by the electric field, allowing its biocompatibility superior to those fueled by traditional toxic solution. Locomotion and morphology of room temperature liquid metal (RTLM) in multiphase fluidic environment and its precise control remain major challenges in the researches and applications of this promising material. Scattered into small droplets and provided with aluminum, autonomous macroscopic Brownian self-propulsion phenomenon of the tiny liquid metal motors can be observed in alkaline solutions [22]. These self-fueled motors can exhibit biomimetic behaviors without the involvement of external energy [7]. In the light of previous studies, a new robotic concept,

Transient State Machine, was also put forward to describe machines that set off free morphological transition for different purposes [20].

Intensive studies have been performed to get command of the motion of liquid metal. Facts prove that liquid metal displays featured behaviors when subjected to electrical field, magnetic field [23,24], graphite substrate [25], ultraviolet-irradiation [26] and electrolyte properties like pH or ionic concentration gradients [27]. Among these, the mechanism of electric field-induced chemical locomotion of conduction objects [3] is more lucid than other factors. Tang et al. [28] had studied liquid metal droplets coated with nanoparticles, namely 'liquid metal marbles', which can be actuated upon the application of electric fields. In addition, they investigated the chaotic advection actuating [29] and pumping [21] performance of the millimeter scale liquid metal droplet under alternating current signal. Yang et al. [30] developed the liquid metal pump in milli/centi-meter scale aided with AC, and demonstrated its bulk and surface motions. Tan et al. revealed the electrical control and stimulation of the moving direction and the velocity of the tiny Al-Ga-In motors [31] as well as the magnetic restriction of their self-propulsion range, termed as the magnetic trap effect [24]. Tang et al. [32] described the usage of modest voltage to direct the continuous flow of liquid metal towards single or multiple directions simultaneously in micro-channels, which could work as a valve. Zhang et al. [33] demonstrated that a self-propelled motor with a nickel cap can be steered by magnetic or electric field, representing a reliable manner for drug delivery. In general, the effect of external fields on the liquid metal seems to be miraculous, especially when the liquid metal is fed with aluminum.

Based on previous studies of liquid metal-aqueous solution multiphase flow, this work focuses on the dynamic behaviors of liquid metal droplets after engulfing a small amount of aluminum, under the effect of electric field. Two non-dimensional parameters are put forward to evaluate the forces exerted on the droplet. Electric field intensity, droplet size, solution concentration and surface roughness are considered crucial factors to affect the morphology and running behavior of tiny liquid metal motors and various comparative experiments are performed. In the process of running, the liquid metal motor is subjected to the viscos dragging between the droplet and the surrounding fluid, the pressure difference on both ends, the friction between the bottom of the droplet and the sink base, and bubble propulsion force. Whether these forces propel or hinder the movement of liquid metal motors are discussed, revealing the moving mechanism of liquid metal in aqueous solutions.

2. Materials and Methods

Liquid metal used in this experiment is $GaIn_{10}$, which contains 90% Gallium (Shanxi Zhaofeng Gallium Co., Ltd., Quanyang, China) and 10% Indium (Zhuzhou Smelter Group Co., Ltd., Zhuzhou, China) in weight. $GaIn_{10}$ and Al are mixed in the ratio of 99:1 by weight. The liquid metal droplets are placed in a 12 cm × 1 cm × 1 cm open-top rectangular glass channel filled with sodium hydroxide solution. Two graphite rods are fixed at both terminals of the channel with a distance of 12 cm. A WYK-605 DC stabilized power supply (Dongfang Group East Co., Ltd., Dongguan, China) is configured to set up an adjustable electrical field. A high speed camera (Canon XF-305, Tokyo, Japan) is mounted above the channel to capture the motion. All the experiments were conducted under room temperature. To figure out the morphology and running behavior of the liquid metal droplet under different conditions, we investigate its performance with varied voltages, droplet sizes, solution concentrations and the friction force.

3. Results and Discussion

3.1. Mechanism Analysis

As the GaIn-Al droplet is placed in a channel filled with alkaline solution, a slow chemical reaction occurs between the alloy and the surrounding solution, producing gallates, $[Ga(OH)_4]^-$, resulting in a negative charge aggregation on the liquid metal droplet surface. The negatively charged surface attracts positive ions of the solution and a nearly homogeneous diffuse layer, electrical double layer

(EDL) [34,35] is formed, which can be modeled as a charged capacitor. As noted, the GaIn-Al alloy possesses high conductivity, the potential through the droplet can be reckoned as uniform. In contrast, the aqueous electrical resistance is not neglected. Once a parallel electric field is implemented at both terminals, a potential gradient is generated along the channel. Thus, the potential difference of the EDL alters along the droplet surface, engendering the charge redistribution on the surface as Figure 1 depicts.

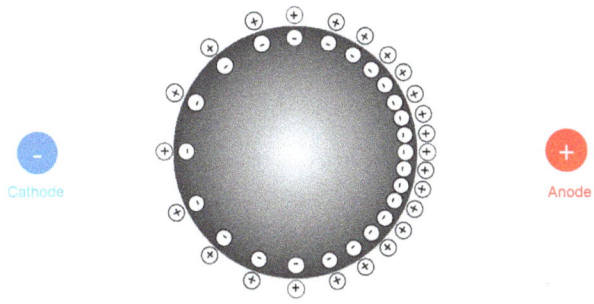

Figure 1. Surface charge redistribution of a liquid metal droplet in the electric field.

The electrical double layer on the surface affects the surface tension between the liquid metal and the solution, which can be described as Lippmann's equation:

$$\gamma = \gamma_0 - \frac{1}{2}c(V - V_0)^2 \tag{1}$$

where γ is the surface tension; γ_0 is the maximum surface tension when $V = 0$; c is the capacitance per unit area of the EDL; V is the electrode potential; V_0 is the potential of zero charge.

Supposing that there is no external electrical field, the EDL is initially charged by q_0, and the initial voltage of the capacitor is $V_0 = q_0/c$. The voltage on the left side of the droplet is $V_L = V - \Delta\varphi_r/2$, and on the right side $V_R = V + \Delta\varphi_r/2$, where $\Delta\varphi_r$ is the potential difference of the external electrical field, φ is the potential in the EDL. It can be inferred that the surface tension on the left side of the liquid metal droplet is larger than that on the right side. Learning from the Young–Laplace's equation, the pressure stress on the two sides of the droplet can be exhibited as:

$$\Delta S = \frac{2\Delta\gamma}{R} = \frac{2q\Delta\varphi}{R} \tag{2}$$

There is a pressure difference between the two sides of the droplet, which generates the propulsive force F_γ for the droplet to move right as Figure 2 presents.

In addition, there are another two driving forces actuating the droplet: the electric force F_e and the bubble propelling force F_b. Hindrance is the force F_v due to flow viscosity. Whether the friction force F_f is an impetus or resistance remains unknown. If the droplet is regarded as a rigid ball, according to the Reynold number, the relative flow of the solution is laminar flow. The viscous force F_v between the solution and the liquid metal droplet can be expressed as [36]:

$$F_v = 6\pi\mu u R \tag{3}$$

Considering gravity G, buoyancy F_{be} from the electrolyte and buoyancy F_{bb} from the bubble generated from the chemical reaction of aluminum and electrolyte on vertical direction, the rolling friction force F_f is demonstrated as

$$F_f = f \cdot (G - F_{be} - F_{bb}) = f \cdot g \cdot \frac{4}{3}\pi R^3 (\rho_m - \rho_a) - f \cdot F_{bb} \tag{4}$$

where f is the rolling friction coefficient between the base of the channel and the droplet, g is the gravity acceleration, ρ_m and ρ_a are the densities of the liquid metal droplet and the electrolyte, $\rho_m = 6.0165 \times 10^3$ kg/m^3 and $\rho_a = 10^3$ kg/m^3, R is the radius of the droplet.

Deriving from the Navier–Stokes equation in the x direction, one has:

$$\rho_a \left(\frac{\partial u}{\partial \tau} + u\frac{\partial u}{\partial x} + v\frac{\partial u}{\partial y} \right) = F_x - \frac{\partial p}{\partial x} + \rho_a v (\frac{\partial^2 u}{\partial x^2} + \frac{\partial^2 u}{\partial y^2}) \tag{5}$$

Regarding the model as steady-state laminar flow, ignoring the velocity in the y direction, the following momentum equation can be obtained:

$$u\frac{\partial u}{\partial x} = \frac{F_e - F_f}{\frac{4}{3}\pi R^3 \rho_m} - \frac{1}{\rho_a}\frac{dp}{dx} + v(\frac{\partial^2 u}{\partial x^2} + \frac{\partial^2 u}{\partial y^2}) \tag{6}$$

Extracting an infinitesimal Δx on the droplet, the pressure difference at two ends is

$$\Delta p = \frac{\Delta S}{\Delta A}\cos\theta = \frac{\Delta S \cos\theta}{2\pi R \sin\theta \Delta x} \tag{7}$$

Figure 3 shows the geometrical meaning of parameters in Equation (7).

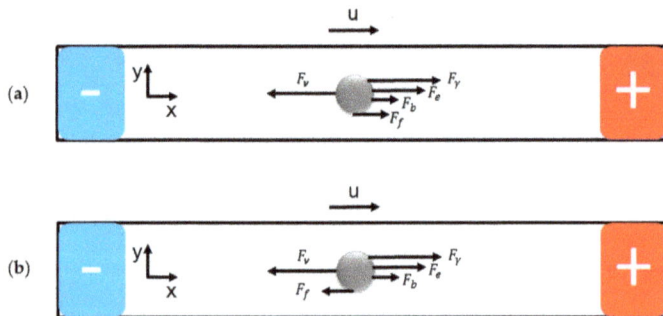

Figure 2. Forces analysis diagram for the locomotive liquid metal droplet in the electrical field when (a) the friction force F_f is an impetus; (b) the friction force F_f is a resistance.

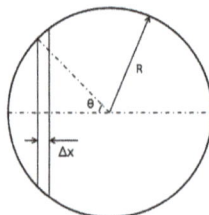

Figure 3. Geometrical meaning of parameters in equations.

Approximately, $\frac{dp}{dx} \approx \frac{\Delta p}{\Delta x} = \frac{2q_0\Delta\varphi}{R\Delta2\pi R\tan\theta(\Delta x)^2}$. Thus,

$$u\frac{\partial u}{\partial x} = \frac{q_0 U}{\frac{4}{3}\pi R^3 \rho_m L} - \frac{gf(\rho_m - \rho_a)}{\rho_m} - \frac{q_0\Delta\varphi}{\rho_a\pi R^2\tan\theta(\Delta x)^2} + \nu\left(\frac{\partial^2 u}{\partial x^2} + \frac{\partial^2 u}{\partial y^2}\right) \tag{8}$$

Taking the diameter of the droplet D as the length scale, u as the velocity scale, the voltage of the external electric field U as the electrical potential scale, Equation (8) can be non-dimensionalized as:

$$\frac{u^2}{D}\left(u^*\frac{\partial u^*}{\partial x^*}\right) = \frac{3q_0 U}{4\pi D^4\rho_m}\frac{1}{(R^*)^3 L^*} - \frac{gf(\rho_m-\rho_a)}{\rho_m} + \frac{q_0 U}{\pi D^4\rho_a}\frac{\varnothing^*}{(R^*)^2\tan\theta(\Delta x^*)^2} + \frac{\nu u}{D^2}\left(\frac{\partial^2 u^*}{\partial x^{*2}} + \frac{\partial^2 u^*}{\partial y^{*2}}\right) \tag{9}$$

where $\varnothing^* = \frac{-\Delta\varphi}{U}$. The parameters with star-superscript are dimensionless.

Divided by $\frac{\nu u}{D^2}$, Equation (9) can be converted into:

$$\frac{uD}{\nu}\left(u^*\frac{\partial u^*}{\partial x^*}\right) = \frac{3q_0 U}{4\pi D^2\nu u\rho_m}\frac{1}{(R^*)^3 L^*} - \frac{gf(\rho_m-\rho_a)D^2}{\rho_m\nu u} + \frac{q_0 U}{\pi D^2\nu u\rho_a}\frac{\varnothing^*}{(R^*)^2\tan\theta(\Delta x^*)^2} + \left(\frac{\partial^2 u^*}{\partial x^{*2}} + \frac{\partial^2 u^*}{\partial y^{*2}}\right) \tag{10}$$

In Equation (10), the first combined quantity $\frac{uD}{\nu}$ is the Reynold number. $\left(u^*\frac{\partial u^*}{\partial x^*}\right)$, $\frac{1}{(R^*)^3 L^*}$, $\frac{\varnothing^*}{(R^*)^2\tan\theta(\Delta x^*)^2}$ and $\left(\frac{\partial^2 u^*}{\partial x^{*2}} + \frac{\partial^2 u^*}{\partial y^{*2}}\right)$ are dimensionless terms, indicating that $\frac{3q_0 U}{4\pi D^2\nu u\rho_m} + \frac{q_0 U}{\pi D^2\nu u\rho_a}$ and $\frac{gf(\rho_m-\rho_a)D^2}{\rho_m\nu u}$ are non-dimensional. Respectively multiplied by the Reynold number $\frac{uD}{\nu}$, two new non-dimensional parameters are obtained as:

$$Ä = \frac{q_0 U}{\pi D\nu^2}\left(\frac{3}{4}\frac{1}{\rho_m} + \frac{1}{\rho_a}\right), \text{ and} \tag{11}$$

$$Ö = \frac{gf(\rho_m - \rho_a)D^3}{\rho_m\nu^2}. \tag{12}$$

Physically, Ä is a metric of the ratio of the forces resulting from the electric field to the fluidic viscous force. The increase of Ä denotes that the forces induced by the electrical field strengthen. Ö is a metric of the ratio of the friction force to the fluidic viscous force.

3.2. The Voltage Effect

The liquid metal droplet is injected statically at the cathode terminal initially. As the direct power supply is switched on, the droplet stretches and accelerates, then moves at a relatively steady speed towards the anode. When it approaches the anode terminal, it abruptly slows down and bounds back. Figure 4 shows the transient velocity of the liquid metal droplet under different voltages. We select the droplet with diameter of 2.5 mm and 0.2 mol/L sodium hydroxide solution from all the options. The applied voltages vary from 2 to 22 V, for water electrolysis intensifies as the voltage rises, creating more variables for the case when the voltage becomes too large. The channel is long and narrow, allowing the electrical field developed by two graphite rods to be approximately uniform. In the uniform electric field, the liquid metal droplet accelerates until the force reaches a balance and the droplet moves at a basically constant velocity. The curves tangle together when the applied voltage is above 10 V. The fluctuations of the curve, representing the oscillation of the motor during motion, can be drawn to the propelling bubbles generated by the liquid metal motor which consumes aluminum in advance, resulting in unpredictable moving direction. Plus, the inevitable adherence of the running droplet to the wall also intensifies the noise.

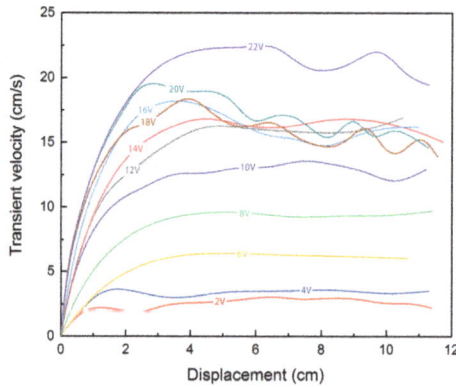

Figure 4. The transient velocity of the liquid metal droplet motor under different voltages.

The variations of the displacement and morphology with time are depicted in Figure 5. With low voltage, the droplet moves in a sphere without undergoing any noticeable deformation. As the voltage rises, both ends of the droplet deform differently, bringing about smaller curvature on the side with larger surface tension. When the voltage reaches 20 V, the deformation intensifies and a distinct pointy tail can be observed on the end opposite to the moving direction. Young–Laplace equation indicates that the surface tension difference is proportional to the potential difference. The increasing voltage magnifies the driving force stemmed from the surface tension difference on the liquid metal droplet. It should be noted that when the external voltage is 2 V or less, the electric force exerted on the liquid metal droplet is insufficient to overcome the resistance and the droplet would stagnate in the channel.

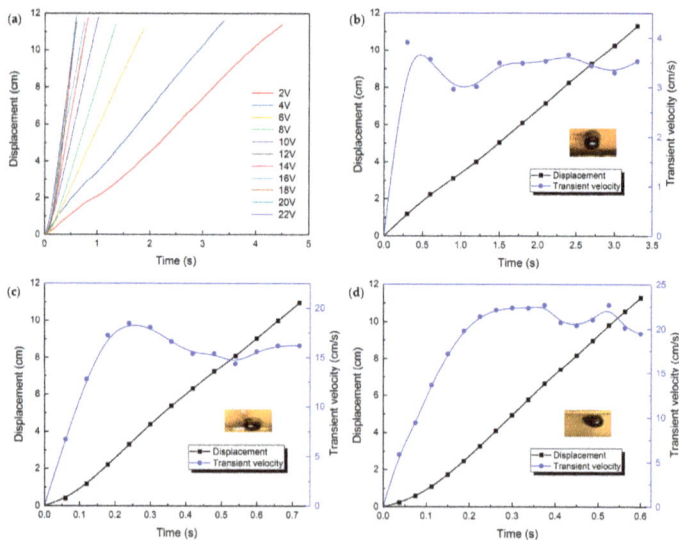

Figure 5. (**a**) Movement of the liquid metal droplet on a smooth surface under the impact of different voltages. The displacement and transient velocity change with time of the liquid metal droplet when the applied voltage is (**b**) 4 V; (**c**) 16 V; (**d**) 22 V, respectively, each with a snapshot of the droplet when it is moving at a constant speed.

3.3. The Size Effect

Figure 6 indicates how the average velocity of the motor varies with the droplet sizes from 1 to 9 mm under 20 V. The electrolyte solution remains 0.2 mol/L sodium hydroxide solution. Originally, the average velocity of the droplet rises with the magnified droplet size. As the diameter approaches 2.5 mm, the average velocity reaches the peak. Continuing to increase the droplet size, the velocity descends reversely. The surface tension driving force is positively correlative to the diameter of the droplet, as Equation (13) [30] shows:

$$F_\gamma = \frac{U}{R_t} \frac{4\pi q_0 \rho_a R^2}{h_1 h_2 - 2\pi R^2/3}.$$

(13)

When the droplet size is small, the surface tension driving force F_γ is too weak to actuate the droplet. Mounting the droplet size, the driving force will ascend and surpass the viscous and frictional drags. However, the viscous force F_v between the solution and the liquid metal droplet increases linearly with the droplet size learned from Equation (2). As the droplet size enlarges, the viscous force gradually takes a dominant role in the net force. As a consequence, the average velocity declines as the droplet size climbs further.

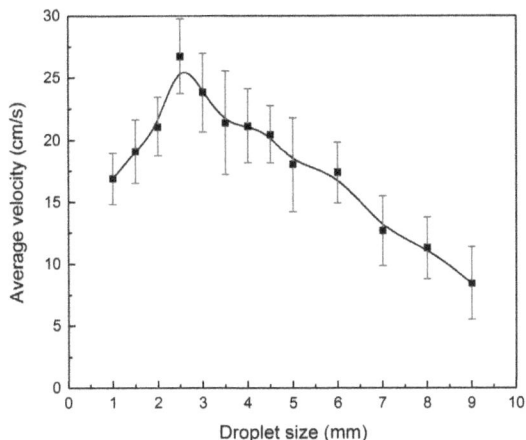

Figure 6. The average velocity of the motor varies with droplet sizes.

3.4. The Concentration Effect

As illustrated in Figure 7, different solution concentrations also influence the locomotion performance of the liquid metal droplet. The diameter of the droplet is 2.5 mm and the applied voltage is 20 V. Surprisingly, the concentration has little impact on the average velocities of the droplets which are already moving at a constant speed but rather on the acceleration process. As can be seen from the curve, the average velocity of the droplet increases slowly at first with the rising concentration and attains the summit at about 0.5 mol/L, followed by a gentle fluctuation. The rising concentration enhances the electrical conductivity of the solution [37] and results in the shrink of the total electrical resistance. On the other hand, the initial EDL charge density q_0 rises as the concentration grows [30], accordingly strengthening the surface tension driving force F_γ. Therefore, the higher the concentration is, the faster the droplet runs. Nevertheless, the viscosity has an exponential relationship [38] with the concentration of the strong electrolyte solution, which would be a counter-balance when the droplet size goes up. Besides, when the concentration attains and exceeds 0.5 mol/L, the EDL charge density saturates. Therefore, the velocity flattens out.

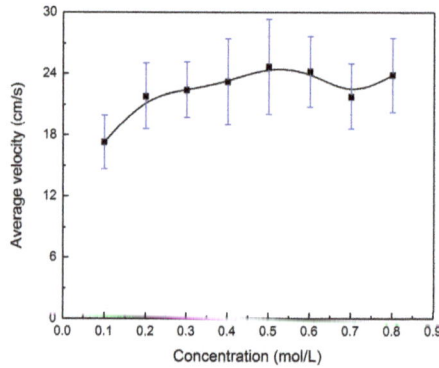

Figure 7. The average velocity of the motor varies with solution concentrations.

3.5. The Friction Effect

At last, after the above study, it still needs to verify that the friction between the bottom of the droplet and the sink base impelling the motion of the liquid metal motor instead of hindering. So we designed a contrast experiment accordingly, replacing the base from smooth glass to ground glass and maintaining other variables. The displacement and transformation tendency is shown in Figure 8. Little difference on deformation under the same voltage is observed.

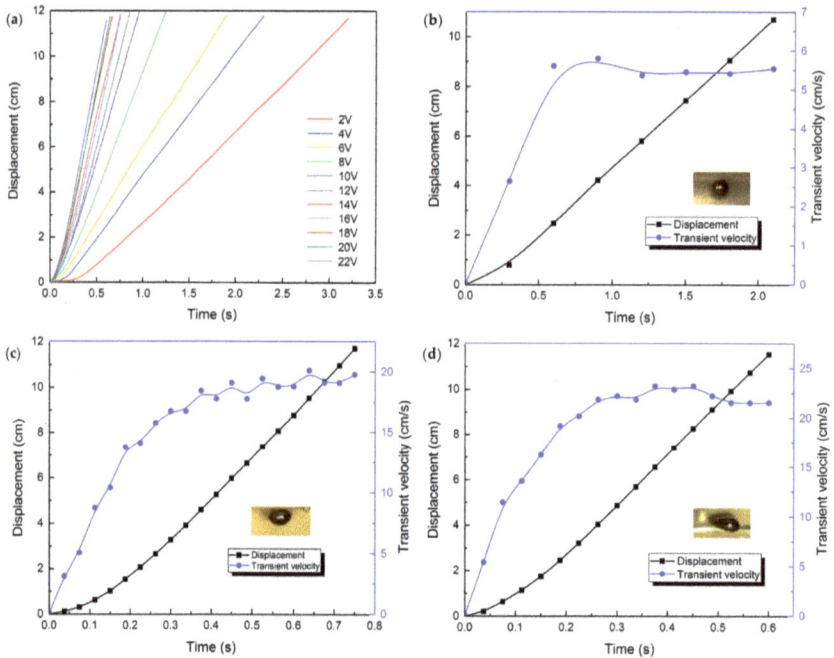

Figure 8. (a) Movement of the liquid metal droplet on a rough surface under the impact of different voltages. The displacement and transient velocity change with time of the liquid metal droplet when the applied voltage is (b) 4 V; (c) 16 V; (d) 22 V, respectively, each with a snapshot of the droplet when it is moving at a constant speed.

Micromachines **2018**, *9*, 192

The comparison on the average velocity of the liquid metal motor between the smooth base and the rough base can be seen from Figure 9. It is obvious that, with the increasing voltage, the average velocity of the liquid metal droplet rises gradually. The upward tendencies for two bases bear a close resemblance to each other, with the curve for the smooth base remaining underneath. The uncertainty grows with the velocity of the liquid metal droplet, on account of the existence of oscillation we mentioned earlier. Hence, it is safe to deduce that the friction given by the base of the groove drives the liquid metal droplet when it is moving towards the anode. Reasonable guess is that the contact of the bottom of the liquid metal and the surface of the groove base is close to the non-slip boundary condition.

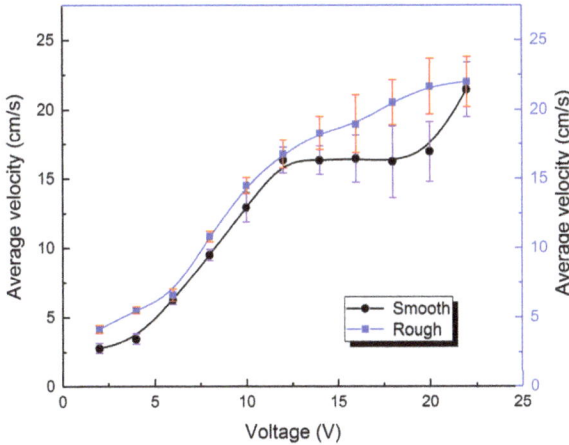

Figure 9. The comparison on average velocities of the liquid metal motor between the case of smooth base and the rough base.

4. Conclusions

This work comprehensively disclosed the motion characteristics of liquid metal motors in the electric field, which is fed with aluminum in advance. Two new non-dimensional parameters, Ä and Ö, are proposed to evaluate the forces exerted on the droplet, representing the ratio of the forces resulting from the electric field to the fluidic viscous force and the ratio of the friction force to the fluidic viscous force respectively. Four essentials, electric field intensity, droplet size, solution concentration and surface roughness, are characterized based on compared experiments. The velocity of the droplet grows with the increasing applied voltage, yet the growth rate reducing. The average velocities of the liquid metal droplet go up and down as the sizes of the droplet increase, with the maximum velocity at the diameter around 2.5 mm. Concentration does not play a leading role in the moving process of the liquid metal motor, as the average velocity grows and flattens with the rising concentration. This work puts forward the idea that the friction between the bottom of the liquid metal and the surface of the groove base propels the droplet, implying the non-slip boundary condition on the liquid metal motor and the channel bottom interface. Further studies could focus on the numerical study of the inner flow field and deformation of the liquid metal motors. Coupled fields such as the combination of the magnetic field, the chemical field or the electric field might also give inspirations in the evolution of the precise control of the liquid metal, partially bringing the legendary science fiction to real life.

Acknowledgments: This work is partially supported by the NSFC Key Project under Grant No. 91748206, Dean's Research Funding of the Chinese Academy of Sciences and the Frontier Project of the Chinese Academy of Sciences.

Micromachines **2018**, *9*, 192

Author Contributions: Yue Sun and Jing Liu conceived the experiments; Yue Sun, Shuo Xu and Sicong Tan designed the experiments; Yue Sun performed the experiments; Shuo Xu developed the mechanism analysis; Yue Sun analyzed the data and wrote the paper.

Conflicts of Interest: The authors declare no conflict of interest.

References

1. Balasubramanian, S.; Kagan, D.; Manesh, K.M.; Calvo-Marzal, P.; Flechsig, G.U.; Wang, J. Thermal modulation of nanomotor movement. *Small* **2009**, *5*, 1569–1574. [CrossRef] [PubMed]
2. Brzoska, J.B.; Brochardwyart, F.; Rondelez, F. Motions of droplets on hydrophobic model surfaces induced by thermal gradients. *Langmuir* **1993**, *9*, 2220–2224. [CrossRef]
3. Loget, G.; Kuhn, A. Electric field-induced chemical locomotion of conducting objects. *Nat. Commun.* **2011**, *2*, 1407–1422. [CrossRef] [PubMed]
4. Calvo-Marzal, P.; Manesh, K.M.; Kagan, D.; Balasubramanian, S.; Cardona, M.; Flechsig, G.U.; Posner, J.; Wang, J. Electrochemically-triggered motion of catalytic nanomotors. *Chem. Commun.* **2009**, *45*, 4509. [CrossRef] [PubMed]
5. Kline, T.R.; Paxton, W.F.; Mallouk, T.E.; Sen, A. Catalytic nanomotors: Remote-controlled autonomous movement of striped metallic nanorods. *Angew. Chem. Int. Ed.* **2010**, *44*, 744–746. [CrossRef] [PubMed]
6. Khalil, I.S.M.; Magdanz, V.; Sanchez, S.; Schmidt, O.G.; Abelmann, L.; Misra, S. Magnetic control of potential microrobotic drug delivery systems: nanoparticles, magnetotactic bacteria and self-propelled microjets. *Eng. Med. Biol. Soc.* **2013**, 5299–5302. [CrossRef]
7. Lübbe, A.S.; Bergemann, C.; Brock, J.; McClure, D.G. Physiological aspects in magnetic drug-targeting. *J. Magn. Magn. Mater.* **1999**, *194*, 149–155. [CrossRef]
8. Paxton, W.F.; Kistler, K.C.; Olmeda, C.C.; Sen, A.; St. Angelo, S.K.; Cao, Y.; Mallouk, T.E.; Lammert, P.E.; Crespi, V.H. Catalytic nanomotors: Autonomous movement of striped nanorods. *Angew. Chem.* **2004**, *126*, 13424. [CrossRef]
9. Solovev, A.A.; Mei, Y.; Bermúdez Ureña, E.; Huang, G.; Schmidt, O.G. Catalytic microtubular jet engines self-propelled by accumulated gas bubbles. *Small* **2009**, *5*, 1688. [CrossRef] [PubMed]
10. Ismagilov, R.F.; Schwartz, A.; Bowden, N.; Whitesides, G.M. Autonomous movement and self-assembly. *Angew. Chem. Int. Ed.* **2010**, *41*, 652–654. [CrossRef]
11. Garciagradilla, V.; Orozco, J.; Sattayasamitsathit, S.; Soto, F.; Kuralay, F.; Pourazary, A.; Katzenberg, A.; Gao, W.; Shen, Y.F.; Wang, J. Functionalized ultrasound-propelled magnetically guided nanomotors: Toward practical biomedical applications. *ACS Nano* **2013**, *7*, 9232. [CrossRef] [PubMed]
12. Wang, W.; Li, S.; Mair, L.; Ahmed, S.; Huang, T.J.; Mallouk, T.E. Acoustic propulsion of nanorod motors inside living cells. *Angew. Chem.* **2014**, *53*, 3201. [CrossRef] [PubMed]
13. Eelkema, R.; Pollard, M.M.; Vicario, J.; Katsonis, N.; Ramon, B.S.; Bastiaansen, C.W.; Broer, D.J.; Feringa, B.L. Molecular machines: Nanomotor rotates microscale objects. *Nature* **2006**, *440*, 163. [CrossRef] [PubMed]
14. Barrell, M.J.; Campaña, A.G.; Von, D.M.; Geertsema, E.M.; Leigh, D.A. Light-driven transport of a molecular walker in either direction along a molecular track. *Angew. Chem.* **2011**, *50*, 285–290. [CrossRef] [PubMed]
15. Ma, K.; Liu, J. Liquid metal cooling in thermal management of computer chips. *Front. Energy Power Eng. China* **2007**, *1*, 384–402. [CrossRef]
16. Zhang, J.; Sheng, L.; Jin, C.; Liu, J. Liquid metal as connecting or functional recovery channel for the transected sciatic nerve. *arXiv* **2014**, arXiv:1404.5931.
17. Lu, Y.; Hu, Q.; Lin, Y.; Pacardo, D.B.; Wang, C.; Sun, W.; Ligler, F.S.; Dickey, M.D.; Gu, Z. Transformable liquid-metal nanomedicine. *Nat. Commun.* **2015**, *6*. [CrossRef] [PubMed]
18. Wang, Q.; Yu, Y.; Yang, J.; Liu, J. Fast fabrication of flexible functional circuits based on liquid metal dual-trans printing. *Adv. Mater.* **2015**, *27*, 181–190. [CrossRef]
19. Sheng, L.; Zhang, J.; Liu, J. Diverse transformations of liquid metals between different morphologies. *Adv. Mater.* **2014**, *26*, 6036–6042. [CrossRef] [PubMed]
20. Lei, S.; He, Z.; Yao, Y.; Liu, J. Transient state machines: transient state machine enabled from the colliding and coalescence of a swarm of autonomously running liquid metal motors. *Small* **2015**, *11*. [CrossRef]

21. Tang, S.Y.; Khoshmanesh, K.; Sivan, V.; Petersen, P.; O'Mullane, A.P.; Abbott, D.; Mitchell, A.; Kalantar-zadeh, K. Liquid metal enabled pump. *Proc. Natl. Acad. Sci. USA* **2014**, *111*, 3304–3309. [CrossRef] [PubMed]
22. Yuan, B.; Tan, S.; Zhou, Y.; Liu, J. Self-powered macroscopic Brownian motion of spontaneously running liquid metal motors. *Chin. Sci. Bull.* **2015**, *60*, 1203–1210. [CrossRef]
23. Kim, D.; Lee, J.B. Magnetic-field-induced liquid metal droplet manipulation. *J. Korean Phys. Soc.* **2015**, *66*, 282–286. [CrossRef]
24. Tan, S.C.; Gui, H.; Yuan, B.; Liu, J. Magnetic trap effect to restrict motion of self-powered tiny liquid metal motors. *Appl. Phys. Lett.* **2015**, *107*, 13424. [CrossRef]
25. Hu, L.; Wang, L.; Ding, Y.; Zhan, S.; Liu, J. Manipulation of liquid metals on a graphite surface. *Adv. Mater.* **2016**, *28*, 9015. [CrossRef]
26. Tang, X.; Tang, S.Y.; Sivan, V.; Zhang, W.; Mitchell, A.; Kalantar-zadeh, K.; Khoshmanesh, K. Photochemically induced motion of liquid metal marbles. *Appl. Phys. Lett.* **2013**, *103*, 174104. [CrossRef]
27. Zavabeti, A.; Daeneke, T.; Chrimes, A.F.; O'Mullane, A.P.; Ou, J.Z.; Mitchell, A.; Khoshmanesh, K.; Kalantar-Zadeh, K. Ionic imbalance induced self-propulsion of liquid metals. *Nat. Commun.* **2016**, *7*, 12402. [CrossRef] [PubMed]
28. Tang, S.Y.; Sivan, V.; Khoshmanesh, K.; O'Mullane, A.P.; Tang, X.; Gol, B.; Eshtiaghi, N.; Lieder, F.; Petersen, P.; Mitchell, A.; et al. Electrochemically induced actuation of liquid metal marbles. *Nanoscale* **2013**, *5*, 5949–5957. [CrossRef] [PubMed]
29. Tang, S.Y.; Sivan, V.; Petersen, P.; Zhang, W.; Morrison, P.D.; Kalantar-zadeh, K.; Mitchell, A.; Khoshmanseh, K. Liquid metal actuator for inducing chaotic advection. *Adv. Funct. Mater.* **2014**, *24*, 5851–5858. [CrossRef]
30. Yang, X.H.; Tan, S.C.; Yuan, B.; Liu, J. Alternating electric field actuated oscillating behavior of liquid metal and its application. *Sci. China Technol. Sci.* **2016**, *59*, 597–603. [CrossRef]
31. Tan, S.C.; Yuan, B.; Liu, J. Electrical method to control the running direction and speed of self-powered tiny liquid metal motors. *Proc. R. Soc. A Math. Phys. Eng. Sci.* **2015**, *471*, 32–38. [CrossRef]
32. Tang, S.Y.; Lin, Y.; Joshipura, I.D.; Khoshmanesh, K.; Dickey, M.D. Steering liquid metal flow in microchannels using low voltages. *Lab Chip* **2015**, *15*, 3905–3911. [CrossRef] [PubMed]
33. Zhang, J.; Guo, R.; Liu, J. Self-propelled liquid metal motors steered by a magnetic or electrical field for drug delivery. *J. Mater. Chem. B* **2016**, *4*, 5349–5357. [CrossRef]
34. Zhang, J.; Yao, Y.; Sheng, L.; Liu, J. Self-fueled biomimetic liquid metal mollusk. *Adv. Mater.* **2015**, *27*, 2648–2655. [CrossRef] [PubMed]
35. Beni, G.; Hackwood, S.; Jackel, J.L. Continuous electrowetting effect. *Appl. Phys. Lett.* **1982**, *40*, 912–914. [CrossRef]
36. Oertel, H. *Prandtl-Essentials of Fluid Mechanics*; Springer Science & Business Media: Berlin, Germany, 2004.
37. Chen, L.M. Measurement of the relationship between conductivity of salt solution and concentration and temperature. *Res. Explor. Lab.* **2010**, *29*, 39–42.
38. Manzhou, H.; Chengrong, L.; Yingli, W.; Jiafu, F.; Xiancheng, Z.; Linli, L. Study on the viscosity of strong electrolyte solution. *Chemistry* **2011**, *74*, 356–361.

micromachines

MDPI

Article

Hydrophobic Janus Foam Motors: Self-Propulsion and On-The-Fly Oil Absorption

Xiaofeng Li, Fangzhi Mou *, Jingjing Guo, Zhuoyi Deng, Chuanrui Chen, Leilei Xu, Ming Luo and Jianguo Guan *

State Key Laboratory of Advanced Technology for Materials Synthesis and Processing, International School of Materials Science and Engineering, Wuhan University of Technology, Wuhan 430070, China; xiaofegli@whut.edu.cn (X.L.); october@whut.edu.cn (J.G.); zydeng2017@whut.edu.cn (Z.D.); chc034@ucsd.edu (C.C.); xull@whut.edu.cn (L.X.); luoming_2016@whut.edu.cn (M.L.)
* Correspondence: moufz@whut.edu.cn (F.M.); guanjg@whut.edu.cn (J.G.)

Received: 21 December 2017; Accepted: 9 January 2018; Published: 11 January 2018

Abstract: In this work, we for the first time have proposed and fabricated a self-propelled Janus foam motor for on-the-fly oil absorption on water by simply loading camphor/stearic acid (SA) mixture as fuels into one end of the SA-modified polyvinyl alcohol (PVA) foam. The as-fabricated Janus foam motors show an efficient Marangoni effect-based self-propulsion on water for a long lifetime due to the effective inhibition of the rapid release of camphor by the hydrophobic SA in the fuel mixture. Furthermore, they can automatically search, capture, and absorb oil droplets on the fly, and then be spontaneously self-assembled after oil absorption due to the self-propulsion of the motors as well as the attractive capillary interactions between the motors and oil droplets. This facilitates the subsequent collection of the motors from water after the treatment. Since the as-developed Janus foam motors can effectively integrate intriguing behaviors of the self-propulsion, efficient oil capture, and spontaneous self-assembly, they hold great promise for practical applications in water treatment.

Keywords: self-propulsion; oil/water separation; foam; capillary interaction; Marangoni effect

---

1. Introduction

Oil-leakages/spillages from industrial oily wastes or ship accidents bring many toxic compounds to water resources and may cause disastrous consequences to public health and aquatic ecosystems [1–4]. To address these issues, growing efforts have been devoted to the development of new materials for oil absorption and oil/water separation. The practical application of these materials requires them to have a high oil/water separation efficiency and low cost. Conventional low-cost materials for oil removal, including clay, activated carbon, and natural fibrous sorbent usually bring secondary pollutants into water and suffer from low selectivity, low absorption capacity, and long separation times [5–8]. In recent years, superhydrophobic and superoleophilic meshes, membranes, fabrics, nanofibers, etc. have been developed and used as filters for selective oil separation [4,9–12]. Even though they have been recognized to effectively separate oil contaminants from water, they are less practical on oil spills on open water because the contaminated water needs to be collected first for subsequent filtration [13–16].

Commercial sponge and foam are cheap porous materials and are available in our daily lives, and their interconnected 3D skeleton structures are expected to endow them with huge spaces for oil absorption and storage [17]. In fact, without surface modification, the pristine foam/sponge materials have a poor selectivity and may absorb both oil and water. Enhancing the roughness of foam surfaces and reducing foam surface energy are two common strategies to modify foam for the selective oil absorption from polluted water [18], which include methods such as dip coating [19], in-situ chemical reactions [20], vapor deposition [21], and spray coating [22]. Compared with other methods, dip coating is an effective and low-cost method for depositing hydrophobic materials onto substrates. After surface modification, the sponge and foam-based oil/water separation materials are promising for efficient oil/water separation with obvious advantages, such as high oil absorption capacity, low cost, and high selectivity [15]. However, those foams still cannot be directly used to deal with oil contaminants on open water owing to a lack of strategies to manipulate foams to capture oil floating on water. In particular, recent studies have demonstrated that magnetic foams, which are fabricated by incorporating magnetic particles onto porous foams, can be manipulated under an external magnetic field to target specific oil contaminants on water [23–25]. However, the incorporation of magnetic particles on porous foam complicates the fabrication process and raises the cost of the foams. In addition, as magnetic flux density decreases intensively with distance, the magnetically targeted oil removal is hindered in wide-range water treatment.

Self-propelled motors/objects have attracted considerable attentions over the last decade due to their conversion of other forms of energy into autonomous motion [26–29]. They are capable of picking up, transporting, and releasing various cargoes and thus have shown considerable promises for environmental detection and remediation, such as oil removal [30–32]. For instance, by modifying with hydrophobic organic chains on the surface, self-propelled micromotors, and microengines are capable of capturing oil droplets in water [33,34]. However, to power these micromotors and microengines, surfactants or chemical fuel (e.g., H_2O_2) need to be introduced into the aquatic environment, which not only makes the oil separation less cost-effective but also brings secondary pollutants into the environment. In addition, these micromotors and microengines can only capture the suspended oil droplets in water and are less effective to deal with the floating oil contaminants from oil spill and industrial oily wastewater. Alternatively, the self-propelled polysulfone (PSf) capsules loaded with sodium dodecyl sulfate (SDS) have been proposed to repel floating oil droplets through Marangoni effect, and realize the collision and merging of the scattered oil droplets [32]. However, it is difficult for the capsules to gather oil droplets on wide-open water due to the limited distance (several centimeters) of the repulsive interaction. In addition, as they can only translocate or merge oil droplets on water, an additional process is needed to separate oil from contaminated water. Hence, an ideal self-propelled motor for oil absorption is expected to have the capabilities of capturing and realizing in-situ oil/water separation in wide-open water, thus enhancing the overall performance and avoiding excessive redundancy.

In this work, we have proposed a self-propelled Janus foam motor, which is fabricated by simply loading the camphor/stearic acid (SA) mixture as fuels into one end of the SA-modified polyvinyl alcohol (PVA) foam, for on-the-fly oil absorption on water. The Janus foam motor shows an efficient self-propulsion on water with a long lifetime based on Marangoni effect owing to the sustained asymmetric release of camphor. Due to the attractive capillary interactions between motor and oil droplets and that between motors, the Janus foam motors not only can effectively capture and absorb oil droplets on the fly but also exhibit spontaneous self-assembly after the oil absorption, which facilitates the subsequent collection of the Janus foam motors from water after the treatment. Compared to the immobile or magnetically-driven hydrophobic foams [15,23–25], the as-developed Janus foam motors are expected to exhibit a much higher efficiency for oil removal because they can automatically search and capture oil contaminations in a wide range and then self-assemble into aggregates after oil absorption, due to their integrated properties of self-propulsion and hydrophobicity.

2. Materials and Methods

2.1. Fabrication of Janus Foam Motors

Commercial PVA foam was firstly cut into bar-shaped pieces with dimensions of 10 mm × 3 mm × 3 mm (length, width, thickness). The bar-shaped PVA foam (10 mg) was immersed into an ethanol solution with 0.35 mM stearic acid for 3 s and dried in air at room temperature to obtain SA-modified PVA foam. Afterwards, SA (29 mg) and camphor (14.5 mg) were loaded in one end of the SA-modified

PVA foam by immersing it into liquid camphor/SA mixture at 80 °C. After cooling in air at room temperature, the Janus foam motor is obtained. The motors with different camphor mass ratio (r_c) were also fabricated using the same method to investigate the influence of r_c on the speed and lifetime of the motors. To adjust r_c, the weight of camphor was varied in the mixture while keeping SA weight (29 mg) unchanged. The Janus foam motor loaded with pure camphor was prepared by replacing 29 mg of SA into 30 μL of ethanol with other conditions unchanged.

2.2. Characterization of Janus Foam Motors

Scanning electron microscopy (SEM) images were obtained by a Hitachi S-4800 field-emission SEM (Hitachi, Tokyo, Japan). Fourier-transformed infrared (FTIR) spectra were obtained using a Nicolet 6700 FTIR spectrometer (Thermo Fisher Scientific, Waltham, MA, USA) in the range of 400–4000 cm^{-1} with a resolution of 4 cm^{-1}.

2.3. Self-Propulsion and Oil Absorption

Janus foam motors were placed on water in a Petri dish (diameter: 150 mm, depth: 30 mm) with 100 mL water. A high-definition (HD) video recorder was placed over the dish. The video clips were analyzed with Video Spot Tracker V8.0 software (Center for Computer Integrated Systems for Microscopy and Manipulation (CISMM), UNC Chapel Hill, NC, USA).

The paraffin oil was colored with Solvent Blue 14 (Alfa Aesar, Haverhill, MA, USA) for easy observation. Several oil droplets (10 μL) were placed on the surface of the water. Then Janus foam motors were put onto water to capture and absorb oil droplets on the fly. The oil absorption capacity (Q) is determined by weighing the motors before and after oil absorption, and is calculated as follows.

$$Q = \frac{(m_1 - m_0)}{m_0}$$

Here, m_0 and m_1 are the weights of the motors before and after oil absorption, respectively.

3. Results and Discussion

The preparation of the Janus foam motor is shown in Figure 1. At first, a commercially available bar-shaped PVA foam was modified with stearic acid by a facile dipping-coating method. The pristine PVA foam (Figure 1a) shows an interconnected 3D skeleton structure, so it can quickly absorb the stearic acid solution (0.35 mM in ethanol) after dipping in the solution. The stearic acid modified PVA (SA-PVA) foam (Figure 1b) was then obtained after pulling up the foam and drying in air at room temperature to remove the solvent. Secondly, one end of SA-PVA foam was loaded with camphor/SA mixture (camphor mass ratio, r_c = 0.38) by dipping it into liquid camphor/SA mixture at 80 °C. After cooling in air at room temperature, the Janus foam motor was obtained. As shown in Figure 1c, the Janus structure of the foam motor can be clearly observed, in which the upper end (SA end) of the PVA foam in light yellow color is coated with SA, and the lower end (Camphor/SA end) in bright yellow color is loaded with camphor/SA mixture. From close SEM observation, it can be seen that the pristine PVA foam before modification shows a bulk 3D macroporous structures, and the size of the pores is in several hundred micrometers (Figure S1a). After the modification, the macroporous structure remains in the SA end (Figure S1b), while the pores in the other end is completely filled by the camphor/SA mixture (Figure S1c). The uptake of camphor and SA in the Janus foam motor was confirmed by FTIR spectra of the PVA foams before and after the modification (Figure S1d). The characteristic absorption peak at 1073 cm^{-1}, which corresponds to the C=O group in camphor and stearic acid [35], was observed in the PVA foams modified with SA or with camphor/SA mixture, while this peak was absent in the FTIR spectrum of pristine PVA foam.

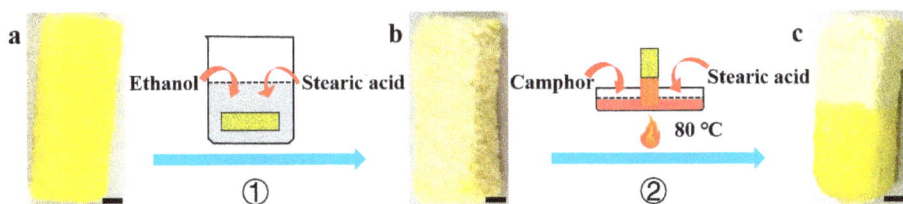

Figure 1. Schematic illustration of the fabrication of the Janus foam motor. At first, the pristine polyvinyl alcohol (PVA) foam (**a**) was modified with stearic acid (SA) by soaking it in SA ethanol solution (step 1) and drying in air to prepare SA-PVA foam (**b**). Then, the Janus foam motor (**c**) was prepared by loading camphor/SA mixture into one end of the SA-PVA foam at 80 °C (step 2). Scale bars: 1 mm.

When the Janus foam motor is put on water, the loaded camphor can be asymmetrically released from the motor. The released camphor is then mainly dissolved in water and adsorbed at the air–water interface with a small mass loss due to its sublimation into the bulk air phase. The dissolved camphor molecules would reduce surface tension (γ) of water around the camphor/SA end of the motor and give rise to a gradient in water surface tension ($\nabla\gamma$), namely a Marangoni stress, across the motor. The Marangoni stress then propels the motor with the SA end forward [36]. The detailed propulsion mechanism of the Janus foam motor based on Marangoni effect is given in Figure 2a. The propulsion of the Janus foam motor under $\nabla\gamma$ can be expressed by the classical Newtonian equation [37]:

$$\rho\ddot{x}_c(t) = \nabla\gamma(u(x_c(t), t)) - \mu\dot{x}_c(t) \tag{1}$$

where $x_c(t)$ denotes the center of mass of the foam motor, $u(x, t)$ is the concentration of the diffused camphor layer, ρ is the surface density of the camphor, and μ is the surface viscosity constant. Figure 2b and Video S1 show the self-propulsion of a typical Janus foam motor on water in a Petri dish (diameter: 15 cm, depth: 3.0 cm) filled with 100 mL of deionized water. It can be seen that the motor can float on water and autonomously move with an initial speed of 24.3 mm/s. The superposition of the translational and rotational motions is frequently observed for the Janus foam motors. This can be explained by the fact that the release of camphor misalign with the horizontal axis symmetry of the motor, generating asymmetric Marangoni stresses and thus a torque rotating the motor [38]. Even though the speed of the Janus foam motor ($r_c = 0.38$) decreased over time, a notable speed of 8.3 mm/s still remained even after 50 min (Figure 2c), indicating the continuous and sustained release of camphor fuel. In contrast, the speed of the Janus foam motor loaded with pure camphor decreased sharply from 32.8 to 0.1 mm/s in 24 min (Figure 2c), revealing the quick exhaustion of camphor fuel. From Equation (1), the propulsion of the Janus foam motor is directly related to the distribution of camphor across the motor. Thus, increasing the mass ratio of camphor in the camphor/SA end would amplify the asymmetric distribution of camphor and enhance the propulsion of the motor. As shown in Figure 2d, the speed of the Janus foam motor increases from 16.6 to 19.0 mm/s with the increasing mass ratio (r_c) of the camphor from 0.17 to 0.38 and becomes stable as r_c is over 0.38. The stable speed can be attributed to the fact that the surface-tension difference between the front SA end and the rear camphor/SA end becomes stable when r_c is over 0.38 due to the saturated adsorption of camphor molecules at the water–air interface [39].

The lifetime of the self-propelled motors is of great significance to its various applications, such as cargo transportation, water treatment, and sensing. [40–44]. In this work, we elaborately loaded the camphor fuel with stearic acid, which could greatly modulate the release behavior of camphor in water due to its hydrophobic nature. In this way, rather than the quick release of the pure camphor in water due to its water solubility (1.2 g/L), camphor in the camphor/SA mixture tended to be sustainably released in a long period, endowing the Janus foam motor with a long lifetime. As shown in Figure 2e, as r_c increased from 0.17 to 0.38, the lifetime of the Janus foam motor increased from 136 to 230 min because of the increasing amount of loaded camphor fuel, but it remained unchanged during further increases in the amount of the loaded camphor fuel as r_c increased from 0.38 to 0.5. This implies that SA can inhibit the rapid release of camphor and elongate the lifetime of the motor at a given amount of the loaded camphor fuel. In contrast, the Janus foam motor loaded with pure camphor only showed a short lifetime of 30 min (Figure 2e). Furthermore, the lifetime of the as-developed Janus foam motor was much longer than other motors also based on the Marangoni effect. For instance, the liquid motor consisting of polyvinyl chloride (PVC) dimethylformamide (DMF) solution only shows a maximum lifetime of 23 min due to the quick exhaustion of DMF fuel [45]. From Figure 2d,e, we can conclude that the optimal r_c for the Janus foam motor is 0.38, with which it exhibits a maximum speed of 19.0 mm/s and simultaneously a maximum lifetime of 230 min. Thanks to its excellent motion behaviors, the Janus foam motor can cruise on almost the entire surface of water in the Petri dish (Figure S2a), while the SA-PVA foam without camphor (Figure S2b) shows a negligible motion distance in the same period of time.

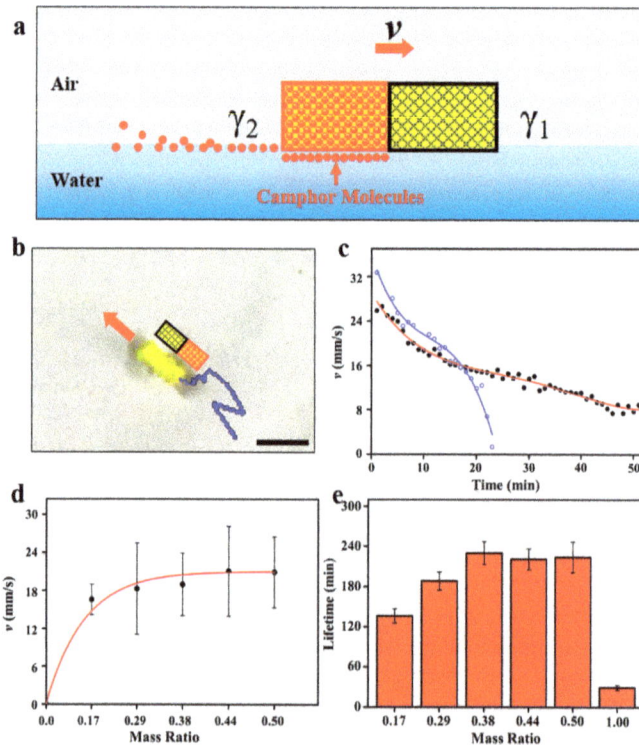

Figure 2. (**a**) Schematic illustration of the propulsion of the Janus foam motor based on Marangoni effect due to the asymmetric release of camphor. The red domain represents the camphor/SA end of the Janus foam motor. (**b**) Motion trajectory of the motor in 1 s. Scale bar: 10 mm. (**c**) The time-dependent speed (v) of the Janus foam motor loaded with the camphor/SA mixture ($r_c = 0.38$) (black dot) and pure camphor (blue circles), respectively. (**d**) The speed (v) of the motor versus mass ratio of camphor in the camphor/SA mixture. (**e**) Lifetime of the motor versus mass ratio of camphor.

The hydrophobicity and oleophilicity of the Janus foam motors is crucial for their performance in oil capture and absorption on water surface. The contact angles (θ) of water on the SA end of the Janus foam motor is measured to be 138° (Figure S3), verifying its hydrophobicity. As shown in Figure S4a–c, when a drop of paraffin oil–water mixture was dropped onto the SA end of the motor, the oil was quickly absorbed by the foam, leaving only water on its surface. The oil–water separation can be explained by the fact that paraffin oil (33.0 mN/m) [46] and stearic acid (24 mN/m) [47] have a similar surface energy, and paraffin oil can easily wet the surface of SA-PVA foam, while water is strongly repelled due to its high surface tension (72.7 mN/m) [48].

Combining the motion behavior and the oleophilicity of the SA end, the Janus foam motor can perform an "on-the-fly" collection of oil on water surfaces through capillary interaction between the motor and oil. As illustrated in Figure 3a and Video S2, with the efficient self-propulsion on water, the Janus foam motor could autonomously approach (0 s), capture (6.35 s), and transport (12 s) the oil droplet (paraffin oil) on the water surface, showing an oil absorption capacity of 0.6 g/g. It is worth noting that, when the motor moves to the area in the vicinity of the oil droplet (typically in 9 mm), the immobile oil droplet would move towards the motor (red trajectory in Figure 3a,b) and then be absorbed, indicating the obvious attractive interaction between the motor and the oil droplet. Due to the hydrophobic nature of the oil droplet and the Janus foam motor, they create negative menisci at the water–air interface [49–51]. The formation of menisci increases the interfacial area and raises the energy of the system [52]. When these two hydrophobic objects approach each other closely enough, the menisci of the two objects overlap and interact, leading to the attractive interaction between them [51]. Then, the motor and oil droplet move toward each other spontaneously on the basis of the minimization of the interfacial free energy of the liquid–air interface [53,54]. Figure 3c shows the detailed mechanism of the attractive interaction between the self-propelled Janus foam motor and the oil droplet. The attractive interaction between the Janus foam motor and the oil droplets facilitates the motor for oil capture and absorption.

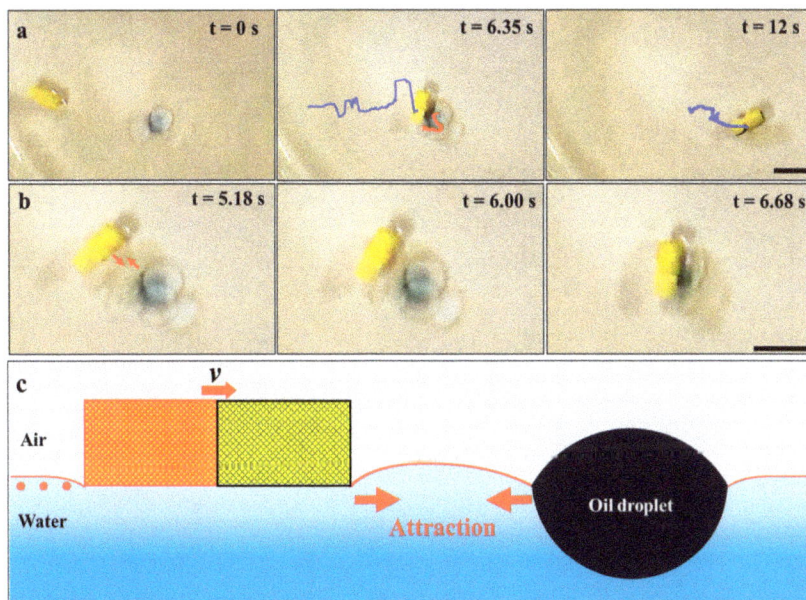

Figure 3. (**a**) Time-lapse images of the oil capture by the Janus foam motor. Red and blue curves represent the trajectories of the oil droplet and Janus foam motors, respectively, suggesting the autonomously approaching, capturing, and transportation of oil droplets by the Janus foam motor. (**b**) Close observation of the attraction between the Janus foam motor and an oil droplet. (**c**) Schematic demonstration of the attractive capillary interaction between the Janus foam motor and oil droplet. Scale bars: 10 mm.

Even though the passive foam materials have demonstrated high removal efficiency and capacity towards the continuous oil layer or film on water [13], it is difficult to remove scattered oil droplets in a wide area as they are immobile if no external driving force is applied. As verified by Figure S2, the Janus foam motor can cruise on a wide water surface and has a much wider treating range than the immobile SA-PVA foam in the same period of time (10 min). Hence, the Janus foam motors developed in this work could capture and absorb scattered oil droplets in a wide area by self-propulsion. As shown in Figure 4 and Video S3, the Janus foam motor can effectively capture and absorb three scattered oil droplets on the water surface in 50 s (Figure 4a), while the immobile SA-PVA foam fails to capture any oil droplets (Figure 4d) because the distance between the oil droplet and the foam is beyond the capillary length (about several millimeters). These different behaviors of oil capture reveal that the Janus foam motor exhibits a much higher oil absorption efficiency than that of the immobile foam materials. Several groups have developed magnetic foams for directed oil removal [23–25]. These magnetic foams can be guided by external magnetic field to target specific oil contaminants on water. However, the incorporation of magnetic particles on porous foam complicates the fabrication process and raises the cost of the foams. In addition, as magnetic flux density decreases intensively with distance, the magnetically targeted oil removal is hindered in wide-range water treatment.

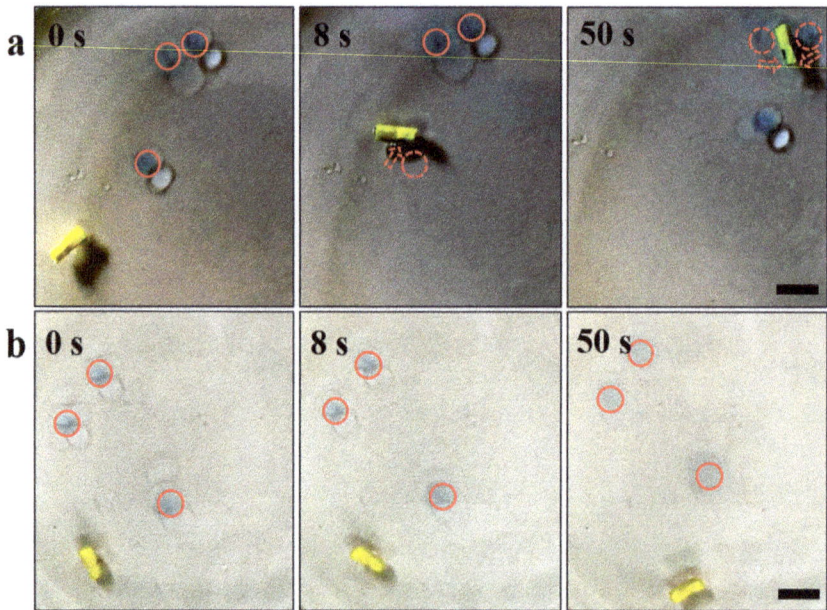

Figure 4. The capture of oil droplets by (**a**) a Janus foam motor and (**b**) an immobile SA-PVA foam at 0, 8, and 50 s, respectively, revealing that the Janus foam motor exhibits a much higher oil absorption efficiency than that of the immobile foam. Scale bars: 10 mm.

The collection of the oil absorbers after the oil absorption is an important process of post treatment in oil remediation. It has been reported that magnetic oil absorbers, such as magnetic superhydrophobic/superoleophilic particles and magnetic foams, can be effectively collected from the treated water by magnetic separation [23]. However, the magnetic separation of oil absorbers on open water is a high energy-consuming process and can only be applied in a short range, so it may be limited in large-scale application. Alternatively, the as-developed Janus foam motors exhibit a spontaneous self-assembly behavior, facilitating their subsequent collection and separation after the oil absorption, as shown in Figure 5a and Video S4. At first, Janus foam motors 1 and 2 cruise on water for oil collecting separately. They then merge into a dimer where they collide with each other. The formed dimer can continuously move on water and grasp the Janus foam motor 3 in its path, forming into a trimer. It is reasonable to speculate that the moving trimer is able to further collect additional motors when they emerge in its motion path and finally grow into a large aggregate. The Janus foam motors with the absorbed oil are expected to show a much faster assembly than those without oil absorption, as evidenced by the results shown in Figure 5b,c and Video S5. It can be seen that two oil-absorbed foams with a distance of 17 mm could merge into a dimer in 7 s, while two oil-free foams (13 mm in distance) take over 30 s to assemble. The enhanced assembly of the oil-absorbed foams is ascribed to the elongated interaction distance between the motors owing to the diffusive spreading of the captured oil on water [55]. The spontaneous assembly of the motors is attributed to the attractive capillary interactions between them, as shown in Figure 5d. For the practical application of the Janus form motor in water treatment, it should be noted that camphor, which is toxic in large doses, does not accumulate in the environment due to their ready metabolism by many bacteria [56–58].

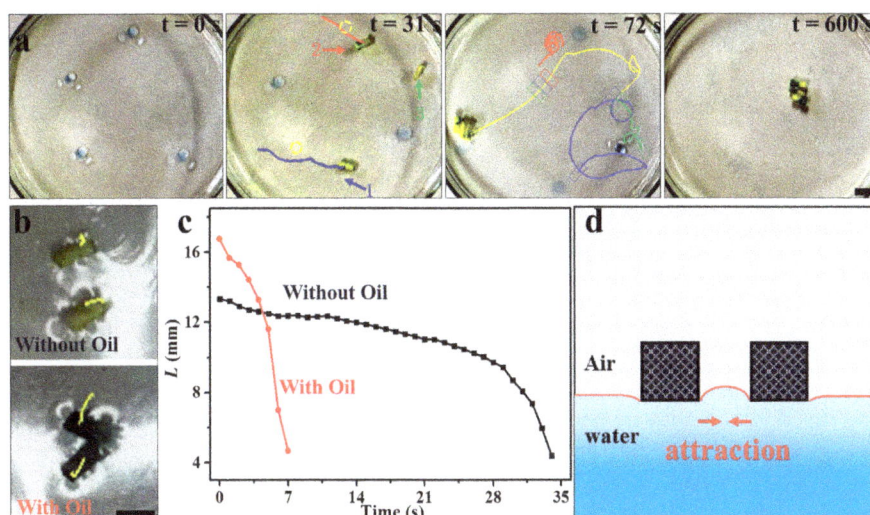

Figure 5. (**a**) Self-assembly of the Janus foam motors. (**b**) Trajectories of SA-PVA foams without oil and those with absorbed oil, reflecting a much faster self-assembly for the motors after oil absorption. (**c**) The distance (L) between the foams versus time. (**d**) Schematic demonstration of the self-assembly of the motors under capillary interactions. Scale bars: 10 mm.

4. Conclusions

In summary, we have demonstrated a novel design of a Janus foam motor and its capability for on-the-fly oil absorption. The Janus foam motor was fabricated by simply loading camphor/SA mixture into one end of the SA-modified PVA foam. It demonstrated efficient self-propulsion on water with a maximum speed of 24.3 mm/s and a long lifetime up to 230 min. The self-propulsion of the Janus foam motor stemmed from the sustained asymmetric release of camphor from the motor, generating Marangoni propulsion. Due to the hydrophobic nature of the surface of the Janus foam motors, they not only can automatically search, capture, and absorb oil droplets on the fly but also can spontaneously merge into large aggregates on water after the oil absorption under the attractive capillary interactions between them. As the as-developed Janus foam motors effectively integrate three intriguing behaviors, including self-propulsion, efficient oil capture, and the spontaneous self-assembly, they may hold great promise for applications in water treatment.

Supplementary Materials: The following are available online at www.mdpi.com/2072-666X/9/1/23/s1, Figure S1: SEM images and FTIR spectra of the pristine PVA foam, PVA foam modified with SA and the Janus foam motor loaded with camphor; Figure S2: Trajectories of a typical Janus foam motor and an SA-PVA foam in 10 min; Figure S3: The contact angle of water on the surface of the SA end of the Janus foam motor; Figure S4: The oil/water separation by the Janus foam motor; Video S1: Self-propulsion of a Janus foam motor on water; Video S2: On-the-fly Oil capture and absorption by a Janus foam motor; Video S3: Oil capture and absorption by a Janus foam motor and an immobile SA-PVA foam; Video S4: Self-assembly of Janus foam motors; Video S5: Self-assembly process of two SA-PVA foams before and after oil absorption.

Acknowledgments: This work was supported by the National Natural Science Foundation of China (No. 21474078, No. 51303144, No. 21705123 and No. 51521001), the Top Talents Lead Cultivation Project and Natural Science Foundation of Hubei Province (No. 2015CFA003), the Yellow Crane Talents Plan of the Wuhan Municipal Government, and the Fundamental Research Funds for the Central Universities (WUT: No. 2016III009 and No. 2015III060).

Author Contributions: Fangzhi Mou and Jianguo Guan conceived and designed the experiments. Xiaofeng Li, Jingjing Guo, and Zhuoyi Deng performed the experiments; Xiaofeng Li, Fangzhi Mou, and Zhuoyi Deng analyzed the data; Xiaofeng Li and Fangzhi Mou wrote the paper. Zhuoyi Deng, Chuanrui Chen, Leilei Xu, Ming Luo, and Jianguo Guan provided critical comments and revised the manuscript.

Conflicts of Interest: The authors declare no competing financial interests.

References

1. Dalton, T.; Jin, D. Extent and frequency of vessel oil spills in US marine protected areas. *Mar. Pollut. Bull.* **2010**, *60*, 1939–1945. [CrossRef] [PubMed]
2. Dubansky, B.; Whitehead, A.; Miller, J.T.; Rice, C.D.; Galvez, F. Multitissue molecular, genomic, and developmental effects of the deepwater horizon oil spill on resident gulf killifish (*fundulus grandis*). *Environ. Sci. Technol.* **2013**, *47*, 5074–5082. [CrossRef] [PubMed]
3. Short, J. Long-term effects of crude oil on developing fish: lessons from the exxon valdez oil spill. *Energy Source* **2003**, *25*, 509–517. [CrossRef]
4. Lee, C.H.; Tiwari, B.; Zhang, D.; Yap, Y.K. Water purification: Oil-water separation by nanotechnology and environmental concerns. *Environ. Sci. Nano* **2017**, *4*, 514–525. [CrossRef]
5. Choi, H.M.; Cloud, R.M. Natural sorbents in oil spill cleanup. *Environ. Sci. Technol.* **1992**, *26*, 772–776. [CrossRef]
6. Sun, H.; Li, A.; Zhu, Z.; Liang, W.; Zhao, X.; La, P.; Deng, W. Superhydrophobic activated carbon-coated sponges for separation and absorption. *ChemSusChem* **2013**, *6*, 1057–1062. [CrossRef] [PubMed]
7. Wahi, R.; Chuah, L.A.; Choong, T.S.Y.; Ngaini, Z.; Nourouzi, M.M. Oil removal from aqueous state by natural fibrous sorbent: An overview. *Sep. Purif. Technol.* **2013**, *113*, 51–63. [CrossRef]
8. Wu, L.; Zhang, J.; Li, B.; Wang, A. Magnetically driven super durable superhydrophobic polyester materials for oil/water separation. *Polym. Chem.* **2014**, *5*, 2382–2390. [CrossRef]
9. Cortese, B.; Caschera, D.; Federici, F.; Ingo, G.M.; Gigli, G. Superhydrophobic fabrics for oil-water separation through a diamond like carbon (DLC) coating. *J. Mater. Chem. A* **2014**, *2*, 6781–6789. [CrossRef]
10. Wu, J.; Wang, N.; Wang, L.; Dong, H.; Zhao, Y.; Jiang, L. Electrospun porous structure fibrous film with high oil adsorption capacity. *ACS Appl. Mater. Interfaces* **2012**, *4*, 3207–3212. [CrossRef] [PubMed]
11. Crick, C.R.; Gibbins, J.A.; Parkin, I.P. Superhydrophobic polymer-coated copper-mesh membranes for highly efficient oil-water separation. *J. Mater. Chem. A* **2013**, *1*, 5943–5948. [CrossRef]
12. Padaki, M.; Surya Murali, R.; Abdullah, M.S.; Misdan, N.; Moslehyani, A.; Kassim, M.A.; Hilal, N.; Ismail, A.F. Membrane technology enhancement in oil-water separation. A review. *Desalination* **2015**, *357*, 197–207. [CrossRef]
13. Ge, J.; Zhao, H.-Y.; Zhu, H.-W.; Huang, J.; Shi, L.-A.; Yu, S.-H. Advanced sorbents for oil-spill cleanup: recent advances and future perspectives. *Adv. Mater.* **2016**, *28*, 10459–10490. [CrossRef] [PubMed]
14. Xue, Z.; Cao, Y.; Liu, N.; Feng, L.; Jiang, L. Special wettable materials for oil/water separation. *J. Mater. Chem. A* **2014**, *2*, 2445–2460. [CrossRef]
15. Pinto, J.; Heredia-Guerrero, J.A.; Athanassiou, A.; Fragouli, D. Reusable nanocomposite-coated polyurethane foams for the remediation of oil spills. *Int. J. Environ. Sci. Technol.* **2017**, *14*, 2055–2066. [CrossRef]
16. Gupta, R.K.; Dunderdale, G.J.; England, M.W.; Hozumi, A. Oil/water separation techniques: A review of recent progresses and future directions. *J. Mater. Chem. A* **2017**, *5*, 16025–16058. [CrossRef]

17. Wang, B.; Liang, W.; Guo, Z.; Liu, W. Biomimetic super-lyophobic and super-lyophilic materials applied for oil/water separation: A new strategy beyond nature. *Chem. Soc. Rev.* **2015**, *44*, 336–361. [CrossRef] [PubMed]

18. Wang, S.; Liu, K.; Yao, X.; Jiang, L. Bioinspired surfaces with superwettability: New insight on theory, design, and applications. *Chem. Rev.* **2015**, *115*, 8230–8293. [CrossRef] [PubMed]

19. Dong, X.; Chen, J.; Ma, Y.; Wang, J.; Chan-Park, M.B.; Liu, X.; Wang, L.; Huang, W.; Chen, P. Superhydrophobic and superoleophilic hybrid foam of graphene and carbon nanotube for selective removal of oils or organic solvents from the surface of water. *Chem. Commun.* **2012**, *48*, 10660–10662. [CrossRef] [PubMed]

20. Wang, B.; Li, J.; Wang, G.; Liang, W.; Zhang, Y.; Shi, L.; Guo, Z.; Liu, W. Methodology for robust superhydrophobic fabrics and sponges from in situ growth of transition metal/metal oxide nanocrystals with thiol modification and their applications in oil/water separation. *ACS Appl. Mater. Interfaces* **2013**, *5*, 1827–1839. [CrossRef] [PubMed]

21. Yin, J.; Li, X.; Zhou, J.; Guo, W. Ultralight three-dimensional boron nitride foam with ultralow permittivity and superelasticity. *Nano Lett.* **2013**, *13*, 3232–3236. [CrossRef] [PubMed]

22. Richardson, J.J.; Björnmalm, M.; Caruso, F. Technology-driven layer-by-layer assembly of nanofilms. *Science* **2015**, *348*, aaa2491. [CrossRef] [PubMed]

23. Calcagnile, P.; Fragouli, D.; Bayer, I.S.; Anyfantis, G.C.; Martiradonna, L.; Cozzoli, P.D.; Cingolani, R.; Athanassiou, A. Magnetically driven floating foams for the removal of oil contaminants from water. *ACS Nano* **2012**, *6*, 5413–5419. [CrossRef] [PubMed]

24. Cheng, M.; Ju, G.; Jiang, C.; Zhang, Y.; Shi, F. Magnetically directed clean-up of underwater oil spills through a functionally integrated device. *J. Mater. Chem. A* **2013**, *1*, 13411–13416. [CrossRef]

25. Liu, L.; Lei, J.; Li, L.; Zhang, R.; Mi, N.; Chen, H.; Huang, D.; Li, N. A facile method to fabricate the superhydrophobic magnetic sponge for oil-water separation. *Mater. Lett.* **2017**, *195*, 66–70. [CrossRef]

26. Guix, M.; Mayorga-Martinez, C.C.; Merkoci, A. Nano/micromotors in (bio)chemical science applications. *Chem. Rev.* **2014**, *114*, 6285–6322. [CrossRef] [PubMed]

27. Zhao, G.; Pumera, M. Macroscopic self-propelled objects. *Chem. Asian J.* **2012**, *7*, 1994–2002. [CrossRef] [PubMed]

28. Gao, W.; Wang, J. The environmental impact of micro/nanomachines: A review. *ACS Nano* **2014**, *8*, 3170–3180. [CrossRef] [PubMed]

29. Xu, L.; Mou, F.; Gong, H.; Luo, M.; Guan, J. Light-driven micro/nanomotors: From fundamentals to applications. *Chem. Soc. Rev.* **2017**, *46*, 6905–6926. [CrossRef] [PubMed]

30. Mou, F.; Kong, L.; Chen, C.; Chen, Z.; Xu, L.; Guan, J. Light-controlled propulsion, aggregation and separation of water-fuelled TiO_2/Pt Janus submicromotors and their "on-the-fly" photocatalytic activities. *Nanoscale* **2016**, *8*, 4976–4983. [CrossRef] [PubMed]

31. Seah, T.H.; Zhao, G.; Pumera, M. Surfactant capsules propel interfacial oil droplets: An environmental cleanup strategy. *ChemPlusChem* **2013**, *78*, 395–397. [CrossRef]

32. Zhao, G.; Seah, T.H.; Pumera, M. External-energy-independent polymer capsule motors and their cooperative behaviors. *Chem. Eur. J.* **2011**, *17*, 12020–12026. [CrossRef] [PubMed]

33. Mou, F.; Pan, D.; Chen, C.; Gao, Y.; Xu, L.; Guan, J. Magnetically modulated pot-like $MnFe_2O_4$ micromotors: Nanoparticle assembly fabrication and their capability for direct oil removal. *Adv. Funct. Mater.* **2015**, *25*, 6173–6181. [CrossRef]

34. Guix, M.; Orozco, J.; García, M.; Gao, W.; Sattayasamitsathit, S.; Merkoçi, A.; Escarpa, A.; Wang, J. Superhydrophobic alkanethiol-coated microsubmarines for effective removal of oil. *ACS Nano* **2012**, *6*, 4445–4451. [CrossRef] [PubMed]

35. Shi, X.; Rosa, R.; Lazzeri, A. On the coating of precipitated calcium carbonate with stearic acid in aqueous medium. *Langmuir* **2010**, *26*, 8474–8482. [CrossRef] [PubMed]

36. Nakata, S.; Kirisaka, J.; Arima, Y.; Ishii, T. Self-motion of a camphanic acid disk on water with different types of surfactants. *J. Phys. Chem. B* **2006**, *110*, 21131–21134. [CrossRef] [PubMed]

37. Hayashima, Y.; Nagayama, M.; Nakata, S. A camphor grain oscillates while breaking symmetry. *J. Phys. Chem. B* **2001**, *105*, 5353–5357. [CrossRef]

38. Edward, B.; Mark, F.; Yelena, B.; Gilad, C.; Viktor, V.; Bernard, P. Superposition of translational and rotational motions under self-propulsion of liquid marbles filled with aqueous solutions of camphor. *Langmuir* **2017**, *33*, 13234–13241.

39. Yuichiro, K.; Shogo, O.; Tomonori, N.; Taro, T.; Masanori, F. Simultaneous measurement of surface tension and its gradient around moving camphor boat on water surface. *Chem. Lett.* **2014**, *43*, 1002–1004.

40. Gao, W.; Pei, A.; Wang, J. Water-driven micromotors. *ACS Nano* **2012**, *6*, 8432–8438. [CrossRef] [PubMed]

41. Mou, F.; Chen, C.; Zhong, Q.; Yin, Y.; Ma, H.; Guan, J. Autonomous motion and temperature-controlled drug delivery of Mg/Pt-poly(*N*-isopropylacrylamide) Janus micromotors driven by simulated body fluid and blood plasma. *ACS Appl. Mater. Interfaces* **2014**, *6*, 9897–9903. [CrossRef] [PubMed]

42. Gao, W.; Feng, X.; Pei, A.; Gu, Y.; Li, J.; Wang, J. Seawater-driven magnesium based Janus micromotors for environmental remediation. *Nanoscale* **2013**, *5*, 4696–4700. [CrossRef] [PubMed]

43. Chen, C.; Mou, F.; Xu, L.; Wang, S.; Guan, J.; Feng, Z.; Wang, Q.; Kong, L.; Li, W.; Wang, J.; et al. Light-steered isotropic semiconductor micromotors. *Adv. Mater.* **2017**, *29*. [CrossRef] [PubMed]

44. Wu, J.; Balasubramanian, S.; Kagan, D.; Manesh, K.M.; Campuzano, S.; Wang, J. Motion-based DNA detection using catalytic nanomotors. *Nat. Commun.* **2010**, *1*, 36. [CrossRef] [PubMed]

45. Wang, L.; Yuan, B.; Lu, J.; Tan, S.; Liu, F.; Yu, L.; He, Z.; Liu, J. Self-propelled and long-time transport motion of PVC particles on a water surface. *Adv. Mater.* **2016**, *28*, 4065–4070. [CrossRef] [PubMed]

46. Jańczuk, B.; Białopiotrowicz, T.; Wójcik, W. The components of surface tension of liquids and their usefulness in determinations of surface free energy of solids. *J. Colloid Interface Sci.* **1989**, *127*, 59–66. [CrossRef]

47. Burnham, N.A.; Dominguez, D.D.; Mowery, R.L.; Colton, R.J. Probing the surface forces of monolayer films with an atomic-force microscope. *Phys. Rev. Lett.* **1990**, *64*, 1931–1934. [CrossRef] [PubMed]

48. Vazquez, G.; Alvarez, E.; Navaza, J.M. Surface tension of alcohol water + water from 20 to 50 °C. *J. Chem. Eng. Data* **1995**, *40*, 611–614. [CrossRef]

49. Bowden, N.; Choi, I.S.; Grzybowski, B.A.; Whitesides, G.M. Mesoscale self-assembly of hexagonal plates using lateral capillary forces: synthesis using the "capillary bond". *J. Am. Chem. Soc.* **1999**, *121*, 5373–5391. [CrossRef]

50. Bowden, N.; Terfort, A.; Carbeck, J.; Whitesides, G.M. Self-assembly of mesoscale objects into ordered two-dimensional arrays. *Science* **1997**, *276*, 233–235. [CrossRef] [PubMed]

51. Ismagilov, R.F.; Schwartz, A.; Bowden, N.; Whitesides, G.M. Autonomous movement and self-assembly. *Angew. Chem. Int. Ed.* **2002**, *41*, 652–654. [CrossRef]

52. Bowden, N.; Oliver, S.R.J.; Whitesides, G.M. Mesoscale self-assembly: Capillary bonds and negative menisci. *J. Phys. Chem. B* **2000**, *104*, 2714–2724. [CrossRef]

53. Kralchevsky, P.A.; Paunov, V.N.; Denkov, N.D.; Ivanov, I.B.; Nagayama, K. Energetical and force approaches to the capillary interactions between particles attached to a liquid-fluid interface. *J. Colloid Interface Sci.* **1993**, *155*, 420–437. [CrossRef]

54. Kralchevsky, P.A.; Paunov, V.N.; Ivanov, I.B.; Nagayama, K. Capillary meniscus interaction between colloidal particles attached to a liquid—Fluid interface. *J. Colloid Interface Sci.* **1992**, *151*, 79–94. [CrossRef]

55. Camp, D.W.; Berg, J.C. The spreading of oil on water in the surface-tension regime. *J. Fluid Mech.* **2006**, *184*, 445–462. [CrossRef]

56. Jones, K.H.; Smith, R.T.; Trudgill, P.W. Diketocamphane enantiomer-specific "Baeyer–Villiger" monooxygenases from camphor-grown Pseudomonas putida ATCC 17453. *Microbiology* **1993**, *139*, 797–805. [CrossRef] [PubMed]

57. Elsa, W. Poisons Information Monograph: Camphor (PIM 095). 1989. Available online: http://www.inchem.org/documents/pims/pharm/camphor.htm (accessed on 11 December 2017).

58. World Health Organization. *International Program on Chemical Safety*; World Health Organization: Geneva, Switzerland, 1984.

micromachines

MDPI

Article

Manipulating Microrobots Using Balanced Magnetic and Buoyancy Forces

Lin Feng [1,2,†], Xiaocong Wu [1,†], Yonggang Jiang [1,*], Deyuan Zhang [1] and Fumihito Arai [3]

1   School of Mechanical Engineering & Automation, Beihang University, Beijing 100191, China;
    linfeng@buaa.edu.cn (L.F.); xiaocongwu@buaa.edu.cn (X.W.); zhangdy@buaa.edu.cn (D.Z.)
2   Beijing Advanced Innovation Center for Biomedical Engineering, Beihang University, Beijing 100083, China
3   Department of Micro-Nano Systems Engineering, Graduate School of Engineering, Nagoya University,
    Nagoya 464-0814, Japan; arai@mech.nagoya-u.ac.jp
*   Correspondence: jiangyg@buaa.edu.cn; Tel.: +86-10-8231-6603
†   These authors contributed equally to this work.

Received: 9 December 2017; Accepted: 24 January 2018; Published: 29 January 2018

Abstract: We present a novel method for the three-dimensional (3D) control of microrobots within a microfluidic chip. The microrobot body contains a hollow space, producing buoyancy that allows it to float in a microfluidic environment. The robot moves in the z direction by balancing magnetic and buoyancy forces. In coordination with the motion of stages in the xy plane, we achieved 3D microrobot control. A microgripper designed to grasp micron-scale objects was attached to the front of the robot, allowing it to hold and deliver micro-objects in three dimensions. The microrobot had four degrees of freedom and generated micronewton-order forces. We demonstrate the microrobot's utility in an experiment in which it grips a 200 μm particle and delivers it in a 3D space.

Keywords: micro-robot; three-dimension manipulation; microfluidic chip

1. Introduction

New technology in biomedical engineering, especially the manipulation of bioparticles (cells, biological tissue, etc.) in three-dimensional (3D) space, has attracted much attention [1]. A number of methods of manipulating bioparticles have been presented in the literature. Noncontact forces have been applied to bioparticle manipulations because they show great merit in terms of flexible actuation. Optical tweezers that use a highly focused laser beam to manipulate bioparticles are optional [2–6]. However, the output force of optical tweezers is not strong enough, especially for manipulating large-sized bioparticles. Dielectrophoresis (DEP), which uses a non-uniform electric field to exert forces on a dielectric particle [7], has been used for manipulating bioparticles [8–10]. However, Gray et al. reported that the electric fields used to rotate cells can cause cell damage [11]. In order to manipulate bioparticles, different kinds of microgrippers have been designed and actuated basing on static electricity [12], electro-thermal expansion [13], mechanical actuation [14,15], and other similar techniques [16–19]. Beyeler et al. [12] built an electrostatic microelectromechanical gripper containing a force sensor that can grip and release microspheres and HeLa cells in a two-dimensional (2D) plane. Colinjivadi et al. [13] presented a polymer "chopstick" gripper consisting of a metal heater layer and demonstrated nanoscale precision along the x, y, and z axes during cell manipulation. Ger et al. [20] presented a cell gripper based on magnetic zig-zag structures, which was actuated by a magnetic field. Wester et al. [15] designed a mechanically actuated microtweezer that allowed for movement in multiple degrees of freedom (DoF). Finally, Chung et al. [21] created a magnetically actuated microrobot capable of manipulating microgels in 3D space.

Bioparticle manipulation often leads to contamination because it is carried out in an open environment. Manipulations are more ideally performed in a closed space, such as inside a microfluidic chip [22]; unfortunately, the previously discussed micromanipulation systems were unable to manipulate particles within such spaces because the systems were too large. For example, the system presented by Chung et al., actuated by magnetic force, was large and complex because the actuator consisted of eight magnetic coils. Moreover, the system's output force was too small to effectively manipulate larger objects, such as oocytes.

To make microrobots more practical and versatile, their size must be reduced for use in a microchip, and output forces must be increased. Applying a noncontact drive is a good method for solving these problems, as it significantly decreases robot size, making them more flexible for microchip use. Magnetic force is one of the most promising non-contact actuation methods for increasing output force and is minimally invasive (with respect to bioparticles). Microrobotic magnetic fields can be generated using a permanent magnet or electromagnetic coils. Electromagnetic coil systems occupy a large space and are more complicated, whereas magnetic fields generated by a permanent magnet are 10–100 times stronger with the same device size [23,24].

A few microrobots are currently capable of working in a 3D space, but their functionality is limited by their size and output force. We previously proposed a microrobotic system that can manipulate cells within a microchip, performing tasks such as the enucleation of bovine oocytes [25], accurate dispensing of single oocytes [26], 3D rotation of a single oocyte [27], and high-accuracy positioning [28]. In addition, we performed work in the area of micro-manipulation in a microchip [29].

In this paper, we propose an innovative method for manipulating microrobots that are actuated by permanent fixed magnets on the manipulator, using magnetic and buoyancy forces. Figure 1 shows a conceptual overview of the microrobot system, where the robot contains a cavity for employing buoyancy in an aquatic environment. A controlled balance between buoyancy and magnetic forces allows for motion in the z direction, which (when combined with the motion of the xy plane) achieves 3D control. The noncontact drive allows the robot to be small enough for use in a microchip, and, using a gripper designed to sit in front of the robot, bioparticles can be easily manipulated. The microgripper is controlled by a stage corresponding to the gripper, a system that introduces opportunities for applications such as delivering objects to a target position or enucleating cells. Manipulation is performed under a microscope, preventing contamination within the enclosed environment. We demonstrate the delivery of a particle with a diameter of approximately 200 µm using this system. We also measured the distance between the two thin beams of the microgripper in various positions to show that it can grip particles of assorted sizes.

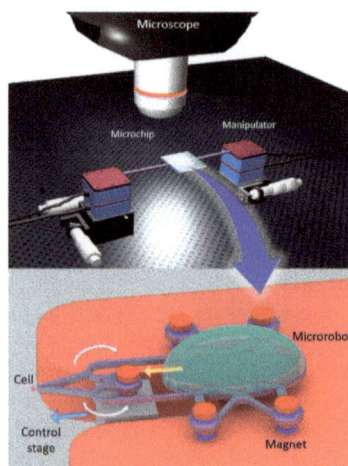

Figure 1. Conceptual overview of the microrobotic system. The robot is actuated by permanent magnets, and works in the aquatic environment of a microfluidic chip.

2. Materials and Methods

2.1. Microrobot Design

In consideration of the aquatic environment of cultured bioparticles, we present a method for 3D actuation in fluid using balanced magnetic and buoyancy forces. Figure 2 shows a microrobot with a cavity in its body for employing appropriate buoyancy force in an aquatic environment. As shown in Figure 2, four magnets permanently fixed to the microrobot's legs play a key role in its 3D control. The magnet fixed to the microgripper controls the gripper. A schematic for the 3D control of the robot is shown in Figure 3.

Figure 3a–d show the process of gripping microscale objects by controlling the microrobot in 3D space. A driven stage is placed under the microrobot to control its 3D motion. The control stage, with a magnet fixed on it, can only move along the x axis on the driven stage. On the tip of the microrobot, a small permanent magnet is assembled in. Under the microfluidic chip, another permanent magnet according to where the magnet position on the tip of the micro-robot is put on the control stage. By controlling the magnet's position on the stage, it is possible to control the opening and closing of the gripper.

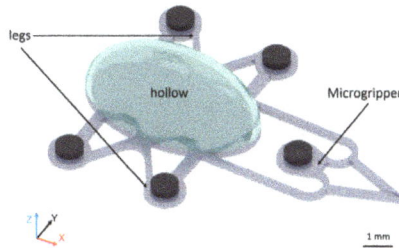

Figure 2. The design of the microrobot. A cavity is placed inside the main body of the robot to provide the buoyancy needed for control.

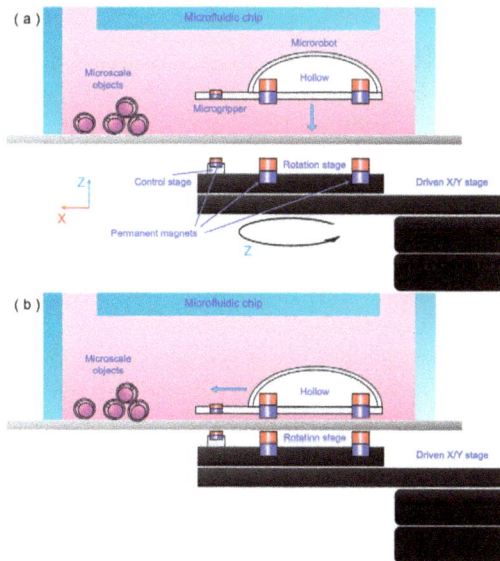

Figure 3. *Cont.*

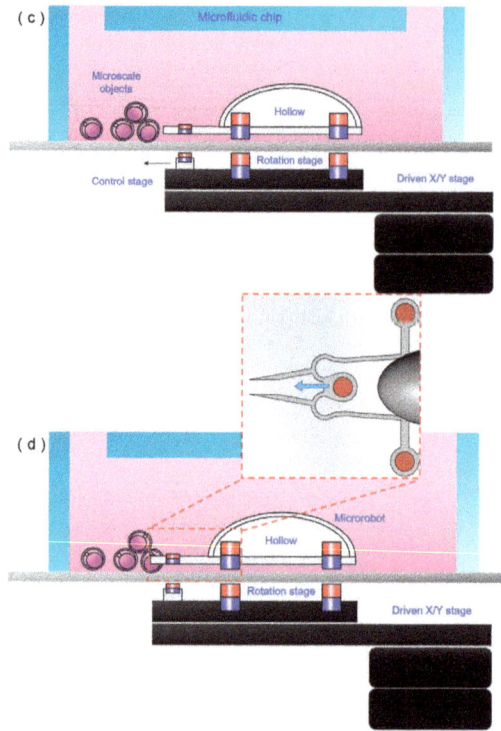

Figure 3. Schematic of the 3D microrobot control. The robot moves towards the stage when the stage moves towards the microfluidic chip, and when the stage moves away from the microfluidic chip, the robot moves away from the stage. (**a,b**) Control the robot to the intended position using the driven stage. (**c,d**) Move the control stage to adjust the state of gripper, to grip the microscale objects.

When the driven stage moves towards the microchip, the distance between the stage and robot decreases, which increases the magnetic force and moves the robot toward the stage, as shown in Figure 3a. On the contrary, when the stage moves away from the microchip, the robot moves away from the stage. Thus, motion along the z axis can be achieved.

The relationship between buoyancy, magnetic force, and weight is a design consideration, so the microrobot's total weight can be calculated by the formulas:

$$G = G_1 + G_2 \tag{1}$$

$$G_1 = \rho_1 g v_1 \tag{2}$$

where G_1 is the weight of the microrobot, G_2 is the weight of the magnets fixed onto the microrobot, ρ_1 is the density of photosensitive resin from which the microrobot is made, and v_1 is the volume of the photosensitive resin.

The microrobot's buoyancy can be calculated by the following formula:

$$F = \rho g v \tag{3}$$

where F is the microrobot's buoyancy, ρ is the density of the liquid (water in this experiment), and v is the microrobot's total volume.

The magnetic force between two cylindrical magnets can be approximated by the formula [30]:

$$F(x) = \frac{\pi \mu_0}{4} M^2 R^4 \left[\frac{1}{x^2} + \frac{1}{(x+2t)^2} - \frac{2}{(x+t)^2} \right] \tag{4}$$

where R is the radius of the cylindrical magnets, t is their height, M is their magnetization, and x is the gap between them.

We experimentally determined the microrobot's weight to be 4.738×10^{-4} N, and its buoyancy to be 6.325×10^{-4} N. As such, without a magnetic force, the robot floats in water. The maximum and minimum values of the magnetic forces are 0.0760 N and 8.59×10^{-3} N, respectively, when the robot is on the top and at the bottom of the liquid. When the robot is on top of the liquid, the magnetic force reaches its minimum, but is still sufficiently large to control the robot's movement, confirming the theoretical feasibility of the actuator method.

2.2. Microgripper Simulation

By moving the control stage, the distance between the magnets in the control stage and the micro-gripper changes, altering the strength of the magnetic field at the microgripper's location. Therefore, the gap state and separation distance of the gripper can be controlled. In our experiments, the component force along the x axis was hard to measure during experiments, as shown in Figure 3, and was experimentally evaluated using the load cell (LVS-5GA, Kyowa Electronic Instruments Co., Ltd. Tokyo, Japan). The maximum component force along the x axis was approximately 0.03 N. We conducted a gripper control simulation, wherein the component force applied to the microgripper was 0.03 N.

Figure 4 shows the simulation wherein a magnetic force of 0.03 N was applied to the microgripper in the +y direction. The gap increased by approximately 123 μm, for a total of 273 μm, which is sufficient for manipulating bioparticles.

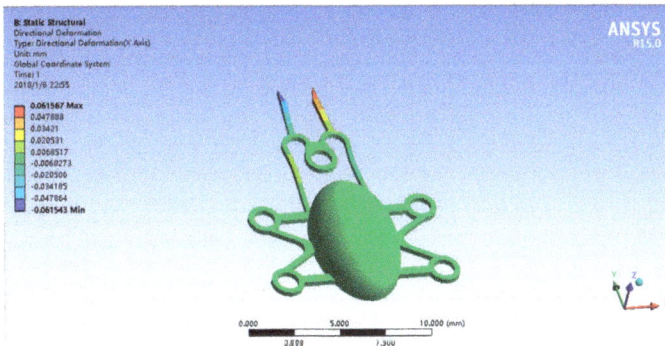

Figure 4. Gripper control simulation. The gap between the two thin beams increases approximately 123 μm compared to the original position; compared to the size of bioparticles, this value is sufficient for manipulation.

2.3. Experimental Setup

Figure 5 shows an overview of the manipulation system, consisting of the observation system and the control system. To observe the manipulation better, the experiment was performed under a camera-attached microscope with an external light source added. The microfluidic chip was fixed on a stage, and the manipulator was operated with a joystick. The equipment mentioned above was fixed to a shock-proof platform to meet the need for high positioning accuracy.

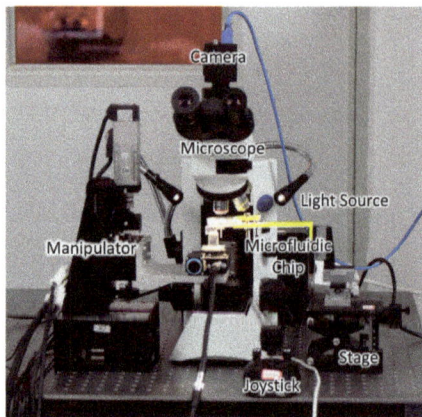

Figure 5. Overview of the manipulation system.

The robot was fabricated via a high-quality 3D printing process (Form2, Formlabs Co., Ltd., MA, USA) using a photosensitive resin (Clear V2, Formlabs Co., Ltd., MA, USA). The microrobot can be 3D-printed directly. However, the hollow inside the robot was full of liquid photosensitive resin due to the liquid manufacturing environment. To empty the hollow chamber, a hole of around 0.6 mm in diameter was left open through which the liquid photosensitive resin could be removed and it was sealed later. The driven and control stages (to which the magnets were fixed) were manufactured by the same 3D printing process. The magnets were fixed to the microrobot, and the stages using a liquid photosensitive resin were fixed by exposure to UV light.

A four-DoF stage (HEIDSTAR Co., Ltd. Fujian, China) was used in the experiment. Its precision is 300 nm, providing enough accuracy compared to the size of the bioparticles and microrobot.

A microfluidic chip consisting of dimethyl siloxane (PDMS, DOW CORNING Co., Ltd. Wiesbaden, Germany) and a glass substrate served as the experiment platform. The cover of the chip was fabricated from PDMS, and the microfluidic chip was filled with water. The robot and bioparticles were encapsulated within the microfluidic chip to protect the particles from contamination. The channel of the microfluidic chip was designed specifically for the requirements of this experiment.

The experiment was observed using a microscope (CX41, OLYMPUS Co., Ltd., Tokyo, Japan) with a mounted camera (GS3-U3-23S6C-C, POINTGREY Co., Ltd., BC, Canada) set above the microfluidic chip.

3. Results

Figure 6 shows the actuation of the microgripper as triggered by the control stage magnet (correspondingly sweeping along the x axis). By moving the control stage magnet along the x axis, its position relative to the micro-gripper magnet changed. A component force along the x axis was produced during this progress. This component force controlled the microgripper's opening and closing. The extent to which the control stage magnet moved determined the magnitude of the component force along the x axis, which in turn changed the distance between the two thin beams of the microgripper, allowing it to grip and release different-sized bioparticles.

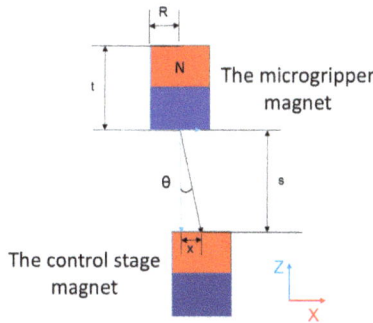

Figure 6. Actuation of the microgripper, triggered by the control stage magnet sweeping along the x axis.

The relationship between the change in the magnets' distance and the gap between the microgripper was examined experimentally, as shown in Figure 7a. The distance between the control stage and microgripper magnet was measured from the center of each magnet on the x axis (as shown) and gripper gap distance D was measured. Figure 7b shows that the rate of increase of D depends on the magnet position (specifically, the value of x). When a microrobot is driven by a permanent magnet beneath the glass substrate, there is a distance in which the microrobot does not follow the drive magnet; we named this distance the "dead band" [31] and it can be seen in the graph. The dead band is caused by the static friction of the microgripper with the glass substrate. This shows that there is a hysteresis caused by the dead band between the gripping phase and release phase in Figure 7b.

As shown in Figure 6, the force along the x axis actuated the microgripper. When angle θ was zero, there was no component force along the x axis, leaving it in its original position. When the control stage magnet moved along the x axis, θ increased and a component force was applied to the gripper along the x axis. The two thin beams on both sides of microgripper were subsequently bent out of shape, causing the gripper to open or close. The size of the opening corresponded to the magnitude of the component force. When the displacement of the control stage magnet was small, the magnetic force decreased only slightly, but θ increased quickly, as shown in Figure 7b. When the displacement was close to zero, the slope of the curve was large, meaning that the gripper's gap increased quickly. With increasing displacement, the increase in the y axis slowed, owing to the geometrically related increase in θ. When displacement exceeded a certain range, the magnetic force decreased rapidly, even though θ continued to increase; this became the most important influence, and the curve's slope decreased. At a specific displacement, the increase in θ and the decrease in magnetic force were balanced—where the slope of the curve was zero and the value of the y axis (gap distance D) was maximized. As shown in Figure 7b, D ranged from 50 to 320 μm, meaning that the microgripper was suitable for manipulating large bioparticles. According to the graph, we can select size x to obtain the corresponding D when manipulating different-sized bioparticles.

Figure 8 shows the result of the experiment wherein the micro-gripper grasped and carried a particle. Figure 8a is an image of the micro-gripper in front of the robot, in its original state. When the corresponding control stage magnet moved in the +x direction, a component force in +x was applied to the microgripper. The applied force stretched the thin beams on both sides of the microgripper out of shape, opening the gripper in preparation to grab the particle. When the opening gap of the tip was large enough to grab the particle, the gripper moved right, and then could close the tip to grab the particle (i.e., the control stage magnet moved in the –x direction).

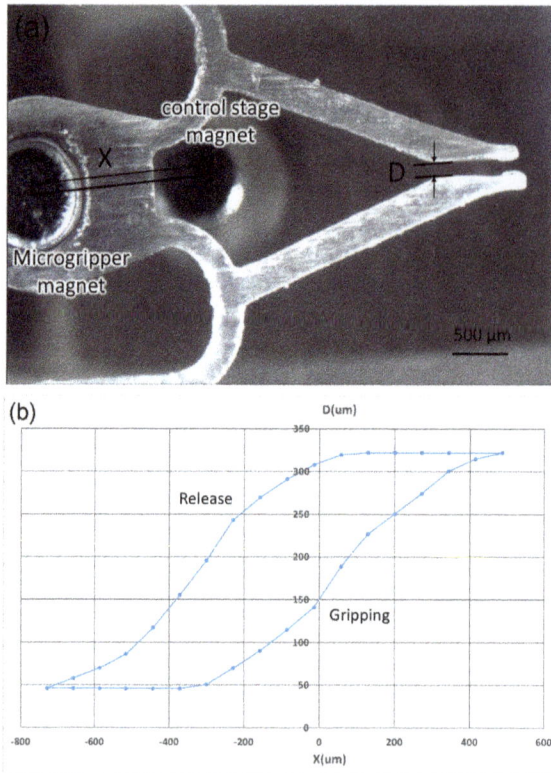

Figure 7. Relationship between magnet movement and the gap between the two thin beams of the microgripper. (**a**) An illustration of the experiment. The distance between the two magnets fixed to the microgripper and control stage is x; D is the gap in the gripper. (**b**) The relationship between the moving of the magnet x and the size of the gripper opening D.

To transport a particle and perform other manipulations, the microrobot must be controlled in 3D space. Motion in the z direction was achieved by balancing the magnetic and buoyancy forces. Motion in the xy plane and rotation around the z axis depended on the motion of the 4-DoF stage. Four other magnets permanently fixed to the microrobot play an important role in 3D control. After moving the robot to the desired position, a reversal of the gripper process opens the gripper and releases the bioparticle.

Figure 9 shows sequential micrographs of the microgripper manipulation experiment. A circular tube with a diameter of 840 μm is located on the left side of the graph in Figure 9, indicating the changing height of the robot in the microfluidic chip. At first, the particle and robot were in their original positions, as shown in Figure 9a. The gripper in front of the microrobot opened to grip the particle by controlling the control stage magnet, as shown in Figure 9b. In Figure 9c, the particle is being gripped. In Figure 9d, the robot and particle changes heights using a balance between magnetic and buoyancy forces. The robot was out of the focal plane; the microrobot tip becomes dimmer in Figure 9d compared to other images. Hence, the microrobot floated and its height in the liquid was controlled by the magnetic force.

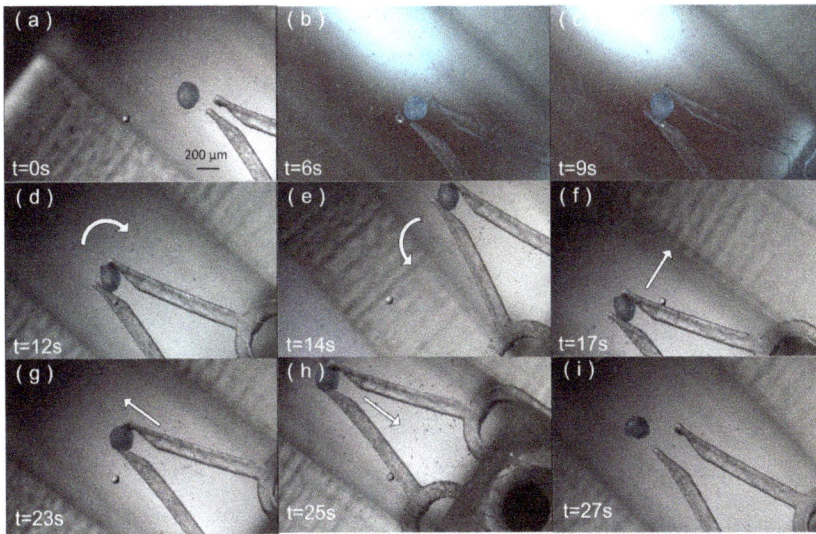

Figure 8. The microgripper grasping and carrying a microparticle. (**a**) The microgripper is in its original state. (**b**) The microgripper ready to be opened. (**c**) The gripper open under action of a magnetic force applied to the gripper. (**d**) The robot is rotated clockwise, carrying the particle. (**e**) The microgripper rotates counter-clockwise. (**f**–**i**) The microgripper carries the micro-particle along a straight line and release the microparticle.

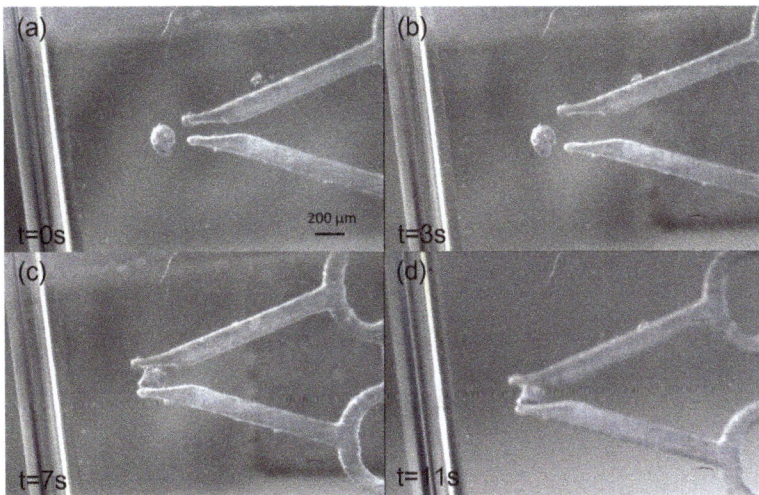

Figure 9. Sequential micrographs of the microgripper manipulation experiment. (**a**) The microgripper is in position to open and grip. (**b**) The microgripper opens under the action of a magnetic component force applied to the gripper. (**c**) The microgripper grips the particle by decreasing the component force. (**d**) The microrobot changes height (it appears dimmer in the image because it has risen in the fluid as a result of changes to the balanced magnetic and buoyancy forces).

4. Discussion

In the first experiment, we examined the relationship between magnet movement and the gap between the two thin beams of the microgripper. The results of this work demonstrate 3D microparticle gripping and transport by innovative actuated microrobots. Figure 7 illustrates this relationship, showing the microrobot's strong adaptability in manipulating a large range of microparticle sizes—from 50 to 320 μm. Furthermore, the demonstration of microparticle grasping and transportation suggests that the microrobot could be successfully controlled by balancing buoyancy and magnetic forces.

5. Conclusions

In this paper, we have described an innovative microrobot manipulation method using balanced magnetic and buoyancy forces in a microfluidic chip. Compared to existing methods, it suffers from less contamination because it functions in a closed space, owing to its small size and noncontact actuation principle. Because of its strong output force, and wide gripping range, it offers the ability to manipulate objects of varying sizes. In addition, the liquid manipulation environment was extremely suitable for bioengineering. In this experiment, we successfully demonstrated that this method is a promising tool for single-cell manipulation. Our future work will strengthen the motion stability on the z axis and add more automation to the microrobot system.

Supplementary Materials: The following are available online at http://www.mdpi.com/2072-666X/9/2/50/s1, Video S1: Microparticle manipulations.

Acknowledgments: This work was supported by the Natural Science Foundation of Beijing (Grant No. 17L20128), and Beihang University, "Zhuobai" project, ZG216S1751.

Author Contributions: L.F., X.W. and Y.J. conceived and designed the experiments; X.W. performed the experiments; L.F., D.Z. and F.A. analyzed the data; F.L., Y.J. and D.Z. contributed reagents/materials/analysis tools; L.F. and X.W. wrote the paper.

References

1. Fujita, H. Recent Progress in Micromachining and Applications to Microactuators. *Jpn. J. Appl. Phys.* **1994**, *33*, 7163–7166. [CrossRef]
2. Bayoudh, S.; Nieminen, T.A.; Heckenberg, N.R. Orientation of biological cells using plane-polarized Gaussian beam optical tweezers 1 Introduction. *J. Mod. Opt.* **2003**, *50*, 1581–1590. [CrossRef]
3. Ichikawa, M.; Kubo, K.; Yoshikawa, K.; Kimura, Y. Tilt control in optical tweezers. *J. Biomed. Opt.* **2008**, *13*, 010503. [CrossRef] [PubMed]
4. Liu, Y.; Yu, M. Investigation of inclined dual-fiber optical tweezers for 3D manipulation and force sensing. *Opt. Express* **2009**, *17*, 13624–13638. [CrossRef] [PubMed]
5. Liang, Y.-L.; Huang, Y.-P.; Lu, Y.-S.; Hou, M.T.; Yeh, J.A. Cell rotation using optoelectronic tweezers. *Biomicrofluidics* **2010**, *4*, 043003. [CrossRef] [PubMed]
6. Onda, K.; Arai, F. Multi-beam bilateral teleoperation of holographic optical tweezers. *Opt. Express* **2012**, *20*, 3633–3641. [CrossRef] [PubMed]
7. Pohl, H.A. Some effects of nonuniform fields on dielectrics. *J. Appl. Phys.* **1958**, *29*, 1182–1188. [CrossRef]
8. Nishioka, M.; Katsura, S.; Hirano, K.; Mizuno, A. Evaluation of cell characteristics by step-wise orientational/nrotation using optoelectrostatic micromanipulation. *IEEE Trans. Ind. Appl.* **1997**, *33*, 1353–1357. [CrossRef]
9. Park, J.; Jung, S.-H.; Kim, Y.H.; Kim, B.; Lee, S.-K.; Ju, B.; Lee, K. An integrated bio cell processor for single embryo cell manipulation. In Proceedings of the IEEE/RSJ International Conference on Intelligent Robots and Systems, Sendai, Japan, 28 September–2 October 2004; pp. 242–247. [CrossRef]

10. Kunikata, R.; Takahashi, Y.; Koide, M.; Itayama, T.; Yasukawa, T.; Shiku, H.; Matsue, T. Three dimensional microelectrode array device integrating multi-channel microfluidics to realize manipulation and characterization of enzyme-immobilized polystyrene beads. *Sens. Actuators B Chem.* **2009**, *141*, 256–262. [CrossRef]

11. Gray, D.S.; Tan, J.L.; Voldman, J.; Chen, C.S. Dielectrophoretic registration of living cells to a microelectrode array. *Biosens. Bioelectron.* **2004**, *19*, 1765–1774. [CrossRef] [PubMed]

12. Beyeler, F.; Neild, A.; Oberti, S.; Bell, D.J.; Sun, Y.; Dual, J.; Nelson, B.J. Monolithically fabricated microgripper with integrated force sensor for manipulating microobjects and biological cells aligned in an ultrasonic field. *J. Microelectromech. Syst.* **2007**, *16*, 7–15. [CrossRef]

13. Colinjivadi, K.S.; Lee, J.-B.; Draper, R. Viable cell handling with high aspect ratio polymer chopstick gripper mounted on a nano precision manipulator. *Microsyst. Technol.* **2008**, *14*, 1627–1633. [CrossRef]

14. Jager, E.W.H. Microrobots for Micrometer-Size Objects in Aqueous Media: Potential Tools for Single-Cell Manipulation. *Science* **2000**, *288*, 2335–2338. [CrossRef] [PubMed]

15. Wester, B.A.; Rajaraman, S.; Ross, J.D.; Laplaca, M.C.; Allen, M.G. Development and characterization of a packaged mechanically actuated microtweezer system. *Sens. Actuators A Phys.* **2011**, *167*, 502–511. [CrossRef]

16. Kawahara, T.; Sugita, M.; Hagiwara, M.; Arai, F.; Kawano, H.; Shihira-Ishikawa, I.; Miyawaki, A. On-chip microrobot for investigating the response of aquatic microorganisms to mechanical stimulation. *Lab Chip* **2013**, *13*, 1070–1078. [CrossRef] [PubMed]

17. Ichikawa, A.; Sakuma, S.; Sugita, M.; Shoda, T.; Tamakoshi, T.; Akagi, S.; Arai, F. On-chip enucleation of an oocyte by untethered microrobots. *J. Micromech. Microeng.* **2014**, *24*, 095004. [CrossRef]

18. Ceylan, H.; Giltinan, J.; Kozielski, K.; Sitti, M.; Münster, R.; Mierka, O.; Turek, S.; Leshansky, A.M.; Fischer, P.; Zhang, L.; et al. Mobile microrobots for bioengineering applications. *Lab Chip* **2017**, *17*, 1705–1724. [CrossRef] [PubMed]

19. Zhang, L.; Peyer, K.E.; Nelson, B.J. Artificial bacterial flagella for micromanipulation. *Lab Chip* **2010**, *10*, 2203–2215. [CrossRef] [PubMed]

20. Ger, T.-R.; Huang, H.-T.; Chen, W.-Y.; Lai, M.-F. Magnetically-controllable zigzag structures as cell microgripper. *Lab Chip* **2013**, *13*, 2364–2369. [CrossRef] [PubMed]

21. Chung, S.E.; Dong, X.; Sitti, M. Three-dimensional heterogeneous assembly of coded microgels using an untethered mobile microgripper. *Lab Chip* **2015**, *15*, 1667–1676. [CrossRef] [PubMed]

22. Castillo, J.; Dimaki, M.; Svendsen, W.E. Manipulation of biological samples using micro and nano techniques. *Integr. Biol.* **2009**, *1*, 30–42. [CrossRef] [PubMed]

23. Cugat, O.; Delamare, J.; Reyne, G. Magnetic Micro-Actuators and Systems (MAGMAS). *IEEE Trans. Magn.* **2003**, *39*, 3607–3612. [CrossRef]

24. Inomata, N.; Mizunuma, T.; Yamanishi, Y.; Arai, F. Omnidirectional actuation of magnetically driven microtool for cutting of oocyte in a chip. *J. Microelectromech. Syst.* **2011**, *20*, 383–388. [CrossRef]

25. Feng, L.; Hagiwara, M.; Ichikawa, A.; Arai, F. On-Chip enucleation of bovine oocytes using microrobot-assisted flow-speed control. *Micromachines* **2013**, *4*, 272–285. [CrossRef]

26. Feng, L.; Sun, Y.; Ohsumi, C.; Arai, F. Accurate dispensing system for single oocytes using air ejection. *Biomicrofluidics* **2013**, *7*, 1–14. [CrossRef] [PubMed]

27. Feng, L.; Di, P.; Arai, F. High-precision motion of magnetic microrobot with ultrasonic levitation for 3-D rotation of single oocyte. *Int. J. Robot. Res.* **2016**, *35*, 1445–1458. [CrossRef]

28. Hagiwara, M.; Kawahara, T.; Yamanishi, Y.; Masuda, T.; Feng, L.; Arai, F. On-chip magnetically actuated robot with ultrasonic vibration for single cell manipulations. *Lab Chip* **2011**, *11*, 2049–2054. [CrossRef] [PubMed]

29. Yamanishi, Y.; Feng, L.; Arai, F. On-demand and Size-controlled Production of emulsion droplets by magnetically driven microtool. In Proceedings of the 2010 IEEE International Conference on Robotics and Automation, Anchorage, AK, USA, 3–7 May 2010; Volume 1, pp. 4094–4099.

30. Vokoun, D.; Beleggia, M.; Heller, L.; Šittner, P. Magnetostatic interactions and forces between cylindrical permanent magnets. *J. Magn. Magn. Mater.* **2009**, *321*, 3758–3763. [CrossRef]
31. Hagiwara, M.; Kawahara, T.; Yamanishi, Y.; Arai, F. Driving method of microtool by horizontally arranged permanent magnets for single cell manipulation. *Appl. Phys. Lett.* **2010**, *97*, 1–4. [CrossRef]

micromachines

MDPI

Article

Pair Interaction of Catalytical Sphere Dimers in Chemically Active Media

Jing-Min Shi [1,†], Ru-Fei Cui [2,†], Jie Xiao [1], Li-Yan Qiao [1,*], Jun-Wen Mao [3] and Jiang-Xing Chen [1,*]

1   Department of Physics, Hangzhou Dianzi University, Hangzhou 310018, China;
    161070042@hdu.edu.cn (J.-M.S.); 151070020@hdu.edu.cn (J.X.)
2   Department of Physics, Zhejiang University, Hangzhou 310027, China; 21636030@zju.edu.cn
3   Department of Physics, Huzhou University, Huzhou 313000, China; jwmao@zjhu.edu.cn
*   Correspondence: qiaoliyan@hdu.edu.cn (L.-Y.Q.); jxchen@hdu.edu.cn (J.-X.C.);
    Tel.: +86-571-8596-5662 (J.-X.C.)
†   These authors contributed equally to this work.

Received: 30 December 2017; Accepted: 11 January 2018; Published: 17 January 2018

Abstract: We study the pair dynamics of two self-propelled sphere dimers in the chemically active medium in which a cubic autocatalytic chemical reaction takes place. Concentration gradient around the dimer, created by reactions occurring on the catalytic sphere surface and responsible for the self-propulsion, is greatly influenced by the chemical activities of the environment. Consequently, the pair dynamics of two dimers mediated by the concentration field are affected. In the particle-based mesoscopic simulation, we combine molecular dynamics (MD) for potential interactions and reactive multiparticle collision dynamics (RMPC) for solvent flow and bulk reactions. Our results indicate three different configurations between a pair of dimers after the collision, i.e., two possible scenarios of bound dimer pairs and one unbound dimer pair. A phase diagram is sketched as a function of the rate coefficients of the environment reactions. Since the pair interactions are the basic elements of larger scale systems, we believe the results may shed light on the understanding of the collective dynamics.

Keywords: pair interaction; catalytically sphere dimer; chemically active medium; multiparticle collision dynamics; phase diagram

---

1. Introduction

Collective dynamics of active particles have attracted great attention in the interdisciplinary fields including biochemistry, materials, and physics in the past decade [1–4]. From the fundamental point, the study on this subject can help us to give insight into the far from equilibrium physics underlying the collective behavior of biological entities. The dynamic of a motor in a many-motor system behaves differently from that of an isolated motor [5,6]. In collections of such motors, the presence of other motors in the surroundings leads to mutual interaction, which would not only change the motion of a single particle but also lead to the emergence of cooperative phenomena [7–9].

From principle individual particle motion to complex multiparticle behavior, the pair interactions set up a bridge as the basic elements of larger scale systems. Thus, studies about pair dynamics are essential for further understanding of collective behavior [10,11]. A great number of interesting phenomena in two or more hydrodynamical collective motors has been observed [12,13]. Since the interactions among individual particles become important, it is common to observe phenomena like clustering and rectification effects [14–16]. In a recent example, the bimetallic rods or dimers would form pairs transiently due to the specific short-range interactions [17]. Wykes et al. found this kind of rods could interact with each other and pair up to form a swimmer or a rotor which exhibit the fundamental form of motility: translation and rotation [18]. Thakur et al. found a variety of bound and unbound states after undergoing a collision of dimer pair [19].

Biochemical environments are generally in and out of equilibrium state [20,21]. Microswimmers often perform tasks in complex environments which are chemically active [22]. The environmental reactions may change the nonequilibrium concentration gradients that are a central element of the propulsion mechanism. A natural raised interesting question is how the nonequilibrium state affects the interaction between a pair of motors. Obviously, the dynamics in such media will be quite different. For example, the periodic supply of fuel to a motor as a result of autonomous oscillations in the surrounding medium will cause oscillations in the motor velocity [23–25]. In this article, we investigate the dynamics of two chemically powered sphere dimer motors in a chemically active medium. The chemical dimer motors are immersed into a solution full of fuel particles that also take part in the bulk phase reactions. The concentration gradient around the dimer, created by reactions occurring on the catalytic sphere surface and responsible for the self-propulsion, are affected by the bulk environmental reactions. Consequently, the communications between the dimer pair are altered, which results in different configurations of dimer pair. In terms of the particles-based method combined with molecular dynamics (MD) and reactive multiparticle collision dynamics (RMPC), we can adjust the activities of the environment to study the pair dynamics of sphere dimers.

2. Mesoscopic Model and Simulation Method

The self-propelled nanodimers consist of the catalytic (C) and noncatalytic (N) spheres which are linked by a fixed distance R. The dimers are surrounded with the point-like fluid (solvent) particles comprising A and B species. In our system, we have considered two sphere-dimer motors. The catalytic sphere catalyzes the irreversible chemical reaction by converting A (fuel particles) to B (product particles) when A encounters C.

$$A + C \rightarrow B + C, \text{on the dimer motor} \tag{1}$$

This model mimics generic features of synthetic nanorods: catalytic reactions occur on one end and the reaction product interacts differently with the catalytic and noncatalytic ends. The motors based on self-diffusiophoresis had been employed to represent the dynamics of sphere dimers comprising a non-catalytic silica sphere connected to a catalytic platinum sphere in experiment [26].

Also there is autocatalytic reaction occurring in the environments with rate constants k_1 and k_2 [19]

$$B + 2A \underset{k_2}{\overset{k_1}{\rightleftharpoons}} 3A, \text{in the environment} \tag{2}$$

The iodate/arsenous acid system can be accurately modeled by such cubic autocatalysis [27]. Here, we select such a special reaction to construct the complex environment. The reactant or product in Equation (1) is then involved in the cubic autocatalytic reaction in Equation (2) taking place in the bulk phase environment, which forms a reaction network coupled by diffusion. The A and B species interact with the dimer sphere through repulsive Lennard-Jones (LJ) interactions of the form

$$V_{\alpha S} = 4\epsilon_{\alpha S}[(\sigma_S/r)^{12} - (\sigma_S/r)^6 + \frac{1}{4}]\theta(r_c - r) \tag{3}$$

where $\theta(r)$ is the Heaviside function and $r_c = 2^{1/6}\sigma_S$ is the cutoff distance. The notation $V_{\alpha S}$, where $S = C, N$ and $\alpha = A, B$, are used to denote various interactions between solvent and dimer monomers. We take $V_{AC} = V_{BC} = V_{AN}$, which are characterized by the same energy parameter ϵ_A. However, interactions between the N sphere and B molecules, V_{BN}, have a different energy parameter ϵ_B and a distance parameter σ_N. The asymmetric potentials are responsible for the diffusiophoretic mechanism of the self-propulsion. The interaction between these two different sphere dimers, denoted by subscripts 1 and 2, respectively, is also described by the repulsive LJ potentials Equation (3) with an energy parameter ϵ_D and a cutoff distance $r_c = 2^{1/6}\sigma_D$ where $\sigma_D = \sigma_{S_1} + \sigma_{S_2} + \delta_D$ with $\delta_D = 0.2$.

The time evolution of the entire system is carried out using a hybrid MD-RMPC scheme, which combines molecular dynamics (MD) for sphere dimers and reactive multiparticle collision (RMPC) dynamics for the fluid particles [28–30]. The hybrid MD-RMPC dynamics consists of the free streaming step and a collision step. In the streaming step, the system including dimers and solvent molecules are propagated by Newton's equations of motion, which are described by molecular dynamics (MD) through deriving forces from Equation (3) with time interval Δt_{MD}.

There is no net force among solvent particles. Instead, the interaction between the solvent particles is described by multiparticle collisions dynamics (MPC). In the collision step, the system is divided into cubic cells with size $a = 1$ and rotation operators $\hat{\omega}_\alpha$ which are assigned to each cell from some set of rotation operators. After the center-of-mass velocity $\vec{v}_{cm}$ of each cell ξ is calculated from $\vec{v}_{cm} = \sum_{j=1}^{N_c} \vec{v}_j / N_c$, where N_c is the total number of particles in the cell, the post-collision velocity $\vec{v}_i(t + \tau)$ of each particle i within the same cell can be obtained according to the rotation rule

$$\vec{v}_i(t + \tau) = \vec{v}_{cm}(t) + \hat{\omega}_\alpha(\vec{v}_i(t) - \vec{v}_{cm}(t)) \tag{4}$$

Grid shifting was employed to ensure Galilean invariance. The hybrid MD-MPC dynamics includes fluctuations, conserves mass, momentum and energy, and accounts for coupling between the C sphere motion and fluid flows.

The cubic autocatalytic reactions in the bulk phase take place independently in each cell at each MPC collision step Δt_{MPC} [28]. In the cell ξ, the bulk reactions in Equation (2) occur according the following probabilities: the forward reaction with probability $\frac{a_1}{a_0}(1 - e^{-a_0\Delta t_{MPC}})$; the backward reaction with probability $\frac{a_2}{a_0}(1 - e^{-a_0\Delta t_{MPC}})$; and no reaction with probability $e^{-a_0\Delta t_{MPC}}$, where $a_1 = k_1 N_B N_A(N_A - 1)$, $a_2 = k_2 N_A(N_A - 1)(N_A - 2)$, and $a_0 = a_1 + a_2$ with N_A (N_B) the total number of A (B) particles in cell ξ.

In our simulations, all quantities are reported in dimensionless LJ units based on energy ϵ, mass m and distance σ parameters: $r/\sigma \to r$, $t(\epsilon/m\sigma^2)^{1/2} \to t$, and $k_BT/\epsilon \to T$. The simulation box of the system is divided into $50 \times 50 \times 50$ cells, and the average number density in each cell is $n_0 \simeq 10$. The MPC rotation angle is fixed at $\alpha = 90°$. The masses for A and B are both fixed at $m = 1$. The system temperature is $k_BT = 1/6$ and the LJ potential parameters are chosen to be $\epsilon_A = 1.0$ and $\epsilon_B = 0.1$. The diameters of the catalytic spheres are $d_C = 4.0$ while the diameters of the noncatalytic spheres are $d_N = 8.0$. The MD time step is $\Delta t_{MD} = 0.01$ and the time step for MPC is $\Delta t_{MPC} = 0.05$. We chose $\epsilon_D = 1.0$, $\delta = 0.8$, and $R = 6.8$.

3. Results and Discussions

Two sphere dimers are initially separated by distance 20.0, and they are targeted to undergo collinear collisions (see the first configuration in Figure 1). Since the fluctuation from the solvent particles, two dimers facing to each other from self-propulsion would not always strictly stay on the same line before the collision. When the motors approach each other and collide, two ultimate possible scenarios are observed: either they would (i) interact and then separate with independent motions or (ii) contact and form a bound pair.

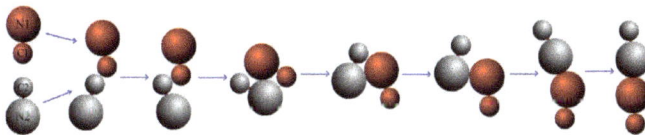

Figure 1. Schematic representation shows the process of formation of a Brownian dimer pair resulted from the collision of two self-propelled nanodimers. The reaction rate coefficient are $k_1 = 0.001$ and $k_2 = 0$.

Since the interactions of dimer pair are mediated by solvent fields, it is necessary to study the general distribution of species A and B resulted from chemical reactions to analyze the problem how the pair forms the post-collision configurations. The autocatalytic chemical reaction $A + C \rightarrow B + C$ converting fuel A to product B, which takes place whenever A particles enter into the boundary layer of the C sphere, may create a gradient concentration field of species A and B around the dimer motors. Since the interaction potentials of species A and B with the N sphere are different, the self-generated chemical gradients of these species give rise to an asymmetric force on the motor directed along its internuclear axis that propels it in solution. In addition, the autocatalytic reaction in the solution,

$$B + 2A \underset{k_2}{\overset{k_1}{\rightleftharpoons}} 3A$$

characterized by the intrinsic rates k_1 and k_2, greatly influences the gradient fields of A and B. Therefore, it is evident that the pair configurations are determined by the environment reactions.

Firstly, we discuss the simple case where the bulk reaction is irreversible, i.e., the rate coefficients $k_2 = 0$. An example illustrating the collision process is plotted in Figure 1. In this case, when the pair approaches each other, they collide firstly and then rotate around each other (see the third and fourth configurations). They keep on moving past each other (the fifth and sixth configurations) until they form a bound pair and move together (the seventh and eighth configurations). The bound-pair may reorient and execute Brownian motion, however, it has no self-propulsion velocity. The stable pair configuration, labeled by "Brownian dimer pair (BP)" is described by the characteristics in Table 1. As shown in Figure 1, the pair is almost linear, which is confirmed by the angles θ_1 and θ_2. One can see the value θ_1 is close to $180°$ while the value of θ_2 is small. The presented distances in Table 1, i.e., $r_{N_1 N_2}$, $r_{C_1 C_2}$, $r_{C_1 N_2}$, and $r_{N_1 C_2}$ all verify the BP configurations.

Table 1. The averaged distance of $r_{N_1 N_2}$, $r_{C_1 C_2}$, $r_{C_1 N_2}$, $r_{N_1 C_2}$ obtained from the bound Brownian dimer pair (BP) and Rotating dimer pair (RP) configurations, respectively. $\theta_1 = \arccos(\hat{r}_{N_2 N_1} \cdot \hat{r}_{C_1 N_1})$, $\theta_2 = \arccos(\hat{r}_{N_2 N_1} \cdot \hat{r}_{C_2 N_2})$, where $\hat{r}_{C_1 N_1}$ (or $\hat{r}_{C_2 N_2}$) and $\hat{r}_{N_2 N_1}$ are the unit vectors pointing from C to N and from N_1 to N_2, respectively.

Configuration	$r_{C_1 C_2}$	$r_{C_1 N_2}$	$r_{N_1 C_2}$	$r_{N_1 N_2}$	θ_1	θ_2
RP	10.938	6.319	6.315	7.119	53.925	126.096
BP	19.937	13.713	13.761	7.227	156.162	22.231

It is the gradient field of B particles that determines the self-propulsion of individual dimer as well as the pair interactions. Thus, it is essential to study its behavior under the influences of reactions in Equations (1) and (2). The forward reaction in Equation (2) requires both sufficient A and B particles while the backward reaction needs only A particles. Since Equation (1) on C sphere surface generates gradient field of B particles shown in Figure 2, the probability of forward reaction in Equation (2) is small in the region too close (bare A particles) or too far (bare B particles) from the C sphere. Thus, the forward reaction often occurs in those radial regions, e.g., around $r = 3$–5 in Figure 2 where profiles of radial distribution of B particles with different k_2 are plotted. To the backward reaction, it frequently appears in the bulk solution full of A particles and a little farther from C sphere. Thus, although Equation (2) is reversible, the forward and backward reactions generally take place in different radial zones. As k_2 is increased slightly, e.g., from $k_2 = 0$ to $k_2 = 0.04$, the forward reaction still plays a major role and it decreases B particles. Therefore, the concentration gradient is getting steeper.

As a consequence of the steeper gradient of B particles, a new bound-pair appears. The process is illustrated in Figure 3. Specifically, when the two dimers approach and collide with each other (the first and second configurations), they are strong attracted mutually (the third and fourth configurations), instead of brushing away in the BP case. The reason is that when the N_2 (N_1) monomer achieves contact with C_1 (C_2), it experiences bigger attraction from the steeper gradient of B particles around C_1 (C_2), which prevents the N_2 moving on. Consequently, a stable configuration is ultimately formed and the bound pair rotates clockwise since the attractive force is pointing from N_2 to C_1 (and N_1 to C_2), which results in a force moment correspondingly. The last four configurations in Figure 3 depict the rotation of the dimer pair. This stable bound state is labeled as "rotating dimer pair" (RP). Compared to the cases in BP state, the value of $r_{C_1C_2}$ in Table 1 is quite small since the C monomers come into contact and assemble. The values of $r_{C_1N_2}$ and $r_{C_2N_1}$ are almost the same here, and they are much smaller than the values in the BP configurations. Since the two N monomers keep on touching, the values of $r_{N_1N_2}$ in BP and RP are approximately equal.

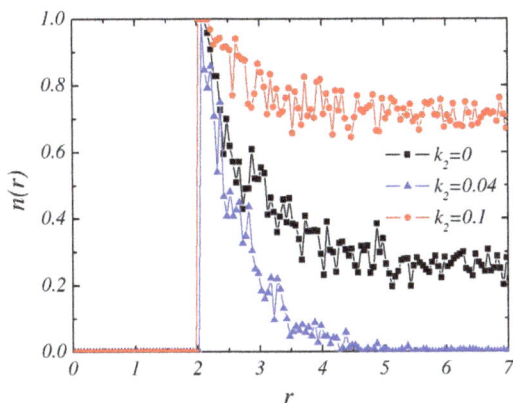

Figure 2. The concentration field of B species as a function of distance r from the center of the catalytic sphere in the steady state with $k_1 = 0.001$.

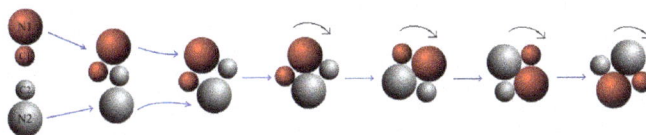

Figure 3. Schematic representation shows the process of formation of a rotating dimer pair resulted from the collision of two self-propelled nanodimers. The reaction rate coefficient are $k_1 = 0.0005$ and $k_2 = 0.02$.

With larger values of $k_2 = 0.1$, the autocatalytic reaction in Equation (2) occurs mainly in a backward direction. The A particles produced by the forward reaction are converted back to B quickly, producing a flat B concentration profile(see Figure 2). So there is very little interaction between these two sphere dimers to keep them bound and hence the pair would move pass each other and "escape" from each other. The state that the two dimers move independently and move away from each other after collision is labeled as "Independent dimer pair" (IP).

Since the two spatiotemporal reactions in Equations (1) and (2) with different response radial regions and diffusions of solvent particles, the dynamics of the system is getting very complex. If k_1 is very small, e.g., $k_1 = 0.0001$, simulation gives us the transition from RP to IP. However, if k_1 is substantially increased to $k_1 = 0.0015$, the transition from BP to IP takes place without RP state when k_2 is increased.

Then, we increase the value of k_1 while $k_2 = 0.06$. If $k_2 = 0$, the RP configuration is observed at very small k_1 (0.0001). Subsequently, the increase of k_1 results in the appearance of BP state. If k_2 is decreased, e.g., $k_2 = 0.02$ and $k_2 = 0.04$, IP, RP, BP successively are obtained when k_1 is enhanced. Figure 4 demonstrates the steady state concentration field of particle B around the catalytic sphere for different values of rate constant k_1 with $k_2 = 0.06$. When $k_1 = 0.0001$, the bulk conversion B to A is pretty slow and then most A particles in the gradient field would become B by the backward reaction, which results in small nonequilibrium B concentration gradient around the dimers. An IP state is observed in the phase diagram. At higher k_1 (0.001), it becomes easier for particles B converting to A in the solution. Especially it is possible to have this bulk reaction near the N sphere and it gives rise to the enhanced decay of density field. The strong concentration field would have the significant effect on the pair interaction of the dimers. The time evolution of the pair shows that they would form RP again. Additionally, those B particles produced by the C catalytic reaction are converted back to A at a fast reaction rate when $k_1 = 0.002$. Less A particles could be supplied near the dimer in this situation and then it yields a flat B concentration profile. From the discussion above, it is confirmed that the configuration of IP is observed in the first and the third cases. As k_2 is increased to large value, such as $k_2 = 0.08$, only IP configurations are observed at any k_1. Different configurations emerge as a result of the concentration gradient of the species in the environment which are influenced by the values of k_1 and k_2. The phase diagram which includes all the three final configurations(i.e., BP, RP and IP) is sketched in Figure 5 in the $k_1 - k_2$ plane.

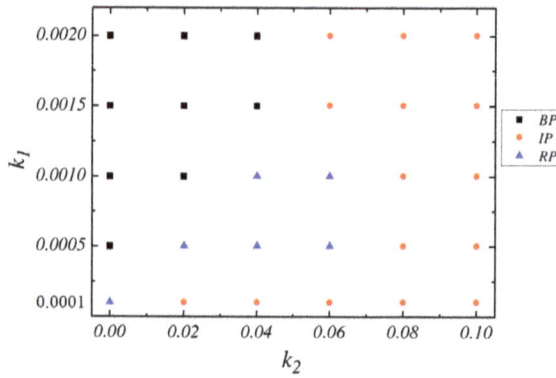

Figure 4. Phase diagram showing the post-collision nature of the dimer pair in the $k_1 - k_2$ plane. Different three regions: two types of bound dimer pairs like Brownian (black square) and rotating dimer pair (blue triangle) and one unbound pair: independently moving dimers (red dot).

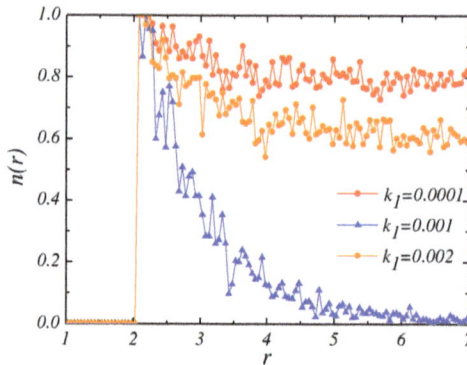

Figure 5. The concentration field of B species as a function of distance r from the center of the catalytic sphere in the steady state with $k_2 = 0.06$.

4. Conclusions

A coarse grain model has been developed to study the pair interaction of catalytical dimer motor propelled by self-diffusiophoresis. Two dimer motors are placed in a chemically active medium where reactions occur at the catalytic monomers and the reactant or product of this reaction is involved in a cubic autocatalytic reaction taking place in the bulk phase environment. The environmental reactions change the non-equilibrium concentration gradients that are a central element of the propulsion mechanism and the consequent pair dynamics. By altering the chemical activities of the medium through changing the intrinsic rates coefficient, the pair shows rich and complex dynamics and forms two bound states, i.e., a rotating dimer pair and Brownian dimer pair. The underlying mechanisms are discussed. A phase diagram describing the dependence of pair configurations on the chemical activities of the environment is presented. The studies in this paper can be extended to other types of motors and more complicated realistic environmental reactions networks.

Acknowledgments: This work was supported by the Natural Science Foundation of Zhejiang Province (No. LR17A050001) and the National Natural Science Foundation of China (No. 11674080, No. 11647118, and No. 11575059).

Author Contributions: Jiang-Xing Chen conceived the study; Jing-Min Shi, Jie Xiao, Ru-Fei Cui, and Li-Yan Qiao performed the simulations; Jing-Min Shi, Li-Yan Qiao and Jun-Wen Mao analyzed the data; Jiang-Xing Chen and Li-Yan Qiao wrote the paper.

Conflicts of Interest: The authors declare no conflict of interest.

References

1. Desai, R.C.; Kapral, R. *Dynamics of Self-Organized and Self-Assembled Structures*; Cambridge University Press: Cambridge, UK, 2009.
2. Oyama, N.; Molina, J.J.; Yamamoto, R. Purely hydrodynamic origin for swarming of swimming particles. *Phys. Rev. E* **2016**, *93*, doi:10.1103/PhysRevE.93.043114.
3. Zöttl, A.; Stark, H. Hydrodynamics determines collective motion and phase behavior of active colloids in quasi-two-dimensional confinement. *Phys. Rev. Lett.* **2014**, *112*, doi:10.1103/PhysRevLett.112.118101.
4. Kapral, R. Perspective: Nanomotors without moving parts that propel themselves in solution. *J. Chem. Phys.* **2013**, *138*, 020901, doi:10.1063/1.4773981.
5. Peruani, F.; Deutsch, A.; Bär, M. Nonequilibrium clustering of self-propelled rods. *Phys. Rev. E* **2006**, *74*, doi:10.1103/PhysRevE.74.030904.
6. Theurkauff, I.; Cottin-Bizonne, C.; Palacci, J.; Ybert, C.; Bocquet, L. Dynamic clustering in active colloidal suspensions with chemical signaling. *Phys. Rev. Lett.* **2012**, *108*, doi:10.1103/PhysRevLett.108.268303.
7. Bechinger, C.; Leonardo, R.D.; Löwen, H.; Reichhardt, C.; Volpe, G.; Volpe, G. Active particles in complex and crowded environments. *Rev. Mod. Phys.* **2016**, *88*, doi:10.1103/RevModPhys.88.045006.
8. Gompper, G.; Ihle, T.; Kroll, D.M.; Winkler, R.G. Multi-Particle Collision Dynamics: A Particle-Based Mesoscale Simulation Approach to the Hydrodynamics of Complex Fluids. In *Advanced Computer Simulation Approaches for Soft Matter Sciences III*; Springer: Berlin/Heidelberg, Germany, 2009; pp. 1–87.
9. Ramaswamy, S. Active matter. *J. Stat. Mech. Theory Exp.* **2017**, *2017*, 054002.
10. Palacci, J.; Sacanna, S.; Steinberg, A.P.; Pine, D.J.; Chaikin, P.M. Living crystals of light-activated colloidal surfers. *Science* **2013**, *339*, 936–940.
11. Bayati, P.; Najafi, A. Dynamics of two interacting active Janus particles. *J. Chem. Phys.* **2016**, *144*, 134901, doi:10.1063/1.4944988.
12. Farzin, M.; Ronasi, K.; Najafi, A. General aspects of hydrodynamic interactions between three-sphere low-Reynolds-number swimmers. *Phys. Rev. E* **2012**, *85*, doi:10.1103/PhysRevE.85.061914.
13. Buttinoni, I.; Bialké, J.; Kümmel, F.; Löwen, H.; Bechinger, C.; Speck, T. Dynamical clustering and phase separation in suspensions of self-propelled colloidal particles. *Phys. Rev. Lett.* **2013**, *110*, doi:10.1103/PhysRevLett.110.238301.
14. Elgeti, J.; Winkler, R.G.; Gompper, G. Physics of microswimmers-single particle motion and collective behavior: A review. *Rep. Prog. Phys.* **2015**, *78*, doi:10.1088/0034-4885/78/5/056601.

15. Chen, X.; Yang, X.; Yang, M.C.; Zhang, H.P. Dynamic clustering in suspension of motile bacteria. *Europhys. Lett.* **2015**, *111*, doi:10.1209/0295-5075/111/54002.

16. Gao, Y.; Mou, F.; Feng, Y.; Che, S.; Li, W.; Xu, L.; Guan, J. Dynamic colloidal molecules maneuvered by light-controlled Janus micromotors. *ACS Appl. Mater. Interfaces* **2017**, *9*, 22704–22712.

17. Wang, W.; Duan, W.; Ahmed, S.; Sen, A.; Mallouk, T.E. From one to many: Dynamic assembly and collective behavior of self-propelled colloidal motors. *Acc. Chem. Res.* **2015**, *48*, 1938–1946.

18. Wykes, M.S.D.; Palacci, J.; Adachi, T.; Ristroph, L.; Zhong, X.; Ward, M.D.; Zhang, J.; Shelley, M.J. Dynamic self-assembly of microscale rotors and swimmers. *Soft Matter* **2016**, *12*, 4584–4589.

19. Thakur, S.; Kapral, R. Dynamics of self-propelled nanomotors in chemically active media. *J. Chem. Phys.* **2011**, *135*, 024509, doi:10.1063/1.3607408.

20. Chen, J.X.; Zhang, H.; Qiao, L.Y.; Liang, H.; Sun, W.G. Interaction of excitable waves emitted from two defects by pulsed electric fields. *Commun. Nonlinear Sci.* **2018**, *54*, 202–209.

21. Chen, J.X.; Xiao, J.; Qiao, L.Y.; Xu, J.R. Dynamics of scroll waves with time-delay propagation in excitable media. *Commun. Nonlinear Sci.* **2018**, *59*, 331–337.

22. Colberg, P.H.; Reigh, S.Y.; Robertson, B.; Kapral, R. Chemistry in motion: Tiny synthetic motors. *Acc. Chem. Res.* **2014**, *47*, 3504–3511.

23. Jülicher, F.; Prost, J. Molecular motors: From individual to collective behavior. *Prog. Theor. Phys. Suppl.* **1998**, *130*, 9–16.

24. Marchetti, M.C.; Joanny, J.F.; Ramaswamy, S.; Liverpool, T.B.; Prost, J.; Rao, M.; Simha, R.A. Hydrodynamics of soft active matter. *Rev. Mod. Phys.* **2013**, *85*, doi:10.1103/RevModPhys.85.1143.

25. Robertson, B.; Kapral, R. Nanomotor dynamics in a chemically oscillating medium. *J. Chem. Phys.* **2015**, *142*, 154902, doi:10.1063/1.4918329.

26. Leonardo, F.V.; Tao, Y.G.; Zacharia, N.S.; Kitaev, V.; Galembeck, F.; Kapral, R.; Ozin, G.A. Catalytic Nanomotors: Self-Propelled Sphere Dimers. *Small* **2010**, *6*, 565–572.

27. Horvath, D.; Petrov, V.; Scott, S.K.; Showalter, K. Instabilities in propagating reaction-diffusion fronts. *J. Chem. Phys.* **1993**, *98*, 6332.

28. Rohlf, K.; Fraser, S.; Kapral, R. Reactive multiparticle collision dynamics. *Comput. Phys. Commun.* **2008**, *179*, 132–139.

29. Echeverria, C.; Kapral, R. Molecular crowding and protein enzymatic dynamics. *Phys. Chem. Chem. Phys.* **2012**, *14*, 6755–6763.

30. Tucci, K.; Kapral, R. Mesoscopic multiparticle collision dynamics of reaction-diffusion fronts. *J. Phys. Chem. B* **2005**, *109*, 21300–21304.

micromachines

MDPI

Review

Controllable Swarming and Assembly of Micro/Nanomachines

Conghui Liu, Tailin Xu *, Li-Ping Xu and Xueji Zhang *

Research Center for Bioengineering and Sensing Technology, School of Chemistry and Biological Engineering, University of Science and Technology Beijing, Beijing 100083, China; conghuiliu@yeah.net (C.L.); xuliping@ustb.edu.cn (L.-P.X.)
* Correspondence: xutailin@ustb.edu.cn (T.X.); zhangxueji@ustb.edu.cn (X.Z.)

Received: 15 November 2017; Accepted: 25 December 2017; Published: 29 December 2017

Abstract: Motion is a common phenomenon in biological processes. Major advances have been made in designing various self-propelled micromachines that harvest different types of energies into mechanical movement to achieve biomedicine and biological applications. Inspired by fascinating self-organization motion of natural creatures, the swarming or assembly of synthetic micro/nanomachines (often referred to micro/nanoswimmers, micro/nanorobots, micro/nanomachines, or micro/nanomotors), are able to mimic these amazing natural systems to help humanity accomplishing complex biological tasks. This review described the fuel induced methods (enzyme, hydrogen peroxide, hydrazine, et al.) and fuel-free induced approaches (electric, ultrasound, light, and magnetic) that led to control the assembly and swarming of synthetic micro/nanomachines. Such behavior is of fundamental importance in improving our understanding of self-assembly processes that are occurring on molecular to macroscopic length scales.

Keywords: collective behavior; micro/nanomotors; swarming behavior

1. Introduction

Collective behavior, involving in cooperative arrangements of a variety of individuals, is a common phenomenon that almost exists on all scales ranging from molecules to galaxies [1–3]. In nature, most of us must have been fascinated by macroscopic flocks of animals (fishes, ants, birds, grasshopper, human, etc.) or microscopic swarming of self-propelled bacteria or cells in the biological world. For example, in the macroscopic world, schools of fish can move in a rather orderly fashion or change direction amazingly abruptly; collective of ants can work together to accomplish complex carrying tasks. In the microscopic world, swarming of bacteria is great and helpful to access new sources of nutrients, to increase the size of the community, and to colonize niches [4]. The collective behavior of swarms often reflects a change of global conditions. The motion of individuals may be affected by "signals" that are transmitted by other individuals, causing spatial–temporal patterns. The study of simple collective behavior of micro/nanoscale particles can help us to better understand more complex collective systems.

Micro/nanomachines are microscale or nanoscale devices (range from dozens of nanometers to hundreds of micrometers) that are capable of converting energy into movements and forces. These synthetic micro/nanomachines can be propelled by various external chemical stimuli (hydrogen peroxide, hydrazine, etc.) and fuel-free stimuli (electric, light, ultrasound, magnetic) on an ultra-small scale [5–24], which are promising to perform complicated biomedical tasks, such as targeted drug delivery, cell manipulation and isolation, noninvasive surgery, bioimaging or biosensing, environmental monitoring and remediation, and in vivo applications [25–30]. In recent years, the swarming or collective behaviors have inspired researchers to design synthetic nanomachines that can cooperate with each other to achieve much complex biological or environmental tasks [31–34]. The ability to regulate the collective behavior of synthetic nanomotors offers considerable promise for creating groups of machines, which can perform cooperative tasks that would be impossible using a single nanomotor [3,5,32].

In the present review, following an approach that is meant to be as much tutorial and non-technical as possible, we present recent progress in controlling the swarming and the assembly of micro/nanomachines developed in our laboratory and by other research groups. We start from the basic natural collective behavior in the aspects of theories and experiments to identify the different features, leading to self-propulsion of microscopic particles in fluid environments, and then summarize the fuel-induced (enzyme, hydrogen peroxide, hydrazine, etc.) or fuel-free induced (electrical, ultrasound, magnetic, light) collective effects and their related biological applications; finally, we outlook the future directions of this field.

2. Collective Behavior in Nature

The macroscopic collective behavior of natural animals, such as a herd of land animals, locusts marching, a flock of birds, a colony of army ants, or a school of fish, is a familiar and beautiful part of the living system (Figure 1a–e) [2,35]. The formation of such collective behaviors may be a consequence of an external stimulus or the local interactions between nearby individuals. Taking a school of fish as an example, each individual complies with its behavior on its position and nearest velocity of neighbors. Hundreds of small silver fish glide in unison, much like a single organism than a collection of individuals. Even not a single fish is lost from the group or suddenly change the direction. Such interactions results in a global collective behavior that may emerge, originating and maintaining the cohesion of the aggregate [35,36].

In the microscopic world, collective behavior of cells and bacteria (Figure 1f) is a common phenomenon that occurs on everybody. For bacteria, many genera of motile bacteria use a mechanism, referred to as 'swarming or collective', to access new sources of nutrients at surface, and to increase the size of the colony [37]. Taking cells into consideration, the migration of cells may respond to the surrounding environment and interacts between other cells through physical contact and soluble substrates, and change their behaviors accordingly.

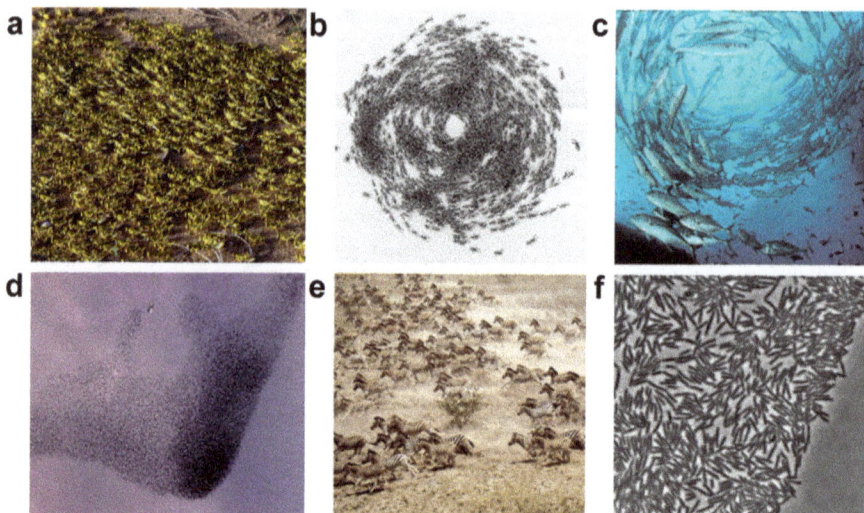

Figure 1. Collective behavior in nature. (**a**) Locusts marching. (**b**) Colony of army ants. (**c**) Fish vortices. (**d**) Thousands of birds producing a fascinating display. (**e**) A herd of zebra. (**a–d** reproduced from [2]) (**f**) Swarming of Escherichia coli (reproduced from [37]).

3. Controlled Swarming and Assembly of Artificial Micromotors

Much like biological systems, micro/nanomachines also respond to external gradients and exhibit schooling and exclusion behaviors. These gradients can be achieved by several approaches, including chemical, electrical, light, ultrasound, and magnetic. Dependent on interparticle attraction and exclusion, and reactions between the gradient and nanomotors, such as catalytic reaction, dielectrophoresis, electromegnetic reaction and etc., controlled swarming and assembly of artificial micromotors can be realized. Such assembly of micro/nanoparticles is of important to fabricate materials with multifunctional properties (e.g., mechanical, conductive, and optical) and to accomplish complex tasks [38–40]. In the following sections, based on the different external gradients, we will discuss the main approaches that can induce the collective behavior of micro/nanomachines and their related applications.

3.1. Chemical Induced Collective Behavior

Micro/nanomotors tend to move preferentially in the direction of externally applied chemical gradient, which is defined as chemotaxis. The chemical induced assembly almost exists in all types of catalytic micro/nanomachines, ranging from a simple spatial geometry, such as Janus spheres, Au-Pt nanowires, or microtubes to complex large units [3,32]. Chemotaxis has been observed in self-propelled unmodified Au microparticles (0.8–1.5 μm) by Joseph Wang's group [41]. As demonstrated in Figure 2a, the reversible swarming of Au microparticles can be achieved by repetitive addition of hydrazine. The possible mechanism of such chemically triggered swarming of Au microparticles was the diffusion of ionic products from the catalytic Au surface. The reaction between hydrazine and H_2O_2 catalyzed by the gold surface generated local amount of anions and cations, including H^+, $N_2H_5^+$, and OH^-, and generated chemical gradients on the nearby Au microparticles, which pulled the particles towards the center of the swarms where the largest electrolyte gradient existed. The same group also reported chemical induced assembly of Ir/SiO_2 Janus nanomotor, as demonstrated in Figure 2b [42]. When compared with previously mentioned assembly of Au microparticle (Figure 2a), the Ir/SiO_2 Janus nanomotor (1.2 μm diameter) displayed a collective swarming behavior in response to their environment autonomously with the sole presence of extreme low levels of hydrazine fuel (only 0.001%). The collective of Janus micromotors (1.2 μm diameter) to propel a six asymmetric teeth (with an external radius of 8 μm) was reported by Leonardo's group, as demonstrated in Figure 2c [43]. The active Janus microparticles were made of platinum half-coated on silica particles (Diameter: 5 μm), and could be propelled in a hydrogen peroxide solution. Self-generated solute gradient was based on dynamical behavior of Janus particles colliding with solid obstacles, the approaching walls particles that were assembled on the surface and oriented their symmetry axis parallel to the wall surface, and finally propelled the six asymmetric teeth.

Self-assembled chemical structures produce molecular aggregates with complex functionalities, which are often inspired by biological systems. While self-assembly has been the focus of intense investigation, the disassembly of these molecular aggregates is much less explored, even though it can lead to temporal control over the aggregate functionality. Sen's group reported a reversible Ag_3PO_4 microparticle system showing collective behaviors and disassembly in (or without) NH_3 medium in Figure 2d [44]. The schooling behavior of Ag_3PO_4 microparticles was based on self-diffusiophoresis. When adding NH_3, the production of large numbers of OH^- resulted in faster diffusion away from the Ag_3PO_4 microparticles than the other ions, and led to exclusionary behavior. When removing NH_3, the reaction was reversed, and schooling behavior was observed. The same group also reported an enzyme pump that nanomotors migrated towards regions of higher substrate concentration. The surface-immobilized enzymes, independent of adenosine triphosphate, were functioned as self-powered micropumps in the presence of their respective substrates. The flow was driven by a gradient in fluid density that was generated by the enzymatic reaction. Such behaviors could autonomously deliver proteins and assemble particles in response to specific chemical stimuli [45,46].

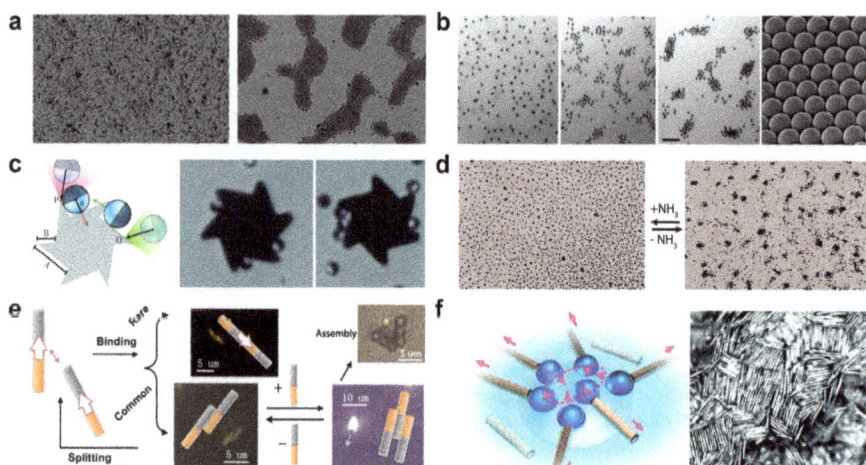

Figure 2. Chemical induced swarming and assembly of micro/nanoparticles. (**a**) Reversible swarm formation of unmodified Au microparticles induced by repetitive hydrazine additions (reproduced from [41]). (**b**) The swarming behavior of the Ir-based micromotors (reproduced from [42]). (**c**) Self-Assembly of micromachining systems powered by Janus micromotors (reproduced from [43]). (**d**) Transition between exclusion and schooling behaviors based on self-diffusiophoresis (reproduced from [44]). (**e**) Catalytically powered dynamic assembly of rod-shaped nanomotors and passive tracer particles (reproduced from [47]). (**f**) Self-assembled tubular microengines (reproduced from [48]).

Au-Pt nanowires that were powered by hydrogen peroxide, also show collective reaction when placed close to each other. Wang et al. studied the mechanism of the interaction of two nanowires as shown in Figure 2e [47]. The numerical simulations of interactions between two Au-Pt nanowires in hydrogen peroxide solutions revealed that the attractive and repulsive interactions were generated in the process of catalytic decomposition of hydrogen peroxide. Electrokinetic effects drove the two assembled doublets moving together in the same direction. Such behavior cannot be observed when nanowires composed of only gold or platinum. Schmidt's group studied the collective behavior of catalytic tubular microengines, as sketched in Figure 2f [48]. The catalytic microtubes can attract each other by bubbles at air-liquid interface and can self-assemble into compact patterns due to the meniscus-climbing effect, as shown in Figure 2f right. Such a balanced system offers an intriguing way to study dynamic self-assembly of bubble-propelled microengines.

Chemical gradient provides an approach to study the autonomous collective behavior, and can trigger swarming behavior of Janus spheres, Au-Pt nanowires, or microtubes into complex large units to achieve multifunctional properties. However, it should be noted that most of the aforementioned chemical triggered swarming behaviors are based on the interaction between the micro/nanomachines and the medium, which makes it not easily reversible and controllable. Besides, the swarming time is highly dependent on the reaction speed that makes it much longer than other approaches.

3.2. External Field Induced Collective Behavior

Besides chemotaxis, several external physical stimuli (for example, optical, electric, ultrasound, and magnetic fields) were used to induce the swarming of microparticles. When compared with chemotaxis, external physical stimuli induced assembly has shown promising performance with major advantages of on-demand motion control. The following sections will discuss recent progress that has been made in on-demand triggering the swarming or collective behavior of micromachines in the past decade.

3.2.1. Electric Induced Collective or Assembly Behavior

Electric induced dielectrophoresis, is a phenomenon exerted on dielectric micro/nanoparticles suspended in water solution when the micro/nanoparticles exist in a nonuniform electric field. Such phenomenon can be used to trigger the assembly of cells. As demonstrated in Figure 3a [49], applying alternating current on a patterned SU-8 on indium tin oxide (ITO)-coated glass slides resulted in a spatially patterned electric field, and led to the propelling the cells toward circular gaps, where the high intensity electric field was located and forming the two-dimensional (2D) patterns. Such behavior is greatly helpful in cell functions research. By combining electric manipulation with microfluidic, Chiang et al. demonstrated electro-microfluidic platform for microgel formation and architecture assembly (Figure 3b) [50]. The conceptual electro-microfluidic platform employed four cross-linkable elementary materials to assemble an architecture consisting of 3 × 3 microgels. The formation of patterned microgels was based on the manipulations of suspended particles, liquid droplets, and crosslinked microgels by electrowetting and dielectrophoresis between parallel plates with appropriate electrodes and dielectric and hydrophobic layers.

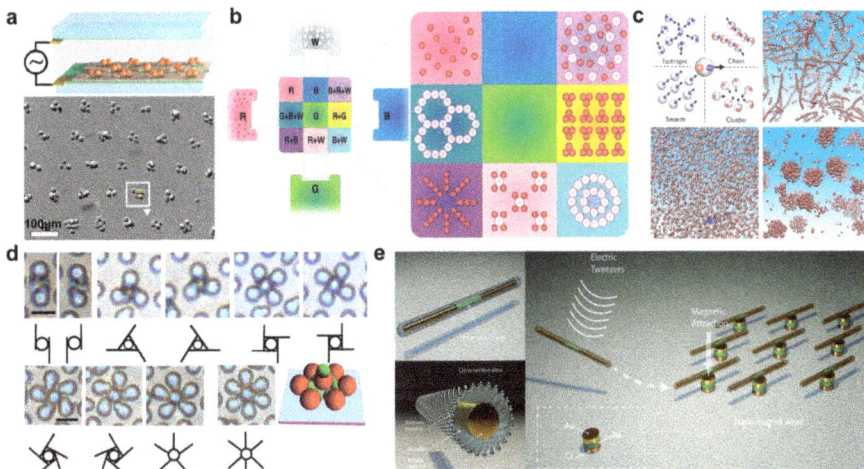

Figure 3. Electric field induced swarming and assembly of micro/nanoparticles. (**a**) Patterned electrode for multiply cell patterns (reproduced from [49]). (**b**) Programmable electromicrofluidic platform for microgel formation and architecture assembly (reproduced from [50]). (**c**) The strategy is to program dual electric charges shifted from the sphere center onto opposing hemispheres (reproduced from [51]). (**d**) Electric-field-induced assembly and propulsion of chiral colloidal clusters (reproduced from [52]). (**e**) Electric tweezers (reproduced from [53]).

Reconfiguring assembly of active Janus particles by electrostatic imbalance was achieved by Yan et al. [51]. A three-dimensional (3D) simulation shown in Figure 3c, the different electric charges on two end of Janus particles could result in dramatic assembly behavior. Janus microswimmers with equal-and-opposite charges on each hemisphere can polymerize into connected chains; if the charge on one hemisphere greatly exceeded that on the other, a torque arose when Janus microswimmer came into proximity. This torque diverted head-repulsive swimmers to avoid head-to-head collisions, and resulted in two-Janus particle alignment, and led to the collective or swarming of particles on a large scale; when tail-repulsive particles in proximity were preferentially positioned their tails apart, they were encouraged to face each other and to form jammed clusters with high local density. The dynamic reversible assembly of Janus microparticles was also reported by Wu's group, as shown in Figure 3d [52]. The geometry of the assembly depended on both the geometry and orientation of

Janus particles. The assembly can be trimer, tetramer, and pentamer, with an increasing number of petals. The author also checked the tetramers formed from dimers with different aspect ratios. When electric field turned off, the Janus particles could disperse. This strategy of electric-controlled collective assembly illustrated more general principles, holding great potential in fabricating 2D or 3D materials.

Combine DC with AC electric fields to manipulate nanowires suspended in a liquid was reported by Fan' group, called electric tweezers [53]. The electric tweezer always consisted of an Au/Ni/Au nanowire as rotor and an Au/Ni/Cr nanodisk as bearing. The Au/Ni/Au nanowire can attach on the patterned Au/Ni/Cr bearing by magnetic interaction, as shown in Figure 3e [54]. The patterned nanowires motor was controllable and can be rotated under a DC and AC electric field at an ultrafast speed of 18,000 r.p.m [55]. Due to the easy-controllability of such manipulation of a nanowire, by modifying the nanowires with a cytokine, such as tumour-necrosis factor-alpha, same group demonstrated a single cell drug delivery [56].

When compared with chemical induced collective behavior, the electric induced swarming behavior shows advantage of on-command controllability. However, it should be noted that the aforementioned electrically triggered patterning micro/nanoparticles always require complex patterned electrodes to generate 2D plane electric field. The swarming effects are highly dependent on the properties of suspended micro/nanoparticles (dielectric or conductive) and the high voltage of electric field may cause damages to cells, which greatly limit their application towards in vivo studies.

3.2.2. Collective or Assembly Behavior of Magnetic Micro/Nanoswimmer

Magnetism is one of the most efficient and effective ways to remotely control the motion of synthetic micro/nanomachines. Several individual groups have introduced ferromagnetic segments (Ni, Co, or Fe) to navigate micro/nanomachines to achieve certain tasks [57–60]. Recently, the magnetic induced assembly of micro/nanomachines attracted much attention for multiply functions [61–63]. As schematized in Figure 4a, movement in pattern (filaments) can be achieved under an applied magnetic field B_x. Such filament was constructed by double-stranded DNA with biotin at each end that can bind the particles together via the specific biotin-streptavidin interaction. The actuation induced a beating pattern that propelled the structure, and the external fields can be adjusted to control the velocity and the direction of motion [64]. Applying magnetic field to assemble colloidal microwheels was achieved by Tasci et al., as shown in Figure 4b [65]. The microwheels can be assembled from individual colloidal under rotating magnetic fields in both B_x and B_y direction of magnetic field at the surface plane. Reversible assembly of wheel-shaped devices can also be achieved, and microwheels can be directed rapidly and precisely along user-defined paths by varying spin frequency and angle relative to the surface.

The schooling behavior of magnetic helical micromotors was investigated by Tottori et al. as shown in Figure 4c [66]. The applied rotating magnetic forces can induce the assembly and disassembly of two swimmers to from a chain configuration. Multiple microswimmers can configure various geometrical configurations including straight, bent chains, and crosses. The speed of the resulting assemblies was highly dependent on their geometrical configurations. Such assembly behavior of individual is greatly helpful for particular designs and configurations. Reversible assembly of magneto-acoustic hybrid nanomotors was reported by Li et al., as demonstrated in Figure 4d [67]. Under the manipulation of both ultrasound and magnetic, the hybrid nanomotors can be regulated into three dramatic forms as follows. Only magnetic field on, directional moving individual nanomotors was observed. Only ultrasound field on, rapid migration of the nanomotors toward the original node position was observed. Both ultrasound and magnetic on, reformation of a dynamic swarm vortex was observed.

The ultimate goal of micro/nanomotors is towards in vivo medical application. Zhao's group reported an approach that accelerated tissue plasminogen activator-mediated thrombolysis by magnetically powered colony nanomotors, as demonstrated in Figure 4e [68]. The Ni-rod nanomotors fabricated by glancing angle deposition approach, and then coated a layer of biocompatible and biodegradable poly (lactic-co-glycolic) aggregates composed of t-PA-coated nanoparticles. Rotating magnetic nanomotors could enhance the mass transport of t-PA molecules at the blood clot interface for local ischemic stroke therapy. Such micro-aggregates can hold the nanoparticles together for enhancing the treatment of thrombolysis. Controlled swimming of a swarm of artificial bacterial flagella (16 μm in length) in Balb-C mouse under 9 mT and 90 Hz was reported by Nelson's group [69]. Due to the florescence of near-infrared probes, whole-body in vivo optical imaging of a swarm of functionalized artificial bacterial flagella was achieved by the magnetically controlled navigation, as shown in Figure 4f. The above mentioned in vivo study of micro/nanomachines hold great potential in future transport in therapeutics and microsurgery.

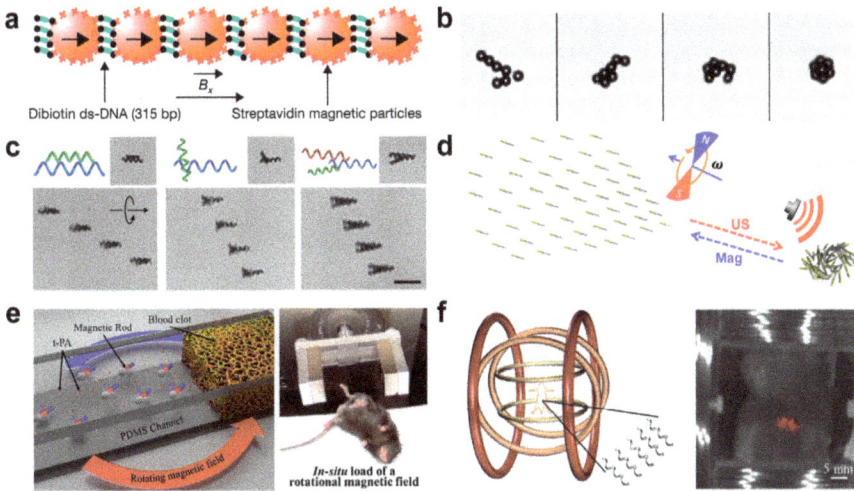

Figure 4. Magnetic field induced swarming and assembly of micro/nanoparticles. (**a**) Magnetic field B_x guided the particles modified with streptavidin to interact with biotin-modified double-stranded DNA and to form filaments via the specific biotin-streptavidin interaction (reproduced from [64]). (**b**) Rotating magnetic field $B_x + B_y$ for colloids assembly (reproduced from [65]). (**c**) Magnetically assisted assembly and disassembly of microhelices (reproduced from [66]). (**d**) Complex spatial-temporal collective behaviors of magneto-acoustic hybrid nanomotors (reproduced from [67]). (**e**) Rotating magnetic nanomotors enhanced the mass transport of t-PA molecules at the blood clot interface for local ischemic stroke therapy (reproduced from [68]). (**f**) Controlled swimming of a swarm of artificial bacterial flagella in the intra peritoneal cavity of a Balb-C mouse (reproduced from [69]).

Magnetism is the most widely used method to remotely trigger and control the swarming behavior of micro/nanoparticles with great biocompability. Although there is still plenty of room for improvement, scientists have already begun to use this approach towards preliminary biological applications. However, such an approach always requires the ferromagnetism of the micro/nanoparticle, which may limit some application prospects.

3.2.3. Ultrasound Induced Collective or Assembly Behavior

Ultrasound is another efficient tool to trigger the assembly or swarming of micro/nanoparticles [70]. Ultrasound manipulation of micro/nanomachines is based on pressure nodes or antinodes, which are responsible for particles migration or collection in certain areas. The using ultrasound field to regulate the swarming behavior of Au-Pt nanomotors hold considerable promise for fabricating man-made nanomachines that mimics the swarming behavior of natural animals. Ultrasound standing waves triggered swarming behavior of Au-Pt nanomotors (200 nm in diameter, 2 μm in length) was reported by Xu et al., as demonstrated in Figure 5a [71]. Without ultrasound field, Au-Pt nanomotors catalyzed decomposition of hydrogen peroxide and displayed autonomous random motion, as shown in Figure 5a left. With ultrasound field, Au-Pt nanomotors fast migrated towards ultrasound pressure nodes and then rapidly formed a swarm, as shown in the right of Figure 5a. The same behavior also reported by Wang et al., as shown in Figure 5b [72]. Transition between aggregated and dispersed states of a group of Au-Ru nanowires (~3 μm long and ~300 nm in diameter) can be found at the bottom of acoustic cell. Ultrasound-triggered reversible swarming of Janus microparticles (2 and 3 μm in diameter) also reported, as demonstrated in Figure 5c [73]. When in a weak acoustic trap, the Janus microswimmers were allowed to explore and reorient freely before reaching the 'ends' of the well. When in a strong acoustic trap, the Janus microswimmers can be fully trapped into a swarm, and then spread diffusively after turning off the ultrasound field. Such ultrasound triggered reversible swarming of micromachines may give insight into the origin of polar order, how and why living organisms align.

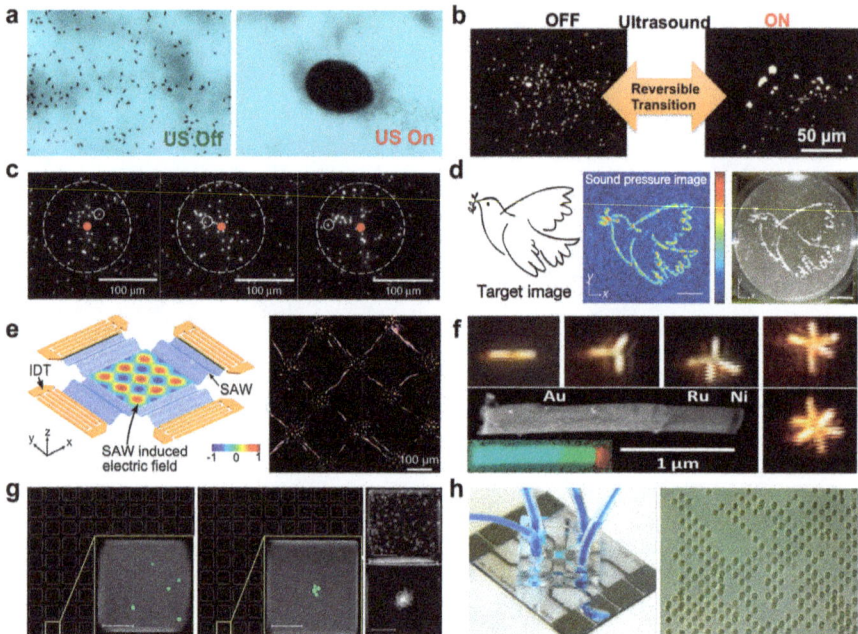

Figure 5. Ultrasound induced swarming and assembly of micro/nanoparticles. (**a**) Ultrasound swarming of Au-Pt catalytic nanowires (reproduced from [71]). (**b**) Reversible swarming behavior of Au-Ru microrods (reproduced from [72]). (**c**) Acoustic trapping of Janus Pt/PS particles (reproduced from [73]). (**d**) Holograms for acoustics (reproduced from [11]). (**e**) Tunable nanowire patterning using standing surface acoustic waves (reproduced from [74]). (**f**) Self-assembly of nanorod motors into geometrically regular multimers and their propulsion by ultrasound (reproduced from [75]). (**g**) Ultrasound-controlled cell aggregation in a multi-well chip (reproduced from [76]). (**h**) Surface acoustic wave for assembly of single cell (reproduced from [77]).

Ultrasound patterned regular geometric graphs have been reported by several groups. Acoustic hologram behavior was investigated, as shown in Figure 5d [11]. The acoustic holograms introduced a monolithic acoustic hologram by using a simple planar transducer driven by a single function generator and amplifier. By modifying the output of a single ultrasonic transducer, a designed 2D phase profile that is generated by the assembled particles can be achieved. Such acoustic induced holograms hold great potential in rapid fabrication of complex sound fields appropriate for super-resolution imaging, selective heating, and personalized medicine. The use of standing surface acoustic waves to trigger nanowires patterning was demonstrated in Figure 5e [74]. Two pairs of parallel interdigital transducers were deposited on a piezoelectric substrate to generate 2D standing surface acoustic waves. With a consistent spacing of half of the wavelength, the polystyrene microspheres were driven to the displacement antinodes, where the nanowires experienced counter-balancing forces from both positive and negative electric charges. Such a hypothesis was also confirmed by adding both silver nanowires and polystyrene microspheres in the 2D standing surface acoustic waves to observe their relative locations. Ultrasound-triggered the self-assembly of nanorod motors into geometrically regular multimers was reported by Mallouk's group [75]. As demonstrated in Figure 5f, Au-Ru-Ni nanowires (300 nm in diameter, 2.0 μm in length) can assemble into geometrically regular dimers, trimers, and higher multimers under a ~4 MHz of ultrasound at the midpoint of a cylindrical ultrasound cell. Such assembly can be propelled autonomously in fluids by excitation with ~4 MHz of ultrasound and exhibit several distinct modes of motion.

Ultrasound tweezers were also used for focusing or patterning of microparticles and cells, as shown in Figure 5g,h [76,77]. Cell lines, 2D cell patterns, and single cell patterns can be achieved by changing the pair numbers of parallel interdigital transducers and using the appropriate frequency. In details, when one pair of parallel interdigital transducers were applied, multiply cell lines can be obtained [78]. As the number of parallel interdigital transducers increased to two pairs, two dimensional ultrasound potential can be achieved, resulting in the generation of a 2D cell patterns [79]. In addition, by adjusting the ratio of λ/D (wavelength/cell diameter), 2D single cell patterns can be obtained [77].

Ultrasound is the fastest and biocompatible approach to trigger the swarming behavior of micro/nanomachines. Theoretically, such an approach could trigger all kinds of micro/nanoparticles without dielectric or magnetic properties. Until now, such approach has achieved some proof-of-concept biological applications, more efforts must be devoted to fabricating more precisely acoustic device to achieve complex in vivo targeted drug delivery studies.

3.2.4. Light Induced Collective or Assembly Behavior

Light is also a powerful tool to study the swarming or collective behavior of the micro/nanomachines. Beside aforementioned electric tweezers and ultrasound tweezers, the use of a strongly focused light beam to precisely trap and manipulate micro/nanoscale objects is called optical tweezers. Micro/nanoscale objects can be polarized by intensity gradients that are generated by a converging beam, and then move toward the highest gradient region of the electric field. Optical tweezers can trap micro/nanoparticles vary from two dimensions to three dimensions, from a single object to multiple objects [80–84]. Light-assisted self-assembly of gold nanoparticle chains has been reported [85]. Optical tweezers patterned highly stable 2D array of Au nanoparticles (diameter: 200 nm) was shown in Figure 6a [86]. The Au nanoparticles can be precisely patterned with a uniform space of 1.1 μm. Such optical tweezers induced 2D pattern is a new kind of template-free and reconfigurable method to reconfigure colloidal crystals.

Ibele et al. demonstrated that silver chloride (AgCl) microparticles interacted with each other at high concentrations and formed schools in presence of UV light, as shown in Figure 6b [87]. AgCl microparticles (1 μm in diameter) randomly scattered in deionized water before UV illumination (Figure 6b left). After 90 s of UV irradiation, asymmetric photodecomposition of AgCl microparticles resulted in electrolyte gradient, and led to the collective behavior of AgCl microparticles (Figure 6b right). Similar behavior can also be observed with titanium dioxide (TiO_2) microparticles (size range 0.2–2.5 μm), as reported by the same group, as shown in Figure 6c [88]. Under UV illumination, the TiO_2 microboats decomposed water and generate electrolyte gradient to trigger the collective behavior of surrounding silica microspheres (2.34 μm), which was unlike AgCl microparticles that photo-decomposed itself to produce the movement.

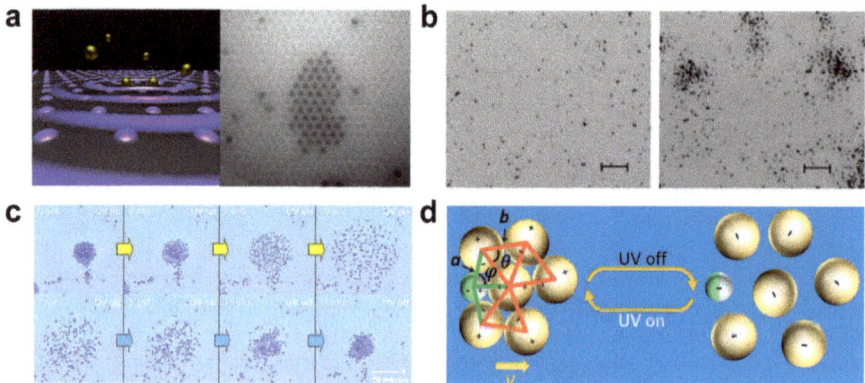

Figure 6. Light induced swarming and assembly of micro/nanoparticles. (**a**) Optical analogue of epitaxial growth to assemble gold nanoparticles (reproduced from [86]). (**b**) Schooling behavior of light-powered AgCl micromotors (reproduced from [87]). (**c**) Titanium-dioxide-based reversible micropump systems (reproduced from [88]). (**d**) TiO$_2$/Pt Janus micromotors with light irradiation for reversible and in-place assembly dissociation of surrounding SiO$_2$ particles (reproduced from [89]).

The light-triggered reversible swarming behavior of Janus microparticles was reported by Palacci et al. [90]. Under the illumination of blue light, such Janus microparticles catalyzed the exothermic chemical decomposition of H$_2$O$_2$, creating chemical gradients, and resulted in assembly into a homogeneous compact pattern. In contrast, the microparticles presented an incompact pattern when the light was off. Guan's group also reported an approach that in-place assembly dissociation in a high accuracy by manipulating TiO$_2$/Pt Janus micromotors with light irradiation (Figure 6d) [89]. Under UV irradiation, TiO$_2$/Pt Janus micromotors generated electron-hole pairs and involved in the oxidation of H$_2$O$_2$, producing a chemical gradient. Such a gradient can result in the assembly of surrounding SiO$_2$ colloidal particles. Visible light-triggered micropump for microparticle assembly was reported by Esplandiu et al. [91]. The micropump was simply fabricated by using standard electron beam lithography and electron beam evaporation, which consisted of 30–50 μm diameter Pt disks on doped silicon wafers. The application of light can lead to photoactivated chemical reactions at the silicon and the metal surfaces, which generated a chemical gradient, and resulted in such collective behavior.

Light has been the most convenient approach to induce the collective or assembly behavior due to its no requirement of complex processes for device fabrication. However, there are also some limitations. For the optical tweezers, the dielectricity of nanoparticles could result in serious local heating problems existing at the optical traps due to the high intensity of the focused laser beam. The photothermal and photocatalytic collective mechanisms always require some special materials (photothermal or photocatalytic properties) that may limit their practical application. Besides, the requirement of a certain degree of transparency in such systems also make it challenging towards clinical studies.

4. The Comparison of Different Approaches

Chemical induced collective behavior is based on the reaction on surface, so it is highly related to the surface properties, regardless of the dimensions and morphology. Chemical-induced assembly is quite simple, however, it does not have the capability of disassembly. For now, it is far away for chemical induced swarming behavior to be a tunable control approach. The response time is a little long, which requires a few minutes. The reaction are always carried out in water solution medium. The adding of extra fuel (H$_2$O$_2$, N$_2$H$_4$, etc.) could be a bit of toxicity towards in vivo studies.

Electric induced collective behavior is based on the interaction between the electric and the microparticle. No special nature and morphology of the particles is required. Electric always require designing the device to be suitable for known particle size. The electric induced assembly behavior could be reversible if the microparticle is alive. The controllability of electric swarming behavior is quite easy, microparticles can be controlled as regular patterns or directional arrangement. The response time is short, which requires only a few seconds. The reaction is often in solution medium, and it is not easy to operate towards in vivo study, because the high voltage may damage the tissue.

Magnetic induced collective behavior is based on the interaction between magnet and ferromagnetic micro/nanoparticles. This mechanism introduce ferromagnetic materials, such as Iron, Cobalt, and Nickel into the micromachines, no special morphology or dimension is required. Magnetic induced collective behavior is a tunable control approach and has the capability of disassembly by controlling the magnetic field. The response time is just a few seconds. The operation can be superfast and can be applied in almost all solution medium without any toxicity.

Ultrasound induced collective behavior is based on the interaction between suspended foreign particles and the applied ultrasound filed. No special nature or morphology or dimension is required. The ultrasound induced collective behavior can be a tunable control and fully reversible by changing the applied acoustic frequency or using active micro/nanoparticles. The response time is just several seconds and can be applied in almost all solution the mediums, without any toxicity. However this approach requires relatively complex fabrication processes to produce specific acoustic waves.

The optical tweezers can only manipulate the micro/nanoscale objects with dimensions smaller than the optical wavelength (few nanometers to hundred nanometers), and most of the morphology of the objects is spherical. It is a fully controllable approach with a short response time, and it can be applied in certain transparent solution medium without any toxicity. For the light-induced chemical reaction collective behavior, the detail is much like the chemical induced collective behavior. The only difference is the requirement of photothermal or photocatalytic material in the micro/nanomachines.

5. Conclusions and Future Directions

Self-organization of natural animals can build complex forms, from multicellular organisms to complex animal structures, such as flocks of birds and schooling of fishes by the interaction of individuals. Although tremendous efforts have been devoted to creating this ability in controlling the swarming or collective behavior of micro/nanomachines by chemical, electrical, magnetic, ultrasound, and light, much challenges must be faced in the design of both programmable and intelligent systems that can operate at microscale or nanoscale. The future direction of the swarming of micro/nanomachines is to create a system that demonstrates programmable self-assembly of thousand-micromachines into complex three-dimensional shapes for certain tasks. With the development of controllable swarming or collective behavior of micro/nanomachines, self-organized systems could be swarm intelligence. Swarm intelligence typically consists of a population of simple units interacting locally with one another and with their environment. The individual units will follow very simple rules, simple control of such a swarm can lead to the emergence of "intelligent" global behavior. This motivates new investigations into advanced collective micro/nanomachines to recover large-scale external damages and to accomplish much complex in vivo or vitro biomedical applications.

Acknowledgments: This research is supported by National Natural Science Foundation of China (21727815, 21475009), Fundamental Research Funds for the Central Universities (FRF-TP-17-066A1).

Author Contributions: Conghui Liu searched literature, collected and analyzed data, completed spreadsheets and pictures, and wrote the paper. Tailin Xu conceived, wrote and revised the paper. Li-Ping Xu and Xueji Zhang wrote and revised the paper.

Conflicts of Interest: The authors declare no conflict of interest.

Micromachines **2018**, *9*, 10

References

1. Poli, R.; Kennedy, J.; Blackwell, T. Particle swarm optimization. *Swarm Intell.* **2007**, *1*, 33–57. [CrossRef]
2. Vicsek, T.; Zafeiris, A. Collective motion. *Phys. Rep.* **2012**, *517*, 71–140. [CrossRef]
3. Zhang, J.; Luijten, E.; Grzybowski, B.A.; Granick, S. Active colloids with collective mobility status and research opportunities. *Chem. Soc. Rev.* **2017**, *46*, 5551–5569. [CrossRef] [PubMed]
4. Kearns, D.B. A field guide to bacterial swarming motility. *Nat. Rev. Microbiol.* **2010**, *8*, 634–644. [CrossRef] [PubMed]
5. Xu, T.; Gao, W.; Xu, L.P.; Zhang, X.; Wang, S. Fuel-Free Synthetic Micro/Nanomachines. *Adv. Mater.* **2017**, *29*, 1603250. [CrossRef] [PubMed]
6. Sanchez, S.; Soler, L.; Katuri, J. Chemically powered micro- and nanomotors. *Angew. Chem. Int. Ed.* **2015**, *54*, 1414–1444. [CrossRef] [PubMed]
7. Li, J.; Rozen, I.; Wang, J. Rocket Science at the Nanoscale. *ACS Nano* **2016**, *10*, 5619–5634. [CrossRef] [PubMed]
8. Wang, H.; Pumera, M. Fabrication of Micro/Nanoscale Motors. *Chem. Rev.* **2015**, *115*, 8704–8735. [CrossRef] [PubMed]
9. Mei, Y.; Solovev, A.A.; Sanchez, S.; Schmidt, O.G. Rolled-up nanotech on polymers: From basic perception to self-propelled catalytic microengines. *Chem. Soc. Rev.* **2011**, *40*, 2109–2119. [CrossRef] [PubMed]
10. Kim, K.; Guo, J.; Xu, X.; Fan, D.L. Recent Progress on Man-Made Inorganic Nanomachines. *Small* **2015**, *11*, 4037–4057. [CrossRef] [PubMed]
11. Melde, K.; Mark, A.G.; Qiu, T.; Fischer, P. Holograms for acoustics. *Nature* **2016**, *537*, 518–522. [CrossRef] [PubMed]
12. Palagi, S.; Mark, A.G.; Reigh, S.Y.; Melde, K.; Qiu, T.; Zeng, H.; Parmeggiani, C.; Martella, D.; Sanchez-Castillo, A.; Kapernaum, N.; et al. Structured light enables biomimetic swimming and versatile locomotion of photoresponsive soft microrobots. *Nat. Mater.* **2016**, *15*, 647–653. [CrossRef] [PubMed]
13. Lin, X.; Wu, Z.; Wu, Y.; Xuan, M.; He, Q. Self-Propelled Micro/Nanomotors Based on Controlled Assembled Architectures. *Adv. Mater.* **2016**, *28*, 1060–1072. [CrossRef] [PubMed]
14. Gao, W.; Wang, J. The environmental impact of micro/nanomachines: A review. *ACS Nano* **2014**, *8*, 3170–3180. [CrossRef] [PubMed]
15. Wu, Z.; Li, T.; Li, J.; Gao, W.; Xu, T.; Christianson, C.; Gao, W.; Galarnyk, M.; He, Q.; Zhang, L.; et al. Turning Erythrocytes into Functional Micromotors. *ACS Nano* **2014**, *8*, 12041–12048. [CrossRef] [PubMed]
16. Xu, T.; Soto, F.; Gao, W.; Garcia-Gradilla, V.; Li, J.; Zhang, X.; Wang, J. Ultrasound-Modulated Bubble Propulsion of Chemically Powered Microengines. *J. Am. Chem. Soc.* **2014**, *136*, 8552–8555. [CrossRef] [PubMed]
17. Wu, Z.; Li, T.; Gao, W.; Xu, T.; Jurado-Sánchez, B.; Li, J.; Gao, W.; He, Q.; Zhang, L.; Wang, J. Cell-Membrane-Coated Synthetic Nanomotors for Effective Biodetoxification. *Adv. Funct. Mater.* **2015**, *25*, 3881–3887. [CrossRef]
18. Li, J.; Thamphiwatana, S.; Liu, W.; Esteban-Fernandez de Avila, B.; Angsantikul, P.; Sandraz, E.; Wang, J.; Xu, T.; Soto, F.; Ramez, V.; et al. Enteric Micromotor Can Selectively Position and Spontaneously Propel in the Gastrointestinal Tract. *ACS Nano* **2016**, *10*, 9536–9542. [CrossRef] [PubMed]
19. Li, T.; Li, J.; Morozov, K.I.; Wu, Z.; Xu, T.; Rozen, I.; Leshansky, A.M.; Li, L.; Wang, J. Highly Efficient Freestyle Magnetic Nanoswimmer. *Nano Lett.* **2017**, *17*, 5092–5098. [CrossRef] [PubMed]
20. Dong, R.; Zhang, Q.; Gao, W.; Pei, A.; Ren, B. Highly Efficient Light-Driven TiO2–Au Janus Micromotors. *ACS Nano* **2016**, *10*, 839–844. [CrossRef] [PubMed]
21. Li, T.; Li, J.; Zhang, H.; Chang, X.; Song, W.; Hu, Y.; Shao, G.; Sandraz, E.; Zhang, G.; Li, L.; et al. Magnetically Propelled Fish-Like Nanoswimmers. *Small* **2016**, *12*, 6098–6105. [CrossRef] [PubMed]
22. Dong, R.; Hu, Y.; Wu, Y.; Gao, W.; Ren, B.; Wang, Q.; Cai, Y. Visible-Light-Driven BiOI-Based Janus Micromotor in Pure Water. *J. Am. Chem. Soc.* **2017**, *139*, 1722–1725. [CrossRef] [PubMed]
23. Wu, Z.; Lin, X.; Wu, Y.; Si, T.; Sun, J.; He, Q. Near-infrared light-triggered "on/off" motion of polymer multilayer rockets. *ACS Nano* **2014**, *8*, 6097–6105. [CrossRef] [PubMed]
24. Wu, Z.; Si, T.; Gao, W.; Lin, X.; Wang, J.; He, Q. Superfast Near-Infrared Light-Driven Polymer Multilayer Rockets. *Small* **2016**, *12*, 577–582. [CrossRef] [PubMed]
25. Guix, M.; Mayorga-Martinez, C.C.; Merkoci, A. Nano/Micromotors in (Bio)chemical Science Applications. *Chem. Rev.* **2014**, *114*, 6285–6322. [CrossRef] [PubMed]

26. Peng, F.; Tu, Y.; Wilson, D.A. Micro/nanomotors towards in vivo application: Cell, tissue and biofluid. *Chem. Soc. Rev.* **2017**, *46*, 5289–5310. [CrossRef] [PubMed]

27. Gao, W.; Wang, J. Synthetic micro/nanomotors in drug delivery. *Nanoscale* **2014**, *6*, 10486–10494. [CrossRef] [PubMed]

28. Nelson, B.J.; Kaliakatsos, I.K.; Abbott, J.J. Microrobots for minimally invasive medicine. *Annu. Rev. Biomed. Eng.* **2010**, *12*, 55–85. [CrossRef] [PubMed]

29. Xu, L.; Mou, F.; Gong, H.; Luo, M.; Guan, J. Light-driven micro/nanomotors: From fundamentals to applications. *Chem. Soc. Rev.* **2017**, *46*, 6905–6926. [CrossRef] [PubMed]

30. Chen, X.-Z.; Hoop, M.; Mushtaq, F.; Siringil, E.; Hu, C.; Nelson, B.J.; Pané, S. Recent developments in magnetically driven micro- and nanorobots. *Appl. Mater. Today* **2017**, *9*, 37–48. [CrossRef]

31. Illien, P.; Golestanian, R.; Sen, A. 'Fuelled' motion: Phoretic motility and collective behaviour of active colloids. *Chem. Soc. Rev.* **2017**, *46*, 5508–5518. [CrossRef] [PubMed]

32. Solovev, A.A.; Sanchez, S.; Schmidt, O.G. Collective behaviour of self-propelled catalytic micromotors. *Nanoscale* **2013**, *5*, 1284–1293. [CrossRef] [PubMed]

33. Shao, J.; Xuan, M.; Zhang, H.; Lin, X.; Wu, Z.; He, Q. Chemotaxis-Guided Hybrid Neutrophil Micromotors for Targeted Drug Transport. *Angew. Chem. Int. Ed.* **2017**, *56*, 12935–12939. [CrossRef] [PubMed]

34. Wu, Y.; Wu, Z.; Lin, X.; He, Q.; Li, J. Autonomous movement of controllable assembled Janus capsule motors. *ACS Nano* **2012**, *6*, 10910–10916. [CrossRef] [PubMed]

35. Sumpter, D.J. The principles of collective animal behaviour. *Philos. Trans. R. Soc. Lond. B Biol. Sci.* **2006**, *361*, 5–22. [CrossRef] [PubMed]

36. Cambui, D.S.; Rosas, A. Density induced transition in a school of fish. *Physica A* **2012**, *391*, 3908–3914. [CrossRef]

37. Swiecicki, J.M.; Sliusarenko, O.; Weibel, D.B. From swimming to swarming: Escherichia coli cell motility in two-dimensions. *Integr. Biol.* **2013**, *5*, 1490–1494. [CrossRef] [PubMed]

38. Ismagilov, R.F.; Schwartz, A.; Bowden, N.; Whitesides, G.M. Autonomous Movement and Self-Assembly. *Angew. Chem. Int. Ed.* **2002**, *41*, 652–654. [CrossRef]

39. Gao, W.; Pei, A.; Feng, X.; Hennessy, C.; Wang, J. Organized self-assembly of Janus micromotors with hydrophobic hemispheres. *J. Am. Chem. Soc.* **2013**, *135*, 998–1001. [CrossRef] [PubMed]

40. Cademartiri, L.; Bishop, K.J. Programmable self-assembly. *Nat. Mater.* **2014**, *14*, 2–9. [CrossRef] [PubMed]

41. Kagan, D.; Balasubramanian, S.; Wang, J. Chemically triggered swarming of gold microparticles. *Angew. Chem. Int. Ed.* **2011**, *50*, 503–506. [CrossRef] [PubMed]

42. Gao, W.; Pei, A.; Dong, R.; Wang, J. Catalytic iridium-based Janus micromotors powered by ultralow levels of chemical fuels. *J. Am. Chem. Soc.* **2014**, *136*, 2276–2279. [CrossRef] [PubMed]

43. Maggi, C.; Simmchen, J.; Saglimbeni, F.; Katuri, J.; Dipalo, M.; De Angelis, F.; Sanchez, S.; Di Leonardo, R. Self-Assembly of Micromachining Systems Powered by Janus Micromotors. *Small* **2016**, *12*, 446–451. [CrossRef] [PubMed]

44. Duan, W.; Liu, R.; Sen, A. Transition between collective behaviors of micromotors in response to different stimuli. *J. Am. Chem. Soc.* **2013**, *135*, 1280–1283. [CrossRef] [PubMed]

45. Sengupta, S.; Patra, D.; Ortiz-Rivera, I.; Agrawal, A.; Shklyaev, S.; Dey, K.K.; Cordova-Figueroa, U.; Mallouk, T.E.; Sen, A. Self-powered enzyme micropumps. *Nat. Chem.* **2014**, *6*, 415–422. [CrossRef] [PubMed]

46. Das, S.; Shklyaev, O.E.; Altemose, A.; Shum, H.; Ortiz-Rivera, I.; Valdez, L.; Mallouk, T.E.; Balazs, A.C.; Sen, A. Harnessing catalytic pumps for directional delivery of microparticles in microchambers. *Nat. Commun.* **2017**, *8*, 14384. [CrossRef] [PubMed]

47. Wang, W.; Duan, W.; Sen, A.; Mallouk, T.E. Catalytically powered dynamic assembly of rod-shaped nanomotors and passive tracer particles. *Proc. Natl. Acad. Sci. USA* **2013**, *110*, 17744–17749. [CrossRef] [PubMed]

48. Solovev, A.A.; Mei, Y.; Schmidt, O.G. Catalytic microstrider at the air-liquid interface. *Adv. Mater.* **2010**, *22*, 4340–4344. [CrossRef] [PubMed]

49. Albrecht, D.R.; Underhill, G.H.; Wassermann, T.B.; Sah, R.L.; Bhatia, S.N. Probing the role of multicellular organization in three-dimensional microenvironments. *Nat. Methods* **2006**, *3*, 369–375. [CrossRef] [PubMed]

50. Chiang, M.Y.; Hsu, Y.W.; Hsieh, H.Y.; Chen, S.Y.; Fan, S.K. Constructing 3D heterogeneous hydrogels from electrically manipulated prepolymer droplets and crosslinked microgels. *Sci. Adv.* **2016**, *2*, e1600964. [CrossRef] [PubMed]

51. Yan, J.; Han, M.; Zhang, J.; Xu, C.; Luijten, E.; Granick, S. Reconfiguring active particles by electrostatic imbalance. *Nat. Mater.* **2016**, *15*, 1095–1099. [CrossRef] [PubMed]

52. Ma, F.; Wang, S.; Wu, D.T.; Wu, N. Electric-field-induced assembly and propulsion of chiral colloidal clusters. *Proc. Natl. Acad. Sci. USA* **2015**, *112*, 6307–6312. [CrossRef] [PubMed]

53. Fan, D.L.; Zhu, F.Q.; Cammarata, R.C.; Chien, C.L. Electric tweezers. *Nano Today* **2011**, *6*, 339–354. [CrossRef]

54. Xu, X.; Kim, K.; Li, H.; Fan, D.L. Ordered arrays of Raman nanosensors for ultrasensitive and location predictable biochemical detection. *Adv. Mater.* **2012**, *24*, 5457–5463. [CrossRef] [PubMed]

55. Kim, K.; Xu, X.; Guo, J.; Fan, D.L. Ultrahigh-speed rotating nanoelectromechanical system devices assembled from nanoscale building blocks. *Nat. Commun.* **2014**, *5*, 3632. [CrossRef] [PubMed]

56. Fan, D.; Yin, Z.; Cheong, R.; Zhu, F.Q.; Cammarata, R.C.; Chien, C.L.; Levchenko, A. Subcellular-resolution delivery of a cytokine through precisely manipulated nanowires. *Nat. Nanotechnol.* **2010**, *5*, 545–551. [CrossRef] [PubMed]

57. Baraban, L.; Makarov, D.; Streubel, R.; Monch, I.; Grimm, D.; Sanchez, S.; Schmidt, O.G. Catalytic Janus motors on microfluidic chip: Deterministic motion for targeted cargo delivery. *ACS Nano* **2012**, *6*, 3383–3389. [CrossRef] [PubMed]

58. Solovev, A.A.; Sanchez, S.; Pumera, M.; Mei, Y.F.; Schmidt, O.G. Magnetic Control of Tubular Catalytic Microbots for the Transport, Assembly, and Delivery of Micro-objects. *Adv. Funct. Mater.* **2010**, *20*, 2430–2435. [CrossRef]

59. Kline, T.R.; Paxton, W.F.; Mallouk, T.E.; Sen, A. Catalytic nanomotors: Remote-controlled autonomous movement of striped metallic nanorods. *Angew. Chem. Int. Ed.* **2005**, *44*, 744–746. [CrossRef] [PubMed]

60. Orozco, J.; Cortes, A.; Cheng, G.; Sattayasamitsathit, S.; Gao, W.; Feng, X.; Shen, Y.; Wang, J. Molecularly imprinted polymer-based catalytic micromotors for selective protein transport. *J. Am. Chem. Soc.* **2013**, *135*, 5336–5339. [CrossRef] [PubMed]

61. Yellen, B.B.; Hovorka, O.; Friedman, G. Arranging matter by magnetic nanoparticle assemblers. *Proc. Natl. Acad. Sci. USA* **2005**, *102*, 8860–8864. [CrossRef] [PubMed]

62. Ahniyaz, A.; Sakamoto, Y.; Bergström, L. Magnetic field-induced assembly of oriented superlattices from maghemite nanocubes. *Proc. Natl. Acad. Sci. USA* **2007**, *104*, 17570–17574. [CrossRef] [PubMed]

63. Erb, R.M.; Son, H.S.; Samanta, B.; Rotello, V.M.; Yellen, B.B. Magnetic assembly of colloidal superstructures with multipole symmetry. *Nature* **2009**, *457*, 999–1002. [CrossRef] [PubMed]

64. Dreyfus, R.; Baudry, J.; Roper, M.L.; Fermigier, M.; Stone, H.A.; Bibette, J. Microscopic artificial swimmers. *Nature* **2005**, *437*, 862–865. [CrossRef] [PubMed]

65. Tasci, T.O.; Herson, P.S.; Neeves, K.B.; Marr, D.W. Surface-enabled propulsion and control of colloidal microwheels. *Nat. Commun.* **2016**, *7*, 10225. [CrossRef] [PubMed]

66. Tottori, S.; Zhang, L.; Peyer, K.E.; Nelson, B.J. Assembly, disassembly, and anomalous propulsion of microscopic helices. *Nano Lett.* **2013**, *13*, 4263–4268. [CrossRef] [PubMed]

67. Li, J.; Li, T.; Xu, T.; Kiristi, M.; Liu, W.; Wu, Z.; Wang, J. Magneto-Acoustic Hybrid Nanomotor. *Nano Lett.* **2015**, *15*, 4814–4821. [CrossRef] [PubMed]

68. Cheng, R.; Huang, W.; Huang, L.; Yang, B.; Mao, L.; Jin, K.; ZhuGe, Q.; Zhao, Y. Acceleration of tissue plasminogen activator-mediated thrombolysis by magnetically powered nanomotors. *ACS Nano* **2014**, *8*, 7746–7754. [CrossRef] [PubMed]

69. Servant, A.; Qiu, F.; Mazza, M.; Kostarelos, K.; Nelson, B.J. Controlled in vivo swimming of a swarm of bacteria-like microrobotic flagella. *Adv. Mater.* **2015**, *27*, 2981–2988. [CrossRef] [PubMed]

70. Xu, T.; Xu, L.-P.; Zhang, X. Ultrasound propulsion of micro/nanomotors. *Appl. Mater. Today* **2017**, *9*, 493–503. [CrossRef]

71. Xu, T.; Soto, F.; Gao, W.; Dong, R.; Garcia-Gradilla, V.; Magana, E.; Zhang, X.; Wang, J. Reversible Swarming and Separation of Self-Propelled Chemically Powered Nanomotors under Acoustic Fields. *J. Am. Chem. Soc.* **2015**, *137*, 2163–2166. [CrossRef] [PubMed]

72. Wang, W.; Duan, W.; Zhang, Z.; Sun, M.; Sen, A.; Mallouk, T.E. A tale of two forces: Simultaneous chemical and acoustic propulsion of bimetallic micromotors. *Chem. Commun.* **2015**, *51*, 1020–1023. [CrossRef] [PubMed]

73. Takatori, S.C.; De Dier, R.; Vermant, J.; Brady, J.F. Acoustic trapping of active matter. *Nat. Commun.* **2016**, *7*, 10694. [CrossRef] [PubMed]

74. Chen, Y.; Ding, X.; Steven Lin, S.C.; Yang, S.; Huang, P.H.; Nama, N.; Zhao, Y.; Nawaz, A.A.; Guo, F.; Wang, W.; et al. Tunable nanowire patterning using standing surface acoustic waves. *ACS Nano* **2013**, *7*, 3306–3314. [CrossRef] [PubMed]
75. Ahmed, S.; Gentekos, D.T.; Fink, C.A.; Mallouk, T.E. Self-assembly of nanorod motors into geometrically regular multimers and their propulsion by ultrasound. *ACS Nano* **2014**, *8*, 11053–11060. [CrossRef] [PubMed]
76. Vanherberghen, B.; Manneberg, O.; Christakou, A.; Frisk, T.; Ohlin, M.; Hertz, H.M.; Onfelt, B.; Wiklund, M. Ultrasound-controlled cell aggregation in a multi-well chip. *Lab Chip* **2010**, *10*, 2727–2732. [CrossRef] [PubMed]
77. Collins, D.J.; Morahan, B.; Garcia-Bustos, J.; Doerig, C.; Plebanski, M.; Neild, A. Two-dimensional single-cell patterning with one cell per well driven by surface acoustic waves. *Nat. Commun.* **2015**, *6*, 8686. [CrossRef] [PubMed]
78. Shi, J.; Ahmed, D.; Mao, X.; Lin, S.C.; Lawit, A.; Huang, T.J. Acoustic tweezers: Patterning cells and microparticles using standing surface acoustic waves (SSAW). *Lab Chip* **2009**, *9*, 2890–2895. [CrossRef] [PubMed]
79. Ding, X.; Lin, S.C.; Kiraly, B.; Yue, H.; Li, S.; Chiang, I.K.; Shi, J.; Benkovic, S.J.; Huang, T.J. On-chip manipulation of single microparticles, cells, and organisms using surface acoustic waves. *Proc. Natl. Acad. Sci. USA* **2012**, *109*, 11105–11109. [CrossRef] [PubMed]
80. Pauzauskie, P.J.; Radenovic, A.; Trepagnier, E.; Shroff, H.; Yang, P.; Liphardt, J. Optical trapping and integration of semiconductor nanowire assemblies in water. *Nat. Mater.* **2006**, *5*, 97–101. [CrossRef] [PubMed]
81. Berthelot, J.; Acimovic, S.S.; Juan, M.L.; Kreuzer, M.P.; Renger, J.; Quidant, R. Three-dimensional manipulation with scanning near-field optical nanotweezers. *Nat. Nanotechnol.* **2014**, *9*, 295–299. [CrossRef] [PubMed]
82. Marago, O.M.; Jones, P.H.; Gucciardi, P.G.; Volpe, G.; Ferrari, A.C. Optical trapping and manipulation of nanostructures. *Nat. Nanotechnol.* **2013**, *8*, 807–819. [CrossRef] [PubMed]
83. Grier, D.G. A revolution in optical manipulation. *Nature* **2003**, *424*, 810–816. [CrossRef] [PubMed]
84. Kim, Y.; Shah, A.A.; Solomon, M.J. Spatially and temporally reconfigurable assembly of colloidal crystals. *Nat. Commun.* **2014**, *5*, 3676. [CrossRef] [PubMed]
85. Jaquay, E.; Martinez, L.J.; Huang, N.; Mejia, C.A.; Sarkar, D.; Povinelli, M.L. Light-assisted, templated self-assembly of gold nanoparticle chains. *Nano Lett.* **2014**, *14*, 5184–5188. [CrossRef] [PubMed]
86. Huang, N.; Martinez, L.J.; Jaquay, E.; Nakano, A.; Povinelli, M.L. Optical Epitaxial Growth of Gold Nanoparticle Arrays. *Nano Lett.* **2015**, *15*, 5841–5845. [CrossRef] [PubMed]
87. Ibele, M.; Mallouk, T.E.; Sen, A. Schooling behavior of light-powered autonomous micromotors in water. *Angew. Chem. Int. Ed.* **2009**, *48*, 3308–3312. [CrossRef] [PubMed]
88. Hong, Y.Y.; Diaz, M.; Cordova-Figueroa, U.M.; Sen, A. Light-Driven Titanium-Dioxide-Based Reversible Microfireworks and Micromotor/Micropump Systems. *Adv. Funct. Mater.* **2010**, *20*, 1568–1576. [CrossRef]
89. Gao, Y.; Mou, F.; Feng, Y.; Che, S.; Li, W.; Xu, L.; Guan, J. Dynamic Colloidal Molecules Maneuvered by Light-Controlled Janus Micromotors. *ACS Appl. Mater. Interfaces* **2017**, *9*, 22704–22712. [CrossRef] [PubMed]
90. Palacci, J.; Sacanna, S.; Steinberg, A.P.; Pine, D.J.; Chaikin, P.M. Living crystals of light-activated colloidal surfers. *Science* **2013**, *339*, 936–940. [CrossRef] [PubMed]
91. Esplandiu, M.J.; Farniya, A.A.; Bachtold, A. Silicon-Based Chemical Motors: An Efficient Pump for Triggering and Guiding Fluid Motion Using Visible Light. *ACS Nano* **2015**, *9*, 11234–11240. [CrossRef] [PubMed]

micromachines

MDPI

Review

Light-Controlled Swarming and Assembly of Colloidal Particles

Jianhua Zhang, Jingjing Guo, Fangzhi Mou *and Jianguo Guan *

State Key Laboratory of Advanced Technology for Materials Synthesis and Processing, International School of Materials Science and Engineering, Wuhan University of Technology, Wuhan 430070, China; zhangjianhua4987@whut.edu.cn (J.Z.); october@whut.edu.cn (J.G.)
* Correspondence: moufz@whut.edu.cn (F.M.); guanjg@whut.edu.cn (J.G.); Tel.: +86-27-8721-8832 (J.G.)

Received: 14 January 2018; Accepted: 11 February 2018; Published: 19 February 2018

Abstract: Swarms and assemblies are ubiquitous in nature and they can perform complex collective behaviors and cooperative functions that they cannot accomplish individually. In response to light, some colloidal particles (CPs), including light active and passive CPs, can mimic their counterparts in nature and organize into complex structures that exhibit collective functions with remote controllability and high temporospatial precision. In this review, we firstly analyze the structural characteristics of swarms and assemblies of CPs and point out that light-controlled swarming and assembly of CPs are generally achieved by constructing light-responsive interactions between CPs. Then, we summarize in detail the recent advances in light-controlled swarming and assembly of CPs based on the interactions arisen from optical forces, photochemical reactions, photothermal effects, and photoisomerizations, as well as their potential applications. In the end, we also envision some challenges and future prospects of light-controlled swarming and assembly of CPs. With the increasing innovations in mechanisms and control strategies with easy operation, low cost, and arbitrary applicability, light-controlled swarming and assembly of CPs may be employed to manufacture programmable materials and reconfigurable robots for cooperative grasping, collective cargo transportation, and micro- and nanoengineering.

Keywords: light control; colloidal particles; swarm; assembly; collective behaviors

1. Introduction

Swarming and assembly represent a process in which multiple entities aggregate together and/or organize into ordered or functional structures through interactions with each other and their environment [1–5]. Swarming and assembly are common phenomena in nature. The examples range from the stacking of atoms and pairing of DNA strands, to the formation of bacterial colonies, schooling of fishes, human crowds, and galaxies [6,7]. Unlike single individuals, the swarms and assemblies in nature may perform complex collective behaviors and cooperative functions. For instance, atoms with different stacking sequences create materials with different properties [8]. To avoid predators, increase their success rate of foraging, and adapt to environmental changes, many animals prefer to face these challenges in the form of swarms and assemblies. As shown in Figure 1A, large schools of fishes change their shape and internal structure to adapt to the surrounding environment [9]. Flocks of wild geese fly in the V formation to decrease the air resistance for migrating, and ant army can transport much larger and heavier food than their own bodies (Figure 1B,C) [10,11]. Similarly, wolves usually act in packs to perform many strategies for hunting and use the group power to defend themselves (Figure 1D) [12]. Some living organisms can communicate with each other by specific modes, like secreting chemicals and making sound waves, to aggregate, but in many cases they form aggregates spontaneously without a centralized control. Drawing inspiration from nature, a thousand-robot swarm could implement a programmable self-assembly [13]. Figure 1E demonstrates that these robots are located in a precise mode to collect information from their neighbors and fulfil a task cooperatively. The more robots there are, the more difficult the tasks that can be accomplished. Although some individuals inevitably commit errors, they can recover from the feedbacks sent by their neighbors when there is a large population of individuals. Therefore, swarms and assemblies represent a living way for biological individuals to face the challenges of a severe living environment and they are the guarantee for robots to accomplish complex and cooperative tasks.

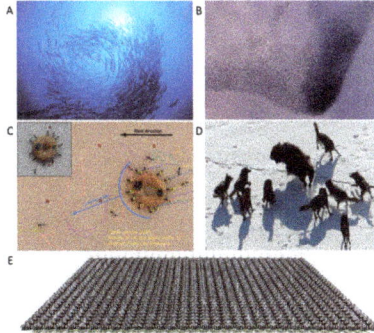

Figure 1. Biological and nonbiological swarms and assemblies. (**A**) A school of fishes (reproduced from [9]); (**B**) a flock of wild geese migrating (reproduced from [10]); (**C**) a group of ants carrying food (reproduced from [11]); (**D**) a pack of wolfs hunting a bison (reproduced from [12]); (**E**) a 2^{10}-robot swarm (reproduced from [13], reprinted with permission from AAAS).

Colloidal particles (CPs), including active and passive CPs, can serve as constituents to form swarms and assemblies. Active CPs, also defined as self-propelled particles or micro- and nanomotors, are CPs that can autonomously move by converting surrounding energies into their own kinetic energies [14–17], while passive CPs, differentiated from active CPs, are those that can migrate only under external forces. Both active and passive CPs are covered in this review. By mimicking their counterparts in nature, such as atoms, molecules, cells, and animals, they can be organized into complex structures in a controlled manner and exhibit cooperative functions [1,18–26]. The equilibrium assembly of CPs, which evolves the building blocks into stable, ordered structures as the system approaches equilibrium, has been studied for decades [27]. The non-equilibrium (or dynamic) swarming and assembly of CPs is promising to create biomimetic, reconfigurable, and "intelligent" materials, which are able to reversibly transform, disassemble, and even move in response to external stimuli [4,23]. Recently, a variety of strategies have been developed to realize the dynamic swarming and assembly of CPs by using different stimuli, such as chemical gradients and external fields (light, magnetic, electric, and ultrasound fields). Chemical gradients can trigger the motions and interparticle interactions of CPs by diffusiophoresis, leading to their swarming and assembly. For example, the spontaneous schooling of AgCl, Ag_3PO_4, TiO_2-SiO_2, and gold (Au) micro- and nanoparticles was demonstrated in the self-generated chemical gradient fields [28–31]. With a chemical gradient produced by a bone crack, negatively charged quantum dots, enzymes, and drug capsules can be dragged to the bone crack for bone crack detection, targeting, and repair [32]. Alternating current electric fields can induce aggregations of CPs with various emergent patterns, such as chains, swarms, and clusters, depending on the electric field frequency [33–35]. Similar to the electric field, the application of a rotating magnetic field (*H*) can maneuver the aggregation of superparamagnetic particles by magnetic dipole–dipole interactions [36]. For instance, Yan et al. reported that magnetic CPs tended to form linear chains under a low-strength rotating *H*, and then these chains dissolved and assembled into swarms once the *H* frequency reached 20 Hz. In addition, ultrasounds can create many pressure nodes, propelling CPs to move together following pressure gradients [37–39]. Light can trigger photoinduced interactions between CPs to form swarms and assemblies. For instance, autonomous TiO_2–Pt micromotors under ultraviolet (UV) irradiation can effectively capture and assemble CPs into dynamic colloidal molecules based on light-controlled electrostatic interactions [40]. On the other hand, CP swarms can perform ample collective behaviors by adjusting to external stimuli. For instance, Ag_3PO_4 microparticle swarms exhibit expansion and contraction with the addition and removal of NH_3, Pt–Au nanowire motor swarms exhibit overall migration under the control of ultrasounds, and the superparamagnetic particle swarms can perform rolling with the cooperative control of ultrasounds and magnetic fields [30,36,41]. Nowadays, the exotic swarming and assembly of CPs motivate intensive endeavors for their applications, such as variable capacitors [42], drug delivery [43,44], biological detection and repair [32,45], targeted diagnosis and therapy [46], magnetic resonance imaging [47], etc.

Among the various external fields to guide the motion of CPs, light is a powerful and versatile external stimulus and has various advantages including remote controllability and high temporospatial resolution. Thus, the light control of CPs has attracted much considerable attention. Back in 1970, Arthur Ashkin invented optical tweezers [48]. He used optical forces induced by lasers to manipulate the motion of micrometre-sized particles and neutral atoms. Hereafter, the optical tweezer technology was frequently used to trap and manipulate CPs in micro- and nanoscale [49–51]. In the meantime, light, acting as an energy source, was widely used to regulate the stop–go motion, speed, and direction of micro- and nanomotors because of its precisely adjustable energy input and direction [52–56]. Lately, vast researches have emerged on the light-controlled swarming and assembly of CPs, which promise the creation of intelligent programmable materials and reconfigurable robots [4,57] but suffer from limitations in arbitrary applicability, precise arrangement of various CPs, enrichment of applications, etc.

Regarding the design strategies, propulsion mechanisms, motion behaviors, and emerging applications of light-driven micro- and nanomotors, we have made a detailed and tutorial review [55]. This review introduces the general principles of light-controlled swarming and assembly of CPs before amply summarizing the recent advances in the field in terms of employing optical forces, photochemical reactions, photothermal effects, and photoisomerizations. Afterwards, the potential applications, challenges, and future prospects of light-controlled swarming and assembly of CPs are also discussed. With the rapidly increasing innovations in mechanisms and strategies with easy operation, low cost, and arbitrary applicability, light-controlled swarming and assembly of CPs may offer new opportunities to develop programmable materials and reconfigurable robots for cooperative grasping, collective cargo transportation, and micro- and nanoengineering. We expect it may provide better horizons to those who wish to participate in this research field and spark the imagination of scientists.

2. General Principles for Light-Controlled Swarming and Assembly of Colloidal Particles (CPs)

Generally, CPs could generate autonomous motion or forced migration under a local or global chemical gradient and external fields (magnetic, electric, thermal, acoustic, flow, and optical fields). Thus, they may come together and form into swarms or assemblies through particle–particle and particle–interface interactions. Even though CP swarms and assemblies are slightly different in their structures, some general principles actually exist in their formation processes.

Swarms (herds, schools, or flocks etc.), including polar and aploar swarms [58,59], usually refer to a group of entities with autonomous motions (Figure 2A), such as microorganisms, insects, animals, or active CPs [25,60]. Individual autonomous particles in the swarm mainly interact with one another via dynamic long-range attraction and short-range repulsion (Figure 2B). The long-range attraction enables the particles to aggregate together, whereas the short-range repulsion guarantees that the particles do not collide with each other [61]. On the other hand, assemblies are hierarchical ordered structures consisting of interacting components (Figure 2A) [4]. Colloidal assemblies, such as colloidal molecules, polymers, and crystals [27,62], can only be formed if the attractive and repulsive interactions between individual CPs are sufficiently balanced in the colloidal system (Figure 2B). Hence, one of the striking features to distinguish swarms and assemblies is the state of the individual particles in the group. The individual CPs in swarms can autonomously move and are not orderly arranged, while those in (dynamic) assemblies gather in an (transient) ordered arrangement and exhibit no relative motions with respect to the assemblies (if there are no external disturbances), because of the dynamic and (transient) balanced interactions, as shown in Figure 2.

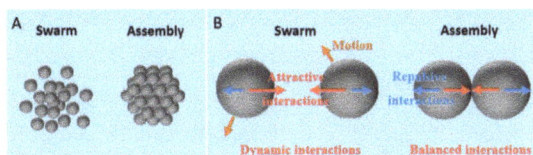

Figure 2. (**A**) Schematic illustration of a swarm and an assembly and (**B**) the state and mutual interactions of the individual particles in the swarm and assembly.

From the above analysis, one common prerequisite to realize the light-controlled swarming and assembly of CPs is to construct light-responsive short-range and long-range interactions to modulate the collective behaviors of CPs. As defined by Wang et al. [26], interactions acting on length scales smaller and comparable to the dimensions of the particles themselves are considered as short-range interactions. Otherwise, they are long-range interactions. The light-controlled short-range interactions consist of the light-controlled Van der Waals interactions (1–10 nm), steric repulsion (1–100 nm), hydrophobic attraction (1–100 nm), electrostatic interactions (1 nm–1 µm), etc. [63,64]. The light-controlled long-range interactions with an acting range from micrometers to millimeters are generally produced from direct optical forces or light–energy conversions, such as photochemical reactions, photothermal conversions, and photoisomerization, which induce diffusiophoresis, thermophoresis, convection, and Marangoni flows to regulate the collective behaviors of CPs [26,63,65–67]. Therefore, the photoresponses of CPs themselves or the environment (liquid media and substrates) are essential to generate the light-controlled short-range and long-range interactions to regulate the swarming and assembly of CPs.

3. Light-Controlled Swarming and Assembly of CPs

Upon light irradiation, the photoactive CPs or environment (liquid media and substrates) respond to it and change their state, which could induce the swarming and assembly of CPs in the following ways. At first, the CPs are subjected to direct optical forces and gather into swarms or assemblies if a highly focused light source is applied for trapping and manipulating the CPs, namely, optical tweezers for the assembly of CPs . Secondly, photoactive CPs, liquid media, and substrates can absorb light energy to produce various photochemical reactions and photophysical effects, such as photocatalytic reactions, photolysis, photothermal conversion, photoisomerization, and so on. In this condition, gradient fields of chemicals or light-induced energies are established around the light-exposed area. Under the local gradient fields, CPs attract or repel their neighbors, thereby causing swarming and assembly of CPs or leading to the exclusion of CPs in the swarms and assemblies formed before light irradiation, respectively. With respect to the interaction nature, light-controlled swarming and assembly of CPs are classified into four categories: optical forces-maneuvered, photochemical reaction-triggered, photothermal effect-induced, and photoisomerization-controlled swarming and assembly of CPs.

3.1. Optical Forces-Maneuvered Swarming and Assembly of CPs

Optical tweezers provide attractive or repulsive optical forces to precisely trap and manipulate micro- and nano-objects using highly focused laser beams. The micro- and nano-objects involve a variety of small matters, such as biological cells and a wide range of CPs. Here, we take the trapping and manipulation of colloidal metal nanoparticles as an example to illustrate the fundamental mechanism of optical tweezers (Figure 3A). Under the irradiation of incident light, the objects can absorb or scatter the photons to create momentum transfer. There are several key forces in this process, including optical gradient forces (the purple arrow) and radiation pressure (the green arrow) resulting from polarization and wavelength-dependent transfer of photon momentum [68]. Moreover, the optical forces can occur between CPs (red arrows). On the other hand, since the laser beams are the highly focused and have a high intensity, they can be used to control just one CP in an extremely tiny irradiation region. Therefore, in theory, various patterns and arrays of CPs can be realized.

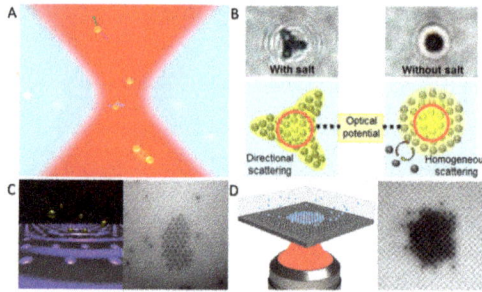

Figure 3. Optical forces-maneuvered assembly of colloidal particles (CPs). (**A**) Schematic illustration of manipulating metal CPs with optical forces (reproduced from [68]); (**B**) assembly of polystyrene (PS) nanoparticles induced by directional and homogeneous light scattering with or without salt (reproduced from [69]); (**C**) optical epitaxial growth of Au nanoparticle arrays (reproduced from [70]); (**D**) light-assisted and templated self-assembly of PS nanoparticles on a photonic crystal slab (reproduced from [71]).

Very recently, Wang and co-workers have demonstrated the dynamic assembly of polystyrene (PS) nanoparticles (diameter: 200 nm) [69]. PS nanoparticles can be assembled into periodic structures with steady states in few minutes. It is intriguing that the scattering directions of the particles can be varied by adding salt, and thus the patterns of the assemblies are different from the situation without salt (Figure 3B). In Figure 3C, Huang et al. have showed the optical epitaxial growth of Au nanoparticle (200 nm in diameter) arrays [70]. Optical forces and optical binding make particles attach to the template, resulting in various particle arrangements. Similarly, Jaquay et al. have illustrated the light-assisted, templated self-assembly of PS nanoparticles (260 nm radius) with a photonic crystal slab. They created arrays of optical traps by a 1.55 μm laser beam, thus the PS nanoparticles could assemble in local regions and disassemble when the laser beam was turned off (Figure 3D) [71].

3.2. Photochemical Reaction-Triggered Swarming and Assembly of CPs

Light can trigger photochemical reactions of photocatalytic and photolytic materials, producing ions or molecules by depleting the reactants. With the diffusion of the produced ions or molecules, chemical gradient fields around CPs are established, which not only can cause CPs to perform self-propulsion [18,72], but also can induce schooling and exclusion of CPs based on diffusiophoresis. If the products of the photochemical reactions are neutral molecules, nonelectrolyte diffusiophoresis governs the motions of CPs, and the electrolyte diffusiophoresis dominates their motions if the products are ions. The velocity (U) of CPs near a substrate in the chemical gradient field of monovalent electrolytes under electrolyte diffusiophoresis contains two contributions, including electrophoresis (the former term) and chemophoresis (the later term), as illustrated in Equation (1) [30].

$$U = \left[\frac{d\ln(C)}{dx}\right]\left[\frac{D_C - D_A}{D_C + D_A}\right]\left[\frac{k_B T}{e}\right]\left[\frac{\varepsilon(\zeta_p - \zeta_w)}{\eta}\right] + \left[\frac{d\ln(C)}{dx}\right]\left[\frac{2\varepsilon k_B^2 T^2}{\eta}\right]\left\{ln\left[1 - \tan h^2\left(\frac{e\zeta_w}{4k_B T}\right)\right] - ln\left[1 - \tan h^2\left(\frac{e\zeta_p}{4k_B T}\right)\right]\right\} \quad (1)$$

Here, $dln(C)/dx$ is the gradient of electrolyte, D_C and D_A represent the diffusivities of the cations and anions, k_B is the Boltzmann constant, T is the temperature, e, ε, and η represent the elementary charge, solution permittivity, and the solution dynamic viscosity, and ζ_P and ζ_W are the zeta potential of the particles and the substrate, respectively. Chemophoresis of CPs is usually negligible unless the M^+ and X^- ions have very similar diffusivities, and electrophoresis usually dominates the electrolyte diffusiophoresis of CPs [73]. As demonstrated in Figure 4A, the M^+ and X^- ions produced from the photochemical reactions diffuse away from the CPs with different rates. Then, the uneven distributed ions induce a local electric field (E). In return, this local E induces the electrophoretic propulsion (electrophoresis or electroosmosis) of the CPs inward or outward, depending on the relative magnitude of ζ_P and ζ_W [26], thus regulating the schooling and exclusion of CPs.

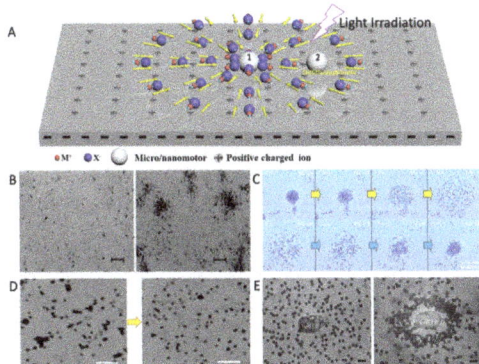

Figure 4. Photochemical reaction-triggered swarming of CPs. (**A**) Schematic illustration of the swarming of CPs based on diffusiophoresis; (**B**) AgCl microparticles before and after ultraviolet (UV) illumination (reproduced from [28]); (**C**) light-controlled reversible expansion–contraction movements of TiO_2–SiO_2 Janus microparticles (reproduced from [29]); (**D**) transition between schooling and exclusion behaviors of Ag_3PO_4 microparticles (reproduced from [30]); (**E**) the diffusiophoretic swarming of amino polystyrene (NH_2-PS) CPs induced by UV light (reproduced from [74]). Scale bars, 20 μm (**B,D**) and 6 μm (**E**).

By employing photochemical reactions, various swarms have been developed based on light-triggered diffusiophoresis of CPs. Sen and co-workers reported the light-induced swarming of AgCl particles (1 μm in diameter), as demonstrated in Figure 4B [28]. In aqueous medium, the photolysis of silver chloride (AgCl) microparticles produces H^+ and Cl^-, as in Equation (2):

$$4AgCl + 2H_2O \xrightarrow{hv,\ Ag^+} 4Ag + 4H^+ + 4Cl^- + O_2 \tag{2}$$

Because of the higher diffusivity of H^+ ions compared to Cl^-, an inward E is established, thereby triggering the inward swarming of AgCl microparticles under electrolyte diffusiophoresis, but the microparticles avoid physical contact because of the short-range repulsive electrostatic interactions between them. In addition, passive SiO_2 particles also swarm towards the AgCl microparticles and exhibit a "predator–prey" behavior due to the long-range attractive diffusiophoretic interactions. The same group also observed a similar diffusiophoretic swarming of SiO_2–TiO_2 Janus particles and Ag_3PO_4 CPs (Figure 4C,D) [29,30]. SiO_2–TiO_2 Janus particles and Ag_3PO_4 CPs showed reversible exclusion–schooling behaviors in response to UV irradiation or ammonia addition, respectively. Furthermore, *N*-hydroxyphthalimide triflate served as a solid photoacid generator which can produce proton and triflate anion that have different diffusion coefficients, ultimately causing the light-induced diffusiophoretic swarming of passive positively charged tracers (NH_2-PS CPs, 2 μm in diameter) (Figure 4E) [74]. There are two categories of swarms (flocks, herds, schools etc.) according to their polarity [58,59]. One category is polar swarms, in which the individual particles move with aligned velocity vectors. Examples include migrating animal herds and migrating colloidal swarms, which exhibit the displacement of the center of mass of the swarm over time [6]. Another category is apolar swarms, whose macroscopic velocity is zero even though the individual particles are polar. Examples of apolar swarms can be found in the patterns of active granular matter and living melanocytes [75,76]. It can be seen that the light-controlled swarms based on diffusiophoresis are apolar.

By utilizing the light-induced diffusiophoresis, light-controlled assembly of CPs can also be realized. As reported by Palacci et al., living crystals could be assembled by light-activated colloidal surfers [77]. Under blue light irradiation, such colloidal surfers photocatalytically decomposed H_2O_2, creating chemical gradients, and resulted in the close packing of the surfers under diffusiophoretic attractions. When the light was turned off, the assembled colloidal crystals dissociated into separated surfers because of the short-range repulsive electrostatic interactions. The assembly and dissociation of the colloidal crystals is reversible and can be swiftly controlled by regulating the light irradiation (Figure 5A). In addition, Mark et al. conducted a research on the interactions between TiO_2–SiO_2 Janus micromotors and passive particles [78]. Under the irradiation of UV light, TiO_2–SiO_2 Janus micromotors can capture and assemble passive particles into ordered structures based on diffusiophoretic effects, as illustrated in Figure 5B. Recently, we have reported that asymmetric redox reactions of H_2O and H_2O_2 on the surface of TiO_2–Pt Janus micromotors make the TiO_2 and Pt ends oppositely charged. Thus, a swimming TiO_2–Pt Janus micromotor under UV irradiation can capture and assemble surrounding micromotors or passive CPs on its surface, resulting in micromotor aggregates or dynamic colloidal molecules, as shown in Figure 5C,D [40,53].

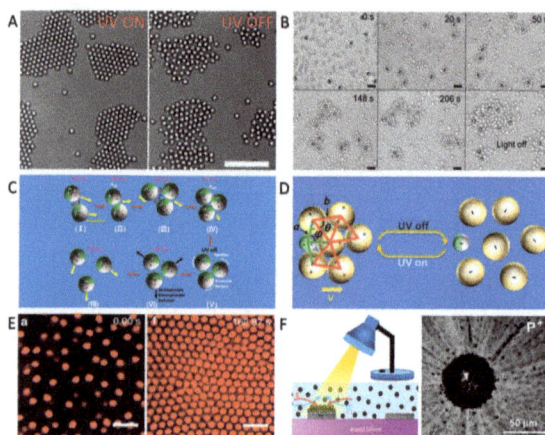

Figure 5. Photochemical reaction-triggered assembly of CPs. (**A–D**) Colloidal assembly based on photochemical reactions of CPs. (**A**) Self-assembly of bimaterial colloid surfers under blue light illumination (reproduced from [77], reprinted with permission from AAAS); (**B**) light-controlled colloidal crystal assembled by TiO_2–SiO_2 Janus micromotors and passive particles (reproduced from [78]); (**C**) schematic diagram of UV light-induced aggregation and separation of the TiO_2–Pt Janus submicromotors (reproduced from [53]); (**D**) schematic diagram of light-controlled assembly and dissociation of a colloidal molecule because of the light-switchable electrostatic interactions between a TiO_2–Pt micromotor (small green-white sphere) and SiO_2 CPs (big yellow spheres) (reproduced from [40]); (**E,F**) colloidal assembly based on photochemical reactions of the substrate; (**E**) colloidal poly (methyl methacrylate) (PMMA) particles concentrating in the region of light irradiation (reproduced from [79]); (**F**) swarming of passive CPs toward the visible light-triggered micropump consisting of a Pt disk on a doped silicon wafer (reproduced from [80]). Scale bars, 10 μm (**A**); and 5 μm (**B,E**).

Apart from the swarming and assembly based on photochemical reactions of the CPs themselves, photochemical reactions of the substrate can also induce the swarming and assembly of CPs. Solomon et al. demonstrated that a local *E* could be established around the light-exposed region of an indium–tin–oxide (ITO) glass substrate because of its photocatalytic reactions. The local *E* then induced the assembly of colloidal poly (methyl methacrylate) (PMMA) stabilized with poly(12-hydroxy-stearic acid) (PHSA) particles in the light-exposed region (Figure 5E) [79]. Complex patterns of particle assemblies could be written on the ITO glass substrate using structured light patterns. Esplandiu et al. revealed the swarming of CPs around a visible light-triggered micropump, which was simply fabricated by depositing Pt disks on a doped silicon wafer. Under light irradiation, the different photocatalytic reactions on Pt and of the silicon surface generated a gradient field of protons, resulting in an inward *E* that drove the swarming of passive CPs under electroosmosis (Figure 5F) [80].

3.3. Photothermal Effect-Induced Swarming and Assembly of CPs

Photothermal materials can absorb light energy and meanwhile convert it into thermal energy and induce a temperature gradient. It has also been manifested that the temperature gradient could manipulate CPs on the basis of thermophoresis. The thermophoretic velocity (*v*) of a single CP can be calculated by Equation (3) [22,67,81,82]:

$$v = -D_T \nabla_T \tag{3}$$

Here, ∇_T is the temperature gradient across the particles, $D_T = S_T D$ is the thermophoretic mobility, nd S_T is Soret coefficient, and D is the diffusion coefficient. Figure 6A shows the thermophoretic propulsion of metal-coated Janus spheres under a local temperature gradient generated from the photothermal conversion [81]. In addition, there are different interactions between those particles depending on the sign of S_T, such as thermophoretic repulsive and thermophoretic attractive interactions. Braun et al. revealed that thermophoretic interactions resulted from the thermophoretic slip flows [67]. As shown in Figure 6B, under the illumination of a laser beam, a temperature gradient arose across the particles and propelled them to the cold region (substrate). Then, a mutual hydrodynamic attraction occurred because of the presence of slip flows, thereby realizing the thermophoretic crystallization of PS CPs. Moreover, Zheng and co-workers realized a light-directed reversible assembly of plasmonic Au nanoparticles (average side length of ~150 nm) by the thermophoretic migration of nanoparticles resulting from their photothermal effect and the associated enhanced local electric field over a plasmonic substrate (Figure 6C) [83]. Because of the high precision and highly controllability of the laser beams, the size and location of assemblies can be controlled precisely to form various patterns. The authors achieved several different dynamic manipulations of selected Au nanotriangle assemblies and demonstrated their patterns transformation. Recently, Zheng and co-workers developed a new strategy to assemble CPs, named opto-thermophoretic assembly. In this strategy, the different rates of thermophoretic migration of cetyltrimethylammonium chloride (CTAC) micelles and Cl⁻ generated a thermoelectric field. This thermoelectric field could then trap and assemble charged CPs. Figure 6D demonstrates various 1D, 2D, and 3D hybrid assemblies of various PS beads and particles based on opto-thermophoresis. The opto-thermophoretic assembly strategy releases the rigorous design rules required by the existing assembly techniques and enriches the structural complexity of the colloidal matter, which will open a new window of opportunities for basic research on matter organization, advanced material design, and applications [84].

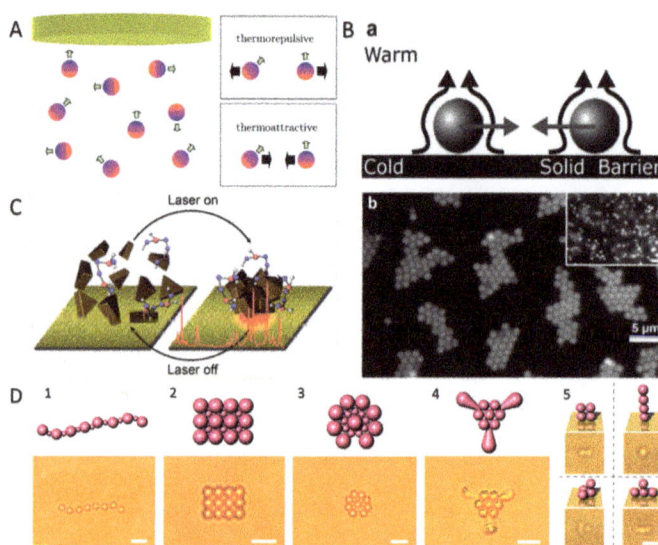

Figure 6. Photothermal effect-induced swarming and assembly of CPs. (**A**) Schematic demonstration of the self-propulsion and mutual interactions of metal-coated Janus spheres produced by the long-ranged temperature profiles due to light irradiation (reproduced from [81]); (**B**) CPs move ballistically to the cold surface in a temperature gradient induced by laser irradiation. The surface deflects the persisting slip flow and leads to mutual hydrodynamic attraction (a), resulting in the thermophoretic crystallization of PS CPs (b). Inset in b: the state of the particles before the presence of the temperature gradient (reproduced from [67]); (**C**) schematic diagram of the assembly and disassembly of positively charged Au nanotriangles with the laser on and off (reproduced from [83]); (**D**) 1D, 2D, and 3D hybrid assembly of 2 and 0.96 μm PS beads and anisotropic PS particles based on opto-thermophoresis; Scale bars, 5 μm (1–4) and 2 μm (5) (reproduced from [84], reprinted with permission from AAAS).

3.4. Photoisomerization-Controlled Swarming and Assembly of CPs

CPs modified with photoactive molecules, such as azobenzenes and spiropyrans, can assemble and disassemble under the control of light, during which the photoactive molecules serve as switches to change the conformations and/or properties of the CPs reversibly [4]. For example, the *trans* isomer of azobenzenes has no dipole moment, while the *cis* isomer, which is obtained under UV irradiation, has a large dipole moment (Figure 7A, (a)). Therefore, UV light can trigger dipole–dipole interactions between azobenzenes and the CPs modified with azobenzenes, inducing the light-controlled assembly of the CPs. When UV is turned off or visible-light irradiation is applied, the CPs disassemble because the dipole moment vanishes along with the *trans* isomerization of azobenzenes. By utilizing the isomerization of azobenzene-terminated ligands on nanoparticles, various metastable colloidal aggregates or crystals were obtained (Figure 7A, (b)) [4,85,86]. On the basis of the azobenzenes switches, Grzybowski et al. realized the aggregation of azobenzenes-functionalized metal nanoparticles (Au or Ag) under UV light through the *cis–trans* isomerization. The aggregation of the nanoparticles caused a color change of the nanoparticles because of the red shifting of the surface plasmonic resonance, allowing the writing of patterns by the structured light (Figure 7B) [87]. As for another class of photoresponsive molecule switches, the spiropyrans, as illustrated in Figure 7C, the opening of the ring gives rise to the merocyanine form and creates positive and negative charges. Thus, CPs decorated with spiropyrans can self-assemble into colloidal aggregates under UV irradiation because of short-range electrostatic interactions and then melt under visible light irradiation [88]. Chen and co-workers showed the assembly and disassembly of amphiphilic Au nanoparticles decorated with hydrophilic poly(ethylene glycol) (PEG) and hydrophobic photoresponsive polymethacrylate (PSPMA) [89]. Under the illumination of UV light, spiropyran units in PSPMA changed into the merocyanine isomer, and Au oligomers formed through π–π stacking and electrostatic attractions (Figure 7D). Similarly, Ren's group prepared SiO$_2$–Pt Janus micromotors with spiropyran moieties on the surface of the SiO$_2$ hemisphere [90]. The functionalized Janus micromotors could exhibit autonomous motion taking hydrogen peroxide as fuel and meanwhile conduct dynamic self-assembly in response to light irradiation. The micromotors assembled into multimers under UV irradiation (365 nm) and melted into mono-motors immediately when UV irradiation was switched to visible light (520 nm) irradiation (Figure 7E). Simultaneously, the photoisomerization of the substrate could also induce the swarming of CPs. Sagues et al. [56] functionalized the substrate with a photosensitive self-assembled azosilane monolayer. When the pear-shaped PS microparticles were dispersed in a nematic liquid crystal confined between a photosensitive and a non-photosensitive plate, they could electrophoretically drive along the local director to form aster and vortex swarms upon application of an alternating current (AC) electric field. One of the most intriguing characteristics is that the colloidal aster and vortex could be interconverted by suitable irradiation procedures because the grafted alkyl-azobenzene chains can be reversibly switched between the *cis* and *trans* isomers (Figure 7F). The formed swarm could also be relocated to a preselected place anywhere within the experimental cell by changing the location of the UV spot.

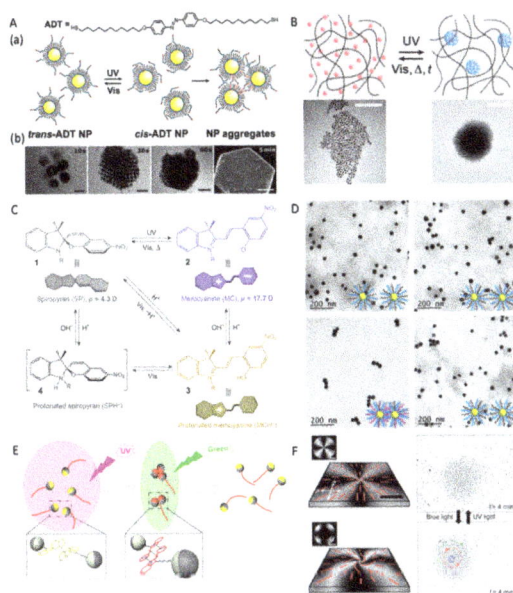

Figure 7. Photoisomerization-controlled assembly of CPs. (**A**) Schematic illustration of an azobenzene switch (a), assembly of nanoparticles coated with azobenzene-terminated ligands under UV light (b) (reproduced from [86]); (**B**) the dynamic aggregation of metal nanoparticles (Au or Ag) functionalized with azobenzenes under UV light, scale bar: 100 nm (reproduced from [87]); (**C**) schematic illustration of a spiropyran switch (reproduced from [88]); (**D**) the assembly and disassembly processes of amphiphilic Au nanoparticle (reproduced from [89]); (**E**) schematic diagram of light-controlled dynamic assembly of SiO$_2$–Pt Janus micromotors with spiropyran moieties attached on SiO$_2$ hemispheres (reproduced from [90]); (**F**) images of a cross-like and of a spiral attraction pattern formed under UV–blue light irradiation, and the corresponding particle aster and vortex after the application of an alternating current (AC) electric field (reproduced from [56]).

4. Applications of Light-Controlled CP Swarms and Assemblies

In nature, biological swarms and assemblies can perform complex and cooperative functions. For instance, ants carry food together and wolf packs hunt a large prey cooperatively [11,12]. In analogy with the biological systems, CP swarms and assemblies have various applications related to the specific properties of individual CPs as well as collective functions attributable to the swarms and assemblies as entities, applicable for drug delivery [43,44], targeted diagnosis and therapy [46], and magnetic resonance imaging [47]. Among them, light-controlled CP swarms and assemblies have enviable advantages owing to their remotely controllability, high temporospatial precision, and non-invasive operation. Up to now, various applications of light-controlled CP swarms and assemblies have been developed.

Firstly, the swarming and assembly can modulate the optical properties of CPs, and thus the light-controlled swarms and assemblies have promising applications in responsive optical devices and photothermal agents. For instance, Grzybowski and co-workers [87] developed an organogel "paper" containing photoresponsive Au or Ag nanoparticles inks, which were fabricated by coating Au or Ag nanoparticles with mixed self-assembled monolayers of dodecylamine (DDA) and photoswitchable azobenzene-terminated thiol. The color of the organogel "paper" experienced dynamic changes (shifting of the surface plasmon resonance) in response to UV irradiation because of the light-controlled assembly–disassembly transition of the photoresponsive Au or Ag nanoparticles. As a result, images and words could be written into the organogel "paper" by UV light, and spontaneously self-erased over time when UV was off (Figure 8A). Alternatively, Klajn et al. created self-erasing patterns in poly(ethylene glycol) gels containing spiropyrans and Au nanoparticles functionalized with 11-mercaptoundecanoic acid by utilizing the photoresponse of the medium instead of that of nanoparticles [91]. Under irradiation of visible blue light, the acidity of the medium increased and it triggered the disassembly of the nanoparticles held together by hydrogen bonds, which thus changed the color of the gel. Furthermore, Gao and co-workers [92] reported that light-triggered aggregation of Au nanoparticles could shift the surface plasmon resonance from the visible to near-infrared region, which not only enhanced photoacoustic imaging, but also improved their effectiveness in photothermal tumor ablation, as shown in Figure 8B. Very recently, we demonstrated that light-controlled TiO_2–Pt micromotors could effectively capture and assemble spherical transparent CPs into dynamic colloidal molecules [40]. The numerical simulation results (Figure 8C) illustrated that the colloidal molecule could act as a swimming microlens array for light manipulation.

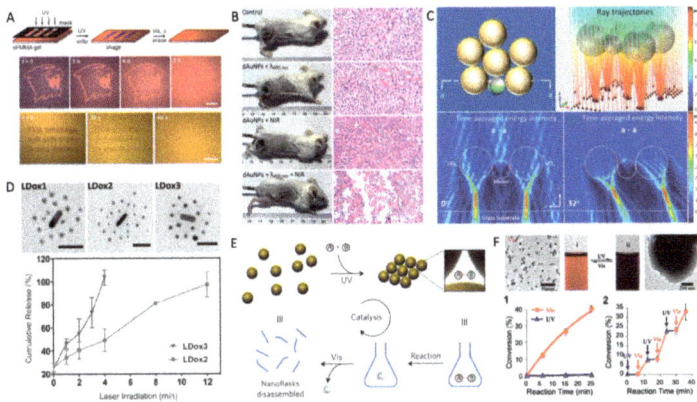

Figure 8. Applications of light-controlled CP swarms and assemblies. (**A**) The writing and erasing processes on an organogel "paper" containing photoresponsive Au or Ag nanoparticles inks under the control of UV light and visible light, with 0.8 s UV exposure through a transparency photomask; images obtained in Au and Ag nanoparticles films; the self-erasing of images in an Au nanoparticles film needs 9 h in daylight and 60 s in an Ag nanoparticles film by exposure to intense (0.3 m·W·cm^{-2}) visible light (reproduced from [87]); (**B**) the in vivo photothermal treatment of malignant tumors by cross-linked Au nanoparticles aggregates (reproduced from [92]); (**C**) numerical simulation results illustrating that the colloidal molecule can act as a microlens array for light manipulation (reproduced from [40]); (**D**) the near-infrared (NIR) light-triggered drug release from core–satellite superstructures; scale bars are 30 nm (reproduced from [93]); (**E**) schematic demonstration of reactions in "dynamic nanoflasks" between assembled azobenzene-coated nanoparticles under UV irradiation, and the subsequent product release due to the disassembly of the cluster under visible light irradiation (reproduced from [94]); (**F**) the light-switchable catalytic activity of photoactive Au nanoparticles; Au nanoparticles decorated with photoactive ligands catalyze a hydrosilylation reaction when in dispersed state, while they have a lose catalytic activity when in aggregated state under UV irradiation (reproduced from [95]).

Secondly, the swarming and assembly of CPs can create inter-particle voids to load drugs or act as nanoreactors. Chan and co-workers designed DNA–nanoparticle-assembled core–satellite superstructures as carriers of therapeutic agents [57,93]. Upon near-infrared light irradiation, the loaded Dox drug was released from the superstructure because of the thermal denaturation of the DNA linkers and the disassembly of the superstructure thanks to the photothermal effect of Au nanorods, as shown in Figure 8D. In addition, Klajn et al. utilized interparticle voids in the nanoparticle assemblies as "dynamic nanoflasks" [4,94]. The "dynamic nanoflasks", which were formed by the assembly of azobenzene-coated nanoparticles under UV irradiation, could selectively trap and concentrate small polar molecules ('A' and 'B' in Figure 8E) to enhance the rates of reactions between them. Once the products ('C' in Figure 8E) were generated, they could be released from the "nanoflasks" when the nanoparticle superstructure was disassembled under visible light irradiation.

Thirdly, the swarming and assembly of CPs can regulate the surface area and the chemical activities of the CPs, thereby creating switchable or tunable catalysts. As shown in Figure 8F, Au nanoparticles covered with azobenzene units and alkyl amine ligands in dispersed state could catalyze reactions such as hydrosilylation. However, once exposed to UV light, the Au nanoparticles assembled into supraspherical aggregates, which significantly lowered the surface area of the nanoparticles and ceased the catalytic reactions. The catalytic activity was switched "on" and "off" by the light-controlled assembly and disassembly of the nanoparticles [4,95]. The aggregation of nanophotocatalysts usually reduces their photocatalytic activities because of the decreasing available surface area and active sites for photocatalytic reactions [96]. However, Sun and co-workers found that TiO$_2$ nanoparticles underwent serious aggregation under UV irradiation because of the shifting of the isoelectric point of the particles, but some aggregates had a superior photocatalytic activity than the dispersed nanoparticles thanks to the facilitated charge separation and transfer between closely contacted nanoparticles [97].

5. Conclusions and Future Prospects

In conclusion, tremendous progress has been made in the light-controlled swarming and assembly of CPs by employing optical forces and light–energy conversions, including photochemical reactions, photothermal effects, and photoisomerizations. However, limitations and challenges remain. For example, optical tweezers can offer a versatile manipulation of CPs for swarming and assembly with an astonishing single particle resolution, but a highly focused light source and specific features (transparency or spherical shape, etc.) of the particles are required [98]. Photochemical reaction-triggered swarming and assembly is usually sensitive to ion concentration in the liquid medium, and photothermal effect-induced swarming and assembly require the photothermal conversion of the CPs or substrate and can only be triggered by light with a specific wavelength. Photoisomerizaion-controlled swarming and assembly are a surface chemistry-based method, posing the problem of applicability to arbitrary particle systems [99]. Furthermore, the so far reported strategies usually aim to the light-controlled swarming and assembly of CPs with the same geometrical or material features, and the precise arrangement of different CPs in a swarm or assembly is of extreme difficulty. In the future, specific attention should be paid to the development of general strategies for light manipulation of CPs regardless of their geometrical and material features. In addition, to obtain programmable materials, it is highly desired to control the precise temporospatial arrangement of different CPs in the swarms and assemblies by light. As for the application aspect, even though some proof-of-concept applications have been demonstrated, including photothermal therapy, drug delivery, pattern writing–self-erasing, switchable–tunable catalysts, etc., some important issues should be addressed before the real applications, such as safety issues in photothermal therapy and drug delivery and cost issues in others. On the other hand, the developed applications are mainly based on light-controlled assembly–disassembly of CPs. Other emergent behaviors, such as collective migrations or dynamic transformations of light-controlled swarms and assemblies, should be taken into consideration in the future. By utilizing the above-mentioned emergent behaviors, it is envisioned that, with elaborate design, CP swarms and assemblies can move to predefined working sites and transform into various user-specified microtools or microdevices, such as micrograspers, microdrillers, microwrenches, microvalves, micropumps, etc. To mimic the complex collective behaviors of biological swarms, such as cooperative carrying, migrating, foraging, nesting, and defending [11,100], it is essential to create "smart", light-controlled, artificial swarms or assemblies comprising active CPs in response to multiple stimuli. With the rapidly increasing innovations in mechanisms and strategies with easy operation, low cost, and arbitrary applicability, light-controlled swarming and assembly of CPs may finally realize intelligent programmable materials and reconfigurable robots for cooperative grasping, collective cargo transportation, microfabrications, etc. [25,26].

Acknowledgments: This work was supported by the National Natural Science Foundation of China (21474078, 51303144, 21705123 and 51521001), the Top Talents Lead Cultivation Project and Natural Science Foundation of Hubei Province (2015CFA003), the Yellow Crane talents plan of Wuhan municipal government, and the Fundamental Research Funds for the Central Universities (WUT: 2016III009 and 2015III060). We thank Xiaofeng Li, Zhuoyi Deng, and Ming You for revising the paper.

Author Contributions: Jianhua Zhang searched, collected, analyzed the literature, completed the pictures, and wrote the paper. Jingjing Guo searched the literature, completed the pictures, and revised the paper. Fangzhi Mou conceived, wrote, and revised the paper. Jianguo Guan conceived and revised the paper.

Conflicts of Interest: The authors declare no conflict of interest.

References

1. Solovev, A.A.; Sanchez, S.; Schmidt, O.G. Collective behaviour of self-propelled catalytic micromotors. *Nanoscale* **2013**, *5*, 1284–1293. [CrossRef] [PubMed]
2. Whitesides, G.M.; Grzybowski, B. Self-assembly at all scales. *Science* **2002**, *295*, 2418–2421. [CrossRef] [PubMed]
3. Jones, M.R.; Mirkin, C.A. Self-assembly gets new direction. *Nature* **2012**, *491*, 42–43. [CrossRef] [PubMed]
4. Grzybowski, B.A.; Fitzner, K.; Paczesny, J.; Granick, S. From dynamic self-assembly to networked chemical systems. *Chem. Soc. Rev.* **2017**, *46*, 5647–5678. [CrossRef] [PubMed]
5. Bricard, A.; Caussin, J.B.; Desreumaux, N.; Dauchot, O.; Bartolo, D. Emergence of macroscopic directed motion in populations of motile colloids. *Nature* **2013**, *503*, 95–98. [CrossRef] [PubMed]
6. Parrish, J.K.; Edelsteinkeshet, L. Complexity, pattern, and evolutionary trade-offs in animal aggregation. *Science* **1999**, *284*, 99–101. [CrossRef] [PubMed]
7. Chen, C.; Liu, S.; Shi, X.-Q.; Chaté, H.; Wu, Y. Weak synchronization and large-scale collective oscillation in dense bacterial suspensions. *Nature* **2017**, *542*, 210–214. [CrossRef] [PubMed]
8. Ashcroft, N.W.; Mermin, N.D. *Solid State Physics*; Academic Press: New York, NY, America, 1976; ISBN 978-0030839931.
9. Lopez, U.; Gautrais, J.; Couzin, I.D.; Theraulaz, G. From behavioural analyses to models of collective motion in fish schools. *Interface Focus* **2012**, *2*, 693–707. [CrossRef] [PubMed]
10. Vicsek, T.; Zafeiris, A. Collective motion. *Phys. Rep.* **2012**, *517*, 71–140. [CrossRef]
11. Gelblum, A.; Pinkoviezky, I.; Fonio, E.; Ghosh, A.; Gov, N.; Feinerman, O. Ant groups optimally amplify the effect of transiently informed individuals. *Nat. Commun.* **2015**, *6*, 7729. [CrossRef] [PubMed]
12. Muro, C.; Escobedo, R.; Spector, L.; Coppinger, R.P. Wolf-pack (*Canis lupus*) hunting strategies emerge from simple rules in computational simulations. *Behav. Process.* **2011**, *88*, 192–197. [CrossRef] [PubMed]
13. Rubenstein, M.; Cornejo, A.; Nagpal, R. Programmable self-assembly in a thousand-robot swarm. *Science* **2014**, *345*, 795–799. [CrossRef] [PubMed]
14. Gao, W.; Feng, X.; Pei, A.; Gu, Y.; Li, J.; Wang, J. Seawater-driven magnesium based Janus micromotors for environmental remediation. *Nanoscale* **2013**, *5*, 4696–4700. [CrossRef] [PubMed]
15. Mou, F.; Chen, C.; Zhong, Q.; Yin, Y.; Ma, H.; Guan, J. Autonomous motion and temperature-controlled drug delivery of Mg/Pt-poly(*N*-isopropylacrylamide) Janus micromotors driven by simulated body fluid and blood plasma. *ACS Appl. Mater. Interfaces* **2014**, *6*, 9897–9903. [CrossRef] [PubMed]
16. Ribeiro, T.; Coutinho, E.; Rodrigues, A.S.; Baleizao, C.; Farinha, J.P.S. Hybrid mesoporous silica nanocarriers with thermovalve-regulated controlled release. *Nanoscale* **2017**, *9*, 13485–13494. [CrossRef] [PubMed]
17. Huergo, M.A.; Maier, C.M.; Castez, M.F.; Vericat, C.; Nedev, S.; Salvarezza, R.C.; Urban, A.S.; Feldmann, J. Optical nanoparticle sorting elucidates synthesis of plasmonic nanotriangles. *ACS Nano* **2016**, *10*, 3614–3621. [CrossRef] [PubMed]
18. Yamamoto, D.; Shioi, A. Self-propelled nano/micromotors with a chemical reaction: Underlying physics and strategies of motion control. *KONA Powder Part. J.* **2015**, *32*, 2–22. [CrossRef]
19. Chen, C.; Mou, F.; Xu, L.; Wang, S.; Guan, J.; Feng, Z.; Wang, Q.; Kong, L.; Li, W.; Wang, J.; et al. Light-steered isotropic semiconductor micromotors. *Adv. Mater.* **2017**, *29*. [CrossRef] [PubMed]
20. Mou, F.; Xu, L.; Ma, H.; Guan, J.; Chen, D.R.; Wang, S. Facile preparation of magnetic gamma-Fe_2O_3/TiO_2 Janus hollow bowls with efficient visible-light photocatalytic activities by asymmetric shrinkage. *Nanoscale* **2012**, *4*, 4650–4657. [CrossRef] [PubMed]
21. Vach, P.J.; Fratzl, P.; Klumpp, S.; Faivre, D. Fast magnetic micropropellers with random shapes. *Nano Lett.* **2015**, *15*, 7064–7070. [CrossRef] [PubMed]
22. Jiang, H.R.; Yoshinaga, N.; Sano, M. Active motion of a Janus particle by self-thermophoresis in a defocused laser beam. *Phys. Rev. Lett.* **2010**, *105*, 268302. [CrossRef] [PubMed]
23. Li, J.; Li, T.; Xu, T.; Kiristi, M.; Liu, W.; Wu, Z.; Wang, J. Magneto-acoustic hybrid nanomotor. *Nano Lett.* **2015**, *15*, 4814–4821. [CrossRef] [PubMed]
24. Guo, J.; Kim, K.; Lei, K.W.; Fan, D.L. Ultra-durable rotary micromotors assembled from nanoentities by electric fields. *Nanoscale* **2015**, *7*, 11363–11370. [CrossRef] [PubMed]
25. Zhang, J.; Luijten, E.; Grzybowski, B.A.; Granick, S. Active colloids with collective mobility status and research opportunities. *Chem. Soc. Rev.* **2017**, *46*, 5551–5569. [CrossRef] [PubMed]

26. Wang, W.; Duan, W.; Ahmed, S.; Sen, A.; Mallouk, T.E. From one to many: Dynamic assembly and collective behavior of self-propelled colloidal motors. *Acc. Chem. Res.* **2015**, *48*, 1938–1946. [CrossRef] [PubMed]

27. Vogel, N.; Retsch, M.; Fustin, C.A.; Del Campo, A.; Jonas, U. Advances in colloidal assembly: The design of structure and hierarchy in two and three dimensions. *Chem. Rev.* **2015**, *115*, 6265–6311. [CrossRef] [PubMed]

28. Ibele, M.; Mallouk, T.E.; Sen, A. Schooling behavior of light-powered autonomous micromotors in water. *Angew. Chem. Int. Ed.* **2009**, *48*, 3308–3312. [CrossRef] [PubMed]

29. Hong, Y.; Diaz, M.; Córdova-Figueroa, U.M.; Sen, A. Light-driven titanium-dioxide-based reversible microfireworks and micromotor/micropump systems. *Adv. Funct. Mater.* **2010**, *20*, 1568–1576. [CrossRef]

30. Duan, W.; Liu, R.; Sen, A. Transition between collective behaviors of micromotors in response to different stimuli. *J. Am. Chem. Soc.* **2013**, *135*, 1280–1283. [CrossRef] [PubMed]

31. Kagan, D.; Balasubramanian, S.; Wang, J. Chemically triggered swarming of gold microparticles. *Angew. Chem. Int. Ed.* **2011**, *50*, 503–506. [CrossRef] [PubMed]

32. Yadav, V.; Freedman, J.D.; Grinstaff, M.; Sen, A. Bone-crack detection, targeting, and repair. *Angew. Chem. Int. Ed.* **2013**, *52*, 10997–11001. [CrossRef] [PubMed]

33. Ma, F.; Wang, S.; Wu, D.T.; Wu, N. Electric-field-induced assembly and propulsion of chiral colloidal clusters. *Proc. Natl. Acad. Sci. USA* **2015**, *112*, 6307–6312. [CrossRef] [PubMed]

34. Yan, J.; Han, M.; Zhang, J.; Xu, C.; Luijten, E.; Granick, S. Reconfiguring active particles by electrostatic imbalance. *Nat. Mater.* **2016**, *15*, 1095–1099. [CrossRef] [PubMed]

35. Albrecht, D.R.; Underhill, G.H.; Wassermann, T.B.; Sah, R.L.; Bhatia, S.N. Probing the role of multicellular organization in three-dimensional microenvironments. *Nat. Methods* **2006**, *3*, 369–375. [CrossRef] [PubMed]

36. Ahmed, D.; Baasch, T.; Blondel, N.; Laubli, N.; Dual, J.; Nelson, B.J. Neutrophil-inspired propulsion in a combined acoustic and magnetic field. *Nat. Commun.* **2017**, *8*, 770. [CrossRef] [PubMed]

37. Ahmed, S.; Gentekos, D.T.; Fink, C.A.; Mallouk, T.E. Self-assembly of nanorod motors into geometrically regular multimers and their propulsion by ultrasound. *ACS Nano* **2014**, *8*, 11053–11060. [CrossRef] [PubMed]

38. Vanherberghen, B.; Manneberg, O.; Christakou, A.; Frisk, T.; Ohlin, M.; Hertz, H.M.; Onfelt, B.; Wiklund, M. Ultrasound-controlled cell aggregation in a multi-well chip. *Lab Chip* **2010**, *10*, 2727–2732. [CrossRef] [PubMed]

39. Wang, W.; Duan, W.; Zhang, Z.; Sun, M.; Sen, A.; Mallouk, T.E. A tale of two forces: Simultaneous chemical and acoustic propulsion of bimetallic micromotors. *Chem. Commun.* **2015**, *51*, 1020–1023. [CrossRef] [PubMed]

40. Gao, Y.; Mou, F.; Feng, Y.; Che, S.; Li, W.; Xu, L.; Guan, J. Dynamic colloidal molecules maneuvered by light-controlled Janus micromotors. *ACS Appl. Mater. Interfaces* **2017**, *9*, 22704–22712. [CrossRef] [PubMed]

41. Xu, T.; Soto, F.; Gao, W.; Dong, R.; Garcia-Gradilla, V.; Magana, E.; Zhang, X.; Wang, J. Reversible swarming and separation of self-propelled chemically powered nanomotors under acoustic fields. *J. Am. Chem. Soc.* **2015**, *137*, 2163–2166. [CrossRef] [PubMed]

42. Lidgi-Guigui, N.; Dablemont, C.; Veautier, D.; Viau, G.; Seneor, P.; Nguyen Van Dau, F.; Mangeney, C.; Vaurès, A.; Deranlot, C.; Friederich, A. Grafted 2D Assembly of colloidal metal nanoparticles for application as a variable capacitor. *Adv. Mater.* **2007**, *19*, 1729–1733. [CrossRef]

43. Ling, D.; Park, W.; Park, S.-J.; Lu, Y.; Kim, K.S.; Hackett, M.J.; Kim, B.H.; Yim, H.; Jeon, Y.S.; Na, K.; et al. Multifunctional tumor pH-sensitive self-assembled nanoparticles for bimodal imaging and treatment of resistant heterogeneous tumors. *J. Am. Chem. Soc.* **2014**, *136*, 5647–5655. [CrossRef] [PubMed]

44. Wang, W.W.; Cheng, D.; Gong, F.M.; Miao, X.M.; Shuai, X.T. Design of multifunctional micelle for tumor-targeted intracellular drug release and fluorescent imaging. *Adv. Mater.* **2012**, *24*, 115–120. [CrossRef] [PubMed]

45. Yadav, V.; Pavlick, R.A.; Meckler, S.M.; Sen, A. Triggered detection and deposition: Toward the repair of microcracks. *Chem. Mater.* **2014**, *26*, 4647–4652. [CrossRef]

46. Gallo, J.; Kamaly, N.; Lavdas, I.; Stevens, E.; Quang-De, N.; Wylezinska-Arridge, M.; Aboagye, E.O.; Long, N.J. CXCR4-targeted and MMP-responsive iron oxide nanoparticles for enhanced magnetic resonance imaging. *Angew. Chem. Int. Ed.* **2014**, *53*, 9550–9554. [CrossRef] [PubMed]

47. Ito, A.; Shinkai, M.; Honda, H.; Kobayashi, T. Medical application of functionalized magnetic nanoparticles. *J. Biosci. Bioeng.* **2005**, *100*, 1–11. [CrossRef] [PubMed]

48. Ashkin, A. Acceleration and Trapping of Particles by Radiation Pressure. *Phys. Rev. Lett.* **1970**, *24*, 156–159. [CrossRef]

49. Pauzauskie, P.J.; Radenovic, A.; Trepagnier, E.; Shroff, H.; Yang, P.D.; Liphardt, J. Optical trapping and integration of semiconductor nanowire assemblies in water. *Nat. Mater.* **2006**, *5*, 97–101. [CrossRef] [PubMed]
50. Chiou, P.Y.; Ohta, A.T.; Wu, M.C. Massively parallel manipulation of single cells and microparticles using optical images. *Nature* **2005**, *436*, 370–372. [CrossRef] [PubMed]
51. Ikin, L.; Carberry, D.M.; Gibson, G.M.; Padgett, M.J.; Miles, M.J. Assembly and force measurement with SPM-like probes in holographic optical tweezers. *New J. Phys.* **2009**, *11*, 023012. [CrossRef]
52. Dong, R.; Zhang, Q.; Gao, W.; Pei, A.; Ren, B. Highly efficient light-driven TiO_2-Au Janus micromotors. *ACS Nano* **2016**, *10*, 839–844. [CrossRef] [PubMed]
53. Mou, F.; Kong, L.; Chen, C.; Chen, Z.; Xu, L.; Guan, J. Light-controlled propulsion, aggregation and separation of water-fuelled TiO_2/Pt Janus submicromotors and their "on-the-fly" photocatalytic activities. *Nanoscale* **2016**, *8*, 4976–4983. [CrossRef] [PubMed]
54. Mou, F.; Li, Y.; Chen, C.; Li, W.; Yin, Y.; Ma, H.; Guan, J. Single-component TiO_2 tubular microengines with motion controlled by light-induced bubbles. *Small* **2015**, *11*, 2564–2570. [CrossRef] [PubMed]
55. Xu, L.; Mou, F.; Gong, H.; Luo, M.; Guan, J. Light-driven micro/nanomotors: From fundamentals to applications. *Chem. Soc. Rev.* **2017**, *46*, 6905–6926. [CrossRef] [PubMed]
56. Hernandez-Navarro, S.; Tierno, P.; Farrera, J.A.; Ignes-Mullol, J.; Sagues, F. Reconfigurable swarms of nematic colloids controlled by photoactivated surface patterns. *Angew. Chem. Int. Ed.* **2014**, *53*, 10696–10700. [CrossRef] [PubMed]
57. Li, F.; Lu, J.; Kong, X.; Hyeon, T.; Ling, D. Dynamic nanoparticle assemblies for biomedical applications. *Adv. Mater.* **2017**, *29*. [CrossRef] [PubMed]
58. Ramaswamy, S. The mechanics and statistics of active matter. *Ann. Rev. Condens. Matter Phys.* **2010**, *1*, 323–345. [CrossRef]
59. Marchetti, M.C.; Joanny, J.F.; Ramaswamy, S.; Liverpool, T.B.; Prost, J.; Rao, M.; Simha, R.A. Hydrodynamics of soft active matter. *Rev. Mod. Phys.* **2013**, *85*, 1143–1189. [CrossRef]
60. Reynolds, C.W. Flocks, herds and schools: A distributed behavioral model. In *ACM SIGGRAPH Computer Graphics*; Association for Computing Machinery (ACM): New York, NY, USA, 1987; Volume 21, pp. 25–34.
61. Juanico, D.E.O. Self-organized pattern formation in a diverse attractive-repulsive swarm. *Europhys. Lett.* **2009**, *86*, 48004. [CrossRef]
62. Li, F.; Josephson, D.P.; Stein, A. Colloidal assembly: The road from particles to colloidal molecules and crystals. *Angew. Chem. Int. Ed.* **2011**, *50*, 360–388. [CrossRef] [PubMed]
63. Grzybowski, B.A.; Wilmer, C.E.; Kim, J.; Browne, K.P.; Bishop, K.J.M. Self-assembly: From crystals to cells. *Soft Matter* **2009**, *5*, 1110–1128. [CrossRef]
64. Bishop, K.J.M.; Wilmer, C.E.; Soh, S.; Grzybowski, B.A. Nanoscale forces and their uses in self-assembly. *Small* **2009**, *5*, 1600–1630. [CrossRef] [PubMed]
65. Illien, P.; Golestanian, R.; Sen, A. 'Fuelled' motion: Phoretic motility and collective behaviour of active colloids. *Chem. Soc. Rev.* **2017**, *46*, 5508–5518. [CrossRef] [PubMed]
66. Palacci, J.; Sacanna, S.; Kim, S.H.; Yi, G.R.; Pine, D.J.; Chaikin, P.M. Light-activated self-propelled colloids. *Philos. Trans. R. Soc. A* **2014**, *372*, 20130372. [CrossRef] [PubMed]
67. Weinert, F.M.; Braun, D. Observation of slip flow in thermophoresis. *Phys. Rev. Lett.* **2008**, *101*, 168301. [CrossRef] [PubMed]
68. Lehmuskero, A.; Johansson, P.; Rubinsztein-Dunlop, H.; Tong, L.M.; Kall, M. Laser trapping of colloidal metal nanoparticles. *ACS Nano* **2015**, *9*, 3453–3469. [CrossRef] [PubMed]
69. Wang, S.F.; Kudo, T.; Yuyama, K.I.; Sugiyama, T.; Masuhara, H. Optically evolved assembly formation in laser trapping of polystyrene nanoparticles at solution surface. *Langmuir* **2016**, *32*, 12488–12496. [CrossRef] [PubMed]
70. Huang, N.; Martinez, L.J.; Jaquay, E.; Nakano, A.; Povinelli, M.L. Optical epitaxial growth of gold nanoparticle arrays. *Nano Lett.* **2015**, *15*, 5841–5845. [CrossRef] [PubMed]
71. Jaquay, E.; Martinez, L.J.; Mejia, C.A.; Povinelli, M.L. Light-assisted, templated self-assembly using a photonic-crystal slab. *Nano Lett.* **2013**, *13*, 2290–2294. [CrossRef] [PubMed]
72. Kline, T.R.; Iwata, J.; Lammert, P.E.; Mallouk, T.E.; Sen, A.; Velegol, D. Catalytically driven colloidal patterning and transport. *J. Phys. Chem. B* **2006**, *110*, 24513–24521. [CrossRef] [PubMed]
73. Wang, W.; Duan, W.; Ahmed, S.; Mallouk, T.E.; Sen, A. Small power: Autonomous nano- and micromotors propelled by self-generated gradients. *Nano Today* **2013**, *8*, 531–554. [CrossRef]

74. Yadav, V.; Zhang, H.; Pavlick, R.; Sen, A. Triggered "on/off" micropumps and colloidal photodiode. *J. Am. Chem. Soc.* **2012**, *134*, 15688–15691. [CrossRef] [PubMed]

75. Narayan, V.; Ramaswamy, S.; Menon, N. Long-lived giant number fluctuations in a swarming granular nematic. *Science* **2007**, *317*, 105–108. [CrossRef] [PubMed]

76. Gruler, H.; Dewald, U.; Eberhardt, M. Nematic liquid crystals formed by living amoeboid cells. *Eur. Phys. J. B Condens. Matter Complex Syst.* **1999**, *11*, 187–192. [CrossRef]

77. Palacci, J.; Sacanna, S.; Steinberg, A.P.; Pine, D.J.; Chaikin, P.M. Living crystals of light-activated colloidal surfers. *Science* **2013**, *339*, 936–940. [CrossRef] [PubMed]

78. Singh, D.P.; Choudhury, U.; Fischer, P.; Mark, A.G. Non-equilibrium assembly of light-activated colloidal mixtures. *Adv. Mater.* **2017**, *29*, 1701328. [CrossRef] [PubMed]

79. Kim, Y.; Shah, A.A.; Solomon, M.J. Spatially and temporally reconfigurable assembly of colloidal crystals. *Nat. Commun.* **2014**, *5*, 3676. [CrossRef] [PubMed]

80. Esplandiu, M.J.; Afshar Farniya, A.; Bachtold, A. Silicon-based chemical motors: An efficient pump for triggering and guiding fluid motion using visible light. *ACS Nano* **2015**, *9*, 11234–11240. [CrossRef] [PubMed]

81. Golestanian, R. Collective behavior of thermally active colloids. *Phys. Rev. Lett.* **2012**, *108*, 038303. [CrossRef] [PubMed]

82. Cohen, J.A.; Golestanian, R. Emergent cometlike swarming of optically driven thermally active colloids. *Phys. Rev. Lett.* **2014**, *112*, 068302. [CrossRef] [PubMed]

83. Lin, L.; Peng, X.; Wang, M.; Scarabelli, L.; Mao, Z.; Liz-Marzan, L.M.; Becker, M.F.; Zheng, Y. Light-directed reversible assembly of plasmonic nanoparticles using plasmon-enhanced thermophoresis. *ACS Nano* **2016**, 9659–9668. [CrossRef] [PubMed]

84. Lin, L.H.; Zhang, J.L.; Peng, X.L.; Wu, Z.L.; Coughlan, A.C.H.; Mao, Z.M.; Bevan, M.A.; Zheng, Y.B. Opto-thermophoretic assembly of colloidal matter. *Sci. Adv.* **2017**, *3*, e1700458. [CrossRef] [PubMed]

85. Klajn, R.; Stoddart, J.F.; Grzybowski, B.A. Nanoparticles functionalised with reversible molecular and supramolecular switches. *Chem. Soc. Rev.* **2010**, *39*, 2203–2237. [CrossRef] [PubMed]

86. Klajn, R.; Bishop, K.J.; Grzybowski, B.A. Light-controlled self-assembly of reversible and irreversible nanoparticle suprastructures. *Proc. Natl. Acad. Sci. USA* **2007**, *104*, 10305–10309. [CrossRef] [PubMed]

87. Klajn, R.; Wesson, P.J.; Bishop, K.J.M.; Grzybowski, B.A. Writing self-erasing images using metastable nanoparticle "Inks". *Angew. Chem. Int. Ed.* **2009**, *48*, 7035–7039. [CrossRef] [PubMed]

88. Fissi, A.; Pieroni, O.; Angelini, N.; Lenci, F. Photoresponsive polypeptides. Photochromic and conformational behavior of spiropyran-containing Poly(l-glutamate)s under acid conditions. *Macromolecules* **1999**, *32*, 7116–7121. [CrossRef]

89. Zhang, L.; Dai, L.; Rong, Y.; Liu, Z.; Tong, D.; Huang, Y.; Chen, T. Light-triggered reversible self-assembly of gold nanoparticle oligomers for tunable SERS. *Langmuir* **2015**, *31*, 1164–1171. [CrossRef] [PubMed]

90. Zhang, Q.; Dong, R.; Chang, X.; Ren, B.; Tong, Z. Spiropyran-decorated SiO$_2$-Pt Janus micromotor: preparation and light-Induced dynamic self-assembly and disassembly. *ACS Appl. Mater. Interfaces* **2015**, *7*, 24585–24591. [CrossRef] [PubMed]

91. Kundu, P.K.; Samanta, D.; Leizrowice, R.; Margulis, B.; Zhao, H.; Borner, M.; Udayabhaskararao, T.; Manna, D.; Klajn, R. Light-controlled self-assembly of non-photoresponsive nanoparticles. *Nat. Chem.* **2015**, *7*, 646–652. [CrossRef] [PubMed]

92. Cheng, X.; Sun, R.; Yin, L.; Chai, Z.; Shi, H.; Gao, M. Light-triggered assembly of gold nanoparticles for photothermal therapy and photoacoustic imaging of tumors in vivo. *Adv. Mater.* **2017**, *29*, 1604894. [CrossRef] [PubMed]

93. Raeesi, V.; Chou, L.Y.; Chan, W.C. Tuning the drug loading and release of DNA-assembled gold-nanorod superstructures. *Adv. Mater.* **2016**, *28*, 8511–8518. [CrossRef] [PubMed]

94. Zhao, H.; Sen, S.; Udayabhaskararao, T.; Sawczyk, M.; Kucanda, K.; Manna, D.; Kundu, P.K.; Lee, J.W.; Kral, P.; Klajn, R. Reversible trapping and reaction acceleration within dynamically self-assembling nanoflasks. *Nat. Nanotechnol.* **2016**, *11*, 82–88. [CrossRef] [PubMed]

95. Wei, Y.; Han, S.; Kim, J.; Soh, S.; Grzybowski, B.A. Photoswitchable catalysis mediated by dynamic aggregation of nanoparticles. *J. Am. Chem. Soc.* **2010**, *132*, 11018–11020. [CrossRef] [PubMed]

96. Stolarczyk, J.K.; Deak, A.; Brougham, D.F. Nanoparticle clusters: Assembly and control over internal order, current capabilities, and future potential. *Adv. Mater.* **2016**, *28*, 5400–5424. [CrossRef] [PubMed]

97. Sun, J.; Guo, L.-H.; Zhang, H.; Zhao, L. UV irradiation induced transformation of TiO_2 nanoparticles in water: aggregation and photoreactivity. *Environ. Sci. Technol.* **2014**, *48*, 11962–11968. [CrossRef] [PubMed]

98. Ashkin, A. Optical trapping and manipulation of neutral particles using lasers. *Proc. Natl. Acad. Sci. USA* **1997**, *94*, 4853–4860. [CrossRef] [PubMed]

99. Zhang, J.; Yan, J.; Granick, S. Directed self-assembly pathways of active colloidal clusters. *Angew. Chem. Int. Ed.* **2016**, *55*, 5166–5169. [CrossRef] [PubMed]

100. Hayakawa, Y.; Furuhashi, S. Group-size distribution of skeins of wild geese. *Phys. Rev. E* **2012**, *86*, 031924. [CrossRef] [PubMed]

MDPI
St. Alban-Anlage 66
4052 Basel
Switzerland
Tel. +41 61 683 77 34
Fax +41 61 302 89 18
www.mdpi.com

Micromachines Editorial Office
E-mail: micromachines@mdpi.com
www.mdpi.com/journal/micromachines

www.ingramcontent.com/pod-product-compliance
Lightning Source LLC
Chambersburg PA
CBHW051858210326
41597CB00033B/5945